Herbicide
BIOASSAYS

Jens C. Streibig
Per Kudsk

CRC Press
Boca Raton Ann Arbor London Tokyo

Library of Congress Cataloging-in-Publication Data

Herbicide bioassays / edited by Jens C. Streibig, Per Kudsk.
p. cm.
Includes bibliographical references and index.
ISBN 0-8493-6603-8
1. Herbicides. 2. Herbicides—Research. 3. Biological assay. I. Streibig, Jens C. II. Kudsk, Per.
SB951.4.H415 1993
632′.954—dc20 92-33598
CIP

Direct all inquiries to CRC Press, Inc., 2000 Corporate Blvd., N. W., Boca Raton, Florida, 33431.

International Standard Book Number 0-8493-6603-8

Library of Congress Card Number 92-33598
Printed in the United States 1 2 3 4 5 6 7 8 9 0
Printed on acid-free paper

PREFACE

Herbicides have become and will certainly continue to be an integrated part of modern crop production. Bioassays have played a vital role in the development of herbicides in assessing activity and selectivity, as well as in the measurements of herbicide residues in soil and other environmental compartments. Previously bioassays using whole plants or plant segments were most common. However, in the recent years they are gradually being replaced by, for example, plant tissue culture systems, *in vitro* systems using isolated enzyme preparations and immunological assays.

It is the sincere hope of the editors and the authors that this book, which to the knowledge of the editors is the first of its kind, will achieve its purpose to be a valuable source of basic scientific information that will help researchers and motivate further good research toward understanding of the biological activity of herbicides in plants and the environment. This book attempts to provide a state of the art overview and could not have been written without the expert cooperation of all the contributors. They are all recognized scientists in their field and we wish to express our thanks to them all for participating in this endeavor.

Jens C. Streibig
Per Kudsk

THE EDITORS

Jens C. Streibig, Ph.D. is Associate Professor in weed science at the Department of Agricultural Sciences, The Royal Veterinary and Agricultural University (RVAU) in Copenhagen. He received his M.Sc. in 1972, his Ph.D. in 1977, and his D.Sc. in 1992 from RVAU. He was Research Officer and Senior Research Fellow at the former Department of Plant Husbandry and Plant Breeding 1977-1982, Research Officer at the Research Center for Plant Protection in 1982 and got the current position as Associate Professor at RVAU in 1983. He was a consultant for a private company in 1986 and a visiting Research Scientist at the Kieth Turnbull Research Institute, Victoria, Australia in 1988. In 1987 he received an honorary reward for professional achievement in weed science.

Since 1980, Dr. Streibig has served as Editor of *Weed Research* and from 1984 as a member of The Referee Group for the *Swedish Journal of Agricultural Research* as well as reviewer for several other scientific journals and publishers of books.

Dr. Streibig's main interests are in herbicide research, weed ecology and crop-weed competition. Within herbicide research it is the use of herbicide bioassay to assess selectivity, effects of herbicide mixtures, adjuvants and measurement of herbicide residues in soil, detection of herbicide-resistant biotypes of weeds and impact of herbicides on weed communities. Within weed ecology his interest is mainly in the area of changes in the weed flora as a result of herbicide use and other agronomic factors. The link between models to assess herbicide phytotoxicity, weed-crop competition and economic thresholds for weed and judicious use of herbicides are also areas of interest.

Dr. Streibig is a member of the European Weed Research Society and the International Weed Science Society. His publications include 23 peer-reviewed scientific journal papers, 2 books, 6 chapters, 34 contributions to conference proceedings, and 11 semi-technical or extension publications.

Per Kudsk, M.Sc. is a Senior Research Scientist at the Department of Weed Control at the Research Centre for Plant Protection in Denmark, a research institute under the Ministry of Agriculture. He received the M.Sc. degree from the Royal Veterinary and Agricultural University and joined the Research Centre for Plant Protection in 1983. His main interests are the influence of adjuvants and formulations, environmental conditions and application techniques on herbicide activity and selectivity. Besides a basic interest in these research areas a further objective of his work is to provide ways of reducing herbicide doses, i.e., reduce the total use of herbicides. Furthermore he is engaged in research work aiming at optimizing the composition and dose of herbicide mixtures. He is a member of the European Weed Research Society and the Weed Science Society of America. His publications include 48 scientific research publications of which 22 have been published internationally in peer-reviewed scientific journals or as invited or contributed papers at international conferences.

CONTRIBUTORS

Z. Amsellem, M.Sc.
Department of Plant Genetics
Weizmann Institute of Science
Rehovot, Israel

Michael R. Bartley, Ph.D.
Weed Science Section
ICI Agrochemicals
Bracknell, Berkshire, England

David V. Clay, B.Sc., A.R.C.S.
Avon Vegetation Research
Nailsea, Bristol, England

Gayle Davidonis, Ph.D.
Plant Physiologist
USDA/ARS
Southern Regional Research Center
New Orleans, Louisianna

Gary Gardner, Ph.D.
Professor, Department Head
Department of Horticultural Science
University of Minnesota
St. Paul, Minnesota

Jerry Green, Ph.D.
Senior Research Biologist
Agricultural Products
E. I. Du Pont de Nemours
Newark, Delaware

Jonathan B. Gressel, Ph.D.
Professor
Department of Plant Genetics
Weizmann Institute of Science
Rehovot, Israel

Petra Günther, Dr.rer.hort.
Weed Research Institute
Federal Biological Research Centre
Braunschweig, Germany

Sigmund Josef Huber, Dr.agr.
Regierungsdirektor
German Patent Office
Munich, Germany

Jens Erik Jensen, Ph.D.
Research Fellow
Department of Agricultural Sciences
The Royal Veterinary and Agricultural University
Copenhagen, Denmark

Daniel A. Kleier, Ph.D.
Research Associate
Agricultural Products
E. I. Du Pont de Nemours
Newark, Delaware

Per Kudsk, M.Sc.
Senior Research Scientist
Department of Weed Control
Research Centre for Plant Protection
Slagelse, Denmark

Wilfried Pestemer, Dr.agr.
Scientific Director
Weed Research Institute
Federal Biological Research Centre
Braunschweig, Germany
Professor
Institute for Phytopathology
University of Hannover
Hannover, Germany

Mats Rudemo, Ph.D.
Professor
Department of Mathematics and Physics
Agricultural University
Copenhagen, Denmark

Robert R. Schmidt, Dr.rer.nat.
Head of Institute of Weed Control
PF-E/H
Bayer AG
Leverkusen, Germany

Yoseph Shaaltiel, Ph.D.
Migal Ltd.
Kiryat Shmona
Israel

Amir Sharon, Ph.D.
Department of Plant Pathology
Cornell University
Ithaca, New York

Jens Carl Streibig, Ph.D.
Associate Professor
Department of Agricultural Sciences
The Royal Veterinary and Agricultural University
Copenhagen, Denmark

Peter Ulvskov, Ph.D.
Senior Scientist
Head, Biotechnology Group
Danish Institute of Plant and Soil Science
Lyngby, Denmark

TABLE OF CONTENTS

Chapter 1

INTRODUCTION

Jens C. Streibig and Per Kudsk

Through the history of agriculture, weeding has been man's lot. For more than 7 millennia human beings have changed the original vegetation by growing crops in monoculture. During most of this period the prevailing weed control methods have been relatively static. Even now agriculture depends on an army of professional hand weeders.

Weeds are pioneers that modify the diversity of agricultural ecosystems by utilizing the environmental potential, especially developed by human beings for crop production. In relation to crops, weeds are successful competitors; they are a nuisance to man's ability to grow crops and they must be controlled to secure yield. Worldwide, they reduce crop growth and yield, block waterways and irrigation systems, and some are even poisonous to man and animals. Without weed control 50 to 100% of certain crops would be lost.

Most changes in control methods have taken place for the last 200 to 300 years, e.g., the introduction of agricultural implements to lessen the manual burden of hand weeding; however, radical changes in the developed world only happened during the last 40 to 50 years where more than 200 chemical compounds have been made commercially available to agriculture.

In 1896 a French wine grower sprayed Bordeaux mixture to control diseases on his vine. He also observed a beneficial side effect in that a weed, *Sinapis arvensis* L., exposed to the spray, died. One year later another French grower discovered the herbicidal activity of sulfuric acid that could selectively control some weeds without injuring the crops. These discoveries were one of the first successful chemical weed control methods. Until recently, sulfuric acid was used as a herbicide in many parts of Europe. Several other inorganic compounds were introduced for a shorter period. The first organic herbicide, DNOC, was patented by George Truffaut and K. Pastac in 1932 for use as a selective herbicide in cereals.

The rapid development of modern chemical substances, experienced for the last 45 years, was initiated by the need for increasing food production during World War II. The shortage of food in the U.K. made agricultural researchers take advantage of previous discoveries of growth substances, auxins, that increase plant growth, primarily by promoting cell elongation. Unknown to each other, workers at Jealott's Hill Research Station and Rothamsted Experimental Station studied the possibilities to boost crop production with the synthetic auxins, chlorinated phenoxyacetic acids. The researchers were capable of realizing when novel results were serendipitous. Both teams reported remarkable herbicidal activity of MCPA. The substances appeared to act like an overdose of auxin on the weeds, but they did not harm the cereal crop.[1] With the return of peace, the original findings were published.[2] In the U.S. work on the physiological activity of 2,4-D and 2,4,5-T already had been published in the early forties.[3]

The phenoxyacetic acids were fundamentally different from the other organic compounds, e.g., DNOC; they were systemic in that they had to be translocated in the plant to exert their action. Soon after the release of MCPA and 2,4-D, 2,4,5-T was released, and a new era of weed control began.[4] Weed science became a science in its own right, rooted in agronomy, plant physiology, and botany.

Within a few years, screening systems to test literally thousands of potential herbicidal chemicals were set up by companies, and they formed the basis for rapid development of new herbicides to control specific weed problems in specific crops. New important herbicides were developed. The *s*-triazines were reported to have unique effects compared with existing

0-8493-6603-8/93/$0.00+$.50

classes of compounds in 1955,[5] and three years before, some substituted ureas were found to be potent inhibitors of the photosynthesis.[6] Until recently about half the commercialized herbicides were photosynthesis inhibitors. In 1971 glyphosate was developed as a herbicide possessing unique properties in controlling particularly perennial grasses and sedges.[7]

Perhaps at no other time in the history of herbicide research has so much knowledge been accumulated about a new group of herbicides in a very short time as that of the sulfonylureas.[8] Only a few years after their discovery, the site of action was found to be inhibition of the enzyme acetolactate synthase that forms a part of the combined pathway responsible for the biosynthesis of valine and isoleucine.[9] The absence of this enzyme in man and other animals helps explain the low toxicity of the sulfonylureas. Because of their specific site of action, sulfonylureas are potential candidates for tailoring crops to fit this chemistry. This specificity has also resulted in unintentionally developed sulfonylurea resistant biotypes of weeds after only a few years use.[10]

The discovery of herbicide resistant biotypes of formerly susceptible species has been taken up by many chemical companies. Concomitantly, the amount of information on mode of action for many groups of herbicides has rapidly increased in the past few years. Many herbicides have only one primary site of action, and for some, the herbicide-resistance gene determinants are single dominant traits, making them amendable to gene transfer techniques. This fact may open up a new productive area of research.

The array of new herbicides has given the farmer a safety factor in growing crops, and has, to a certain extent also freed growers from traditional crop rotation and fallow systems, partly implemented to cope with noxious weeds. Today, the farmer can chemically "weed" unwanted plants on his land, and within limits, grow crops in demand by the consumers without too many crop rotation restrictions.

Chapter 2 reviews the role of bioassays in the development of herbicides, from synthesis by the chemist to introduction to the market, a process that currently requires 5 to 8 years of development work. The development of pesticides is very expensive; depending on the product, the costs are estimated to be between 40 and 50 million $U.S. and can reach up to 95 million $U.S. per compound. The discovery of a new herbicide is chiefly by empirical synthesis and evaluation, but computer-aided molecular modeling may help improve the design and synthesis of new molecules.

Several methods are available for determining the herbicidal properties of a substance. Short term laboratory bioassays using *in situ*, *in vivo*, and *in vitro* tests and greenhouse experiments are used to assess biological effects and selective properties of novel compounds. For laboratory and greenhouse screening of a herbicide, only a few grams are necessary. Further quantities of the chemical are required for field tests to confirm their performance under practical conditions after positive identification of promising leads. Biological testing of efficacy accounts for about 31% of the total costs. Parallel with field studies of a new substance, extensive testing of its toxicology and ecotoxicology properties is necessary. In 1988, the toxicological and ecotoxicological evaluation averaged about 33% of the total research and development costs for herbicides. These high costs are justified to ascertain the safety of producers, applicators, consumers, and the environment. Bioassays should be designed to identify herbicides with different properties; refinement of bioassay methods is therefore indispensable.

Chapter 3 reviews the mathematics of dose-response curves and how to statistically appraise bioassay data. Dose-response curves, commonly used in toxicology and pharmacology, play an important role in the screening assessment of herbicide activity and selectivity.

Statistical methods have been devised and accepted as standard practice in assessing the toxicity of chemicals in many branches of biology. In herbicide research some attempts have been made to use these methods to evaluate herbicides and judge the precision and the accuracy of the results obtained.

Standard textbooks on regressions[11,12] chiefly deal with straight line regression techniques and are thus of only little help in the intrinsic nonlinear situations often encountered in herbicide bioassays. Herbicide research can benefit from the development of modern computers and the rapidly growing statistical theory. The implicit statistical assumption of constant variance and normal distribution of residuals are still important when analyzing bioassays, but today newly developed methods take these problems into account.

Chapter 4 reviews the assessment of herbicide selectivity in the screening for new herbicides. For practical use in crops, we usually talk about selective and nonselective herbicides. Selective herbicides control weeds without seriously affecting the crop. The distinction between selective and nonselective herbicides often is a question of dose; thus, the axiom of Paracelsus still applies:

Was ist das nit Gifft ist:
alle Ding sind Gifft und nichts ohn Gifft.
*Allein die Dosis macht das ein Ding kein Gifft ist.**

Paracelsus (1494–1541)

Paraphrased to our purpose: "a sufficiently high dose of a herbicide will kill almost any plant, whereas a sufficiently low dose will produce no effect."

Selectivity is one of the very fascinating areas of herbicide research. Selectivity is dependent on much more than mere metabolism of compounds in the plant and physical separation of site of application and site of uptake. Environmental conditions and the growth stage of weeds and crops are important issues to better understand why some compounds control weeds but do not harm the crops. About 85% of commercially available herbicides are used in a crop selective manner. This high proportion reflects the importance of developing herbicides that can be used selectively in crops. Consequently, the chemical companies put much effort into the search for novel compounds that are selective in some crops. The screening process is quite comprehensive and requires dose-response curves covering a large range of doses.

Various indices for the assessment of herbicide selectivity can be used under controlled conditions, but the performance of a novel compound in the field is often hard to predict from bioassay studies in greenhouses.

Chapter 5 reviews the role of bioassay in quantitative structure activity relationships (QSAR). If biological assay results are obtained for a series of molecules of sufficient diversity, quantitative relationships can be constructed between assay response and various chemical and physical descriptors of the molecules. Sometimes this is quite simple as one can use plant response from primary screen vs. some easily measured properties of the screened compounds. For more comprehensive QSAR, the choice of bioassay method, be it *in vitro* or *in vivo*, and descriptors of molecules is quite complex, and the methods of evaluation of results often involve multidimensional statistics and graphics. In some instances a detailed image of a herbicide receptor site, including the ability to predict bioactivity of leads, are the aim of QSAR. Good QSAR designs minimize the chance that optimal compounds in an analogue program will be missed.

Chapter 6 reviews the importance of proper formulation of herbicides and use of adjuvants to enhance the efficacy and stability of herbicides. Researchers have tried to surpass the performance of herbicides by formulation techniques, adjuvants, and mixtures. Formulations are used to improve stability, ease mixing with water, and secure efficacy of herbicides under a variety of environmental conditions.

In some instances the formulations and adjuvants are not phytotoxic per se, hence the basic theory of similar, also called parallel, dose-response curves can be applied to appraise the

* "What is not poisonous? All substances are poisonous and no substances are without poison. It is the dosage that makes a substance nonpoisonous."

effect of different formulations and/or adjuvants. The horizontal relative replacement of the dose-response curves for a herbicide with different formulations or adjuvants, in fact, answers the user's question: "How much can I reduce the herbicide dose when adding an adjuvant without losing efficacy?" The effects of adjuvant can be subtle and need proper experimental designs that help to better discern when and sometimes how it pays to use different formulations and/or adjuvants.

Chapter 7 reviews the assessment of the joint action of herbicide mixtures. Herbicide mixtures are commonly used in many crops, and some herbicides are only marketed as mixtures. The study of mixtures may also open up new prospects in the struggle against acquired resistance in weeds.[13]

In weed science literature the terminology and assessment of a herbicide mixture has been confusing in that "additivity of effects" has been confused with "additivity of doses". Additivity of effects at some preset doses is easy to comprehend and calls for ordinary analysis of variance. Additivity of doses is more difficult in that we must know the dose-response curves and the exchange rate between herbicides in the mixture at preset response levels. Additivity of doses answers the question often asked by users of mixtures: "If I mix Herbicide A and B in a certain proportion do I lose or gain efficacy relative to applying the herbicide singly?" This is essentially the same question raised with adjuvants in Chapter 6, and in a way, the reference models for assessing adjuvants and mixtures are the very same.

Chapter 8 reviews the problems with appraisal of the ecotoxicology of herbicides in plant or plant communities. No-observable-effect levels (NOEL) has a central position in that it defines a threshold dose that just does not inflict an effect upon a test organism. Various methods are outlined and the nonlinear dose-response curve plays an important role. As always when dealing with subtle effects, the experimental conditions and the test organisms implicitly define the results, and they cannot always be extrapolated to more natural environments.

Chapter 9 reviews the role of bioassay to assess the effects of herbicide residues. While the other chapters deal with theory, development, and use of bioassay in situations where the herbicide is directly applied to the plant, this chapter deals with the assessment of residues in soil. In a way, bioassay of soil residues can be considered a substitute for qualitative or quantitative analysis of herbicides or their metabolites.

Previously, biological assays used to determine residues in the soils played an important role, but they have gradually been replaced by high tech chemical analysis. Still, bioassay are needed in some situations in which instrumental chemical analysis are unavailable. Poor standardization of assays has often left the impression that assay data are imprecise.

Chapter 10 reviews the use of plant tissue culture in herbicide research. They are used in screening of herbicides and growth regulators, in studing metabolism and the mode of action of herbicides, and in developing herbicide resistant cultures. They can be used to test many different compounds and can be tailored for evaluation of structure activity relationships. Comparison between culture results and whole plant studies can sometimes help identifying putative barriers precluding herbicides to reach their site of action. A point of concern is that all application factors such as media composition and growth stage of the culture must be considered when interpreting the results.

Chapter 11 reviews some of the immunochemical techniques available for herbicide research. They can be used to analyze traces of herbicides in soil, water, urine, and blood. Immunoassays are based on the principles that antibodies to pesticides can be prepared in animals. These antibodies can recognize and attach to certain chemical configurations with exquisite specificity. Small molecules such as herbicides usually are not immunogenic per se, but they can be made so by chemically binding them to large immunogenic proteins. Various types of immunological assay can be used to detect herbicides. Immunological techniques also can be used to detect the distribution of herbicides in the plant tissue.

Chapter 12 reviews the possibilities to biorationally develop herbicide synergists to lower the field rates, to change the selectivity of already developed herbicides, and to lower the adverse effect of either natural or acquired herbicide tolerance/resistance of weeds to certain herbicides. It is a fascinating story about how to understand herbicide action in the plant and how to design synergists for specific purposes. It provides a wealth of information on how to assess synergizing effects from mere visual assessment to sophisticated measurement in prepared enzyme preparations.

Apart from the practical use of putative synergists, the research gives us a good opportunity to better understand compound selectivity by searching for synergists that prevent the detoxification of herbicides or toxic products that they generate. In this context we must know the site of action of the herbicide, the mode of death, and the pathways responsible for detoxification. The full scientific potential of this work requires virtually all sorts of bioassay systems and chemical hardware. *In vitro* assays play an important role in performing highly sensitive prescreening, but many levels of bioassay systems should also be used to find synergists.

REFERENCES

1. **Norman, A. G., Minarik, C. E., and Weintraub, R. L.,** Herbicides. *Annu. Rev. Plant Physiol.,* 1, 141, 1950.
2. **Hance, R. J. and Holly, K.,** *Weed Control Handbook: Principles,* Blackwell Scientific Publications, Oxford, 1990, 57.
3. **King, L. J.,** *Weeds of the World. Biology and Control,* Interscience Publishers, Inc. New York, 1966, 285.
4. **Loos, M. A.,** Phenoxyalkanoic acids, in *Herbicides. Chemistry, Degradation, and Mode of Action,* Kearney, P. C. and Kaufman, D. D., Eds., Marcel Dekker, Inc., New York, 1975, 1.
5. **Esser, H. O., Dupuis, G., Ebert, E., Vogel, C., and Marco, G.J.,** *S*-triazines, in *Herbicides. Chemistry, Degradation, and Mode of Action,* Kearney, P. C. and Kaufman, D. D., Eds., Marcel Dekker, Inc., New York, 1975, 129.
6. **Geissbühler, H., Martin, H., and Voss, G.,** The substituted ureas, in *Herbicides. Chemistry, Degradation, and Mode of Action,* Kearney, P. C. and Kaufman, D. D., Eds., Marcel Dekker, Inc., New York, 1975, 209.
7. **Franz, J.E.,** Discovery, development and chemistry of glyphosate, in *The Herbicide Glyphosate,* Grossbard, E. and Atkinson D., Eds., Butterworths, London, 1985, 3.
8. **Beyer, E.M., Duffy, M.J., Hay, J.V., and Schlueter, D.D.,** Sulfonylureas, in *Herbicides. Chemistry, Degradation and Mode of Action,* Kearney P. C. and Kaufman, D. D., Eds., Marcel Dekker, Inc., New York, 1988, 117.
9. **Ray, T. B.,** Site of action of chlorsulfuron: inhibition of valine and isoleucine biosynthesis in plants, *Plant Physiol.,* 75, 827, 1984.
10. **Saari, L. L., Cotterman, J. C., and Primiani, M. M.,** Mechanism of sulfonylurea herbicide resistance in the broadleaf weed, *Kochia scoparia, Plant Physiol.,* 93, 55, 1990.
11. **Steel, R. G. D. and Torrie, J. H.,** *Principles and Procedures for Statistics,* McGraw-Hill, Inc., New York, 1960.
12. **Snedecor, G. W. and Chochran, W. G.,** *Statistical Methods,* 6th ed., Iowa State University Press, 1967.
13. **Gressel, J.,** Why get resistance? It can be prevented or delayed, in *Herbicide Resistance in Weeds and Crops,* Caseley, J. C., Cussans, G. W., and Atkin, R. K., Eds., Butterworth-Heinemann Ltd., Oxford, 1991, 1.

Chapter 2

DEVELOPMENT OF HERBICIDES — ROLE OF BIOASSAYS

Robert R. Schmidt

TABLE OF CONTENTS

0-8493-6603-8/93/$0.00+$.50

I. INTRODUCTION

A. FROM SYNTHESIS TO MARKETING OF A HERBICIDE

The development of a herbicide, from synthesis by the chemist to the introduction to the market, is a long, involved process that currently requires 5 to 8 years of development work.

The development of pesticides is very expensive; depending on the product, the costs are estimated to be between 80 to 100 million DM and can reach up to 180 million DM per compound.[1] Therefore, a review of the production and investment costs and of the potential market share for the product is essential at an early stage.

For laboratory and greenhouse screening of a herbicide, only a few grams, synthesized by the research chemist, are necessary. The discovery of a new herbicide is traditionally by empirical synthesis and evaluation.[2] Computer-aided molecular modeling can improve the design and synthesis of new molecules.[2]

After a successful screening further quantities of the chemical are required for field tests. The amounts required lie between several hundred grams and several kilograms. Later on, production possibilities must be explored.

B. NEED FOR NEW HERBICIDES

Weeds compete with crop plants for space, water, light, and nutrients. They can cause significant yield reductions and may even result in a total loss of the crop. They often interfere with mechanical harvesting and can also decrease the quality of agricultural products. Some weeds are poisonous to man and animals, while others are transmitters of pathogens and insects.

Even after a long period of successful chemical weed control, there are still biological, ecological, and economic reasons for synthesizing and developing new herbicides. These reasons can be summarized as follows:

- Selective herbicides do not control all weeds, and therefore the unaffected ones will increase in importance.[3]
- Weeds may develop resistant biotypes against herbicides from different classes, (e.g., triazines and more recently sulfonylureas and imidazolinones).[4]
- Changes in agricultural practice cause specific weed problems.[5]
- New concepts in weed control, such as threshold levels or weed suppression rather than eradication, require new types and new modes of action of herbicides.
- Less persistent and less mobile herbicides are needed to avoid groundwater contamination.
- Reduced dosages are required from ecological as well as economic standpoints.

In conclusion, a new herbicide must be adapted to economic and ecological points of views, and it must provide advantages over existing compounds.[6] There is still a need to supply the rapidly growing world population with food; therefore, it is essential that the losses of plant production caused by pests, diseases and weeds be minimized. According to a publication of the Food and Agriculture Organization of the United Nations,[7] the losses may sum up to 40% before harvesting. Minimizing the crop losses can be done only with appropriate methods, including new, efficient, and safe herbicides.

C. REQUIREMENTS FOR THE DEVELOPMENT

After synthesis of a compound, its potential as a herbicide is tested in the greenhouse and laboratory. Bioassays are used to explore the best way of applying the product (e.g., pre- or post-emergence) and to determine its usefulness as a total or selective herbicide. In follow-up field screening trials the performance of the herbicide is checked.

Further field tests in the respective crops and regions and under different climatic conditions are carried out to test the performance under practical conditions. These biological tests are accompanied by extensive toxicological and ecotoxicological studies to provide safety to man, animals, and the environment. In this context the metabolism of the herbicide in plants, animals, and soil is as essential as the research on plant and soil residues and the behavior in the soil.

Due to the fact that, in most cases, the active ingredient per se cannot be applied, appropriate formulations and applications have to be explored. Last but not least, the production possibility and facility and the marketing chances have to be considered.

D. ROLE OF BIOTESTS

Rapid information on the biological performance and the toxicological and ecotoxicological properties of a new candidate are therefore of major importance for the decision to develop a compound or not. In the following sections a survey is given on the necessary tests and the role bioassays play in the development of herbicides.

II. BIOASSAYS FOR DETECTING HERBICIDES

Several methods are available for determining the herbicidal properties of a substance.[8-11] In most cases a compound is applied before seeding, before emergence of the seedling, or after the emergence of the plant.

By application to the seed, compounds that kill the seeds or inhibit germination can be identified.

Applying a substance before the seedling emerges is suitable for detecting herbicides that can be taken up by the root or shoot, and after uptake, interfere with physiological and biochemical processes in the plant.

The third method of application, to the emerged plant, gives the opportunity to find herbicides that destroy the plants immediately after contact with the compound or interfere with physiological processes in the plant after they are taken up by shoot and foliage. All bioassays regarding the detection of herbicides with the above mentioned application methods are dealt with in the section on greenhouse trials. For the detection of specific herbicidal compounds a number of bioassays of short duration can be carried out in the laboratory.

A. LABORATORY BIOASSAYS

1. *In Situ* Tests

Growing plants under open air conditions and using them as bioindicators *in situ* for monitoring environmental stresses and pollution is common in ecotoxicological studies.[12] Physiological processes can be measured without destroying the plants: transpiration, analyzed by porometry;[13,14] photosynthesis and respiration, analyzed by infrared gas analyzers (IRGA).[13]

In contrast to above studies, the detection and discovery of herbicides by *in situ* tests is not common but should nevertheless be mentioned.

Plants growing in the field cannot easily be transferred to the laboratory; therefore, portable laboratory instruments have to be used in the field for measuring specific physiological processes in the plants.

Measuring photosynthesis on whole plants is one of the ways to identify photosynthesis inhibiting herbicides. Nondestructive methods under field conditions are measurements of fluorescence,[15] e.g., with portable fluorometers, and recording of CO_2 uptake, e.g., with portable infrared gas analyzers.[14] Phytotoxicity and herbicidal action can also be measured by nondestructive luminoscopy.[16]

These nondestructive measurements are also suitable tools for detecting herbicide activity with *in vivo* tests.

2. *In Vivo* Tests

The response of a test plant to a compound assumed to inhibit photosynthesis can be detected with rapid biotests. Measurement of photosynthetic CO_2 uptake with IRGA is most common. Whole plants are placed in glass containers through which air is circulating. At the inlet and outlet, the CO_2 content of the air stream is measured by an IRGA. After illumination, the plants take up CO_2 during photosynthesis, and therefore a difference in CO_2 levels can be measured between incoming and outgoing air. Using this method the influence of a herbicide on photosynthetic CO_2 uptake can be measured over several days.

Another method to study *in vivo* the action of a herbicide on photosynthesis is measuring the photosynthetic O_2 evolution of unicellular green algae, e.g., synchronously grown *Chlamydomonas* that can also be used to detect herbicides with the following mechanisms of action:[17]

- Growth inhibition in the light without rapid phytotoxicity (e.g., metribuzin)
- Bleaching without rapid phytotoxicity (e.g., fluridone)
- Rapid phytotoxicity in the light (e.g., diphenylethers such as nitrofen and paraquat)
- Rapid phytotoxicity in the dark (e.g., PCP)
- Direct inhibition of cell division (e.g., amiprophos-methyl and trifluralin)
- Indirect inhibition of cell division (e.g., alachlor)

Furthermore, *Lemna* spp. growing in nutrient or soil solution can be used to find photosynthesis inhibiting herbicides.

Herbicides interfering with germination can be easily detected by Petri dish tests in which seeds of different test plants are placed on a filter paper incubated with water (untreated control) or herbicide solutions of different concentrations. After 3 to 5 days the percentage of germinated seedlings is evaluated; later on (day 8 to 10) the length of the shoot and/or root can also be measured.

A similar test can be carried out with small pots where the seeds are placed in nonadsorptive substrates (e.g., quartz sand or vermiculite). The plants are grown in hydroponic systems using water or nutrient solutions (e.g., Hoagland[18]) containing the herbicide. Compounds that interfere with the biosynthesis of amino acids, e.g., sulfonylureas and imidazolinones, can be easily detected in such tests.

3. *In Vitro* Tests

There are several *in vitro* tests that can be used to detect herbicides. Different tests are required according to the biochemical and physiological mode of action of herbicides.[19-21]

Isolated chloroplasts are commonly employed to test for herbicides that inhibit photosynthesis. The chloroplasts are usually isolated from leaves of spinach or sugar beets, but they can also be obtained from other plants. The isolated chloroplasts can be kept for several days in a refrigerator and may be used when required; the test can be carried out at room temperature. The chloroplasts are illuminated in a reaction mixture containing ferricyanide as an electron acceptor A. In the reaction that occurs in the chloroplast, water is oxidized and O_2 is released (Hill reaction):

$$2H_2O + 2A \xrightarrow{\text{light}} 2AH_2 + O_2 \tag{1}$$

Reduction of the electron acceptor ferricyanide can then be measured spectrophotometrically as a change in absorbance at 420 nm or colorimetrically. Oxygen (O_2) evolution can be measured polarographically by using an oxygen electrode.[22]

Other assays measure the fluorescence or CO_2 uptake with IRGA of leaf discs after incubation with herbicides. Another bioassay for photosynthesis inhibitors is the so called "sink-test", where leaf disks are floating on a phosphate-based medium containing a photosynthesis inhibiting herbicide. The disks sink in a few hours if photosynthesis is inhibited.[23,24] Herbicides directly or indirectly inhibiting the synthesis of chlorophyll can be detected by analyzing the chlorophyll content of cucumber cotyledones incubated in herbicide solutions.[25]

A similar method can be used to find herbicides interfering with carotenoid synthesis ("bleaching"). Plants grown in hydroponic systems containing bleaching herbicides are extracted and analyzed for their content of different carotenoides. Cell-free preparations of carotenoid biosynthesis[26] are also available, but unlike *in vitro* bioassays, they are not as common.

Recently a test was published in which plants are extracted and their primary and secondary metabolites are analyzed by capillary gas chromatography. The profiles of herbicide treated plants are compared to those of untreated plants. Contrasting profiles can be obtained for different herbicidal groups.[27] At the moment, the validity of such a bioassay for screening herbicides cannot be estimated.

Herbicides acting on cell membrane permeability can be rapidly detected by measuring the increase of electrolytic conductivity of the medium in which leaf discs of treated plants are incubated.[28]

Root regeneration of mono- and dicotyledonous plant cuttings, e.g., oat and soybeans, has been employed as a test system to allow the classification of different graminicides like acetamides, (thiol)carbamates, aryl-phenoxy-propionic acids, cyclohexandiones, and benzyl-ethers.[29]

On an enzymatic level, biotests have also been published for the routine testing of compounds inhibiting the biosynthesis of amino acids. Enzyme preparations of acetolactatesynthase (ALS) (= acetohydroxyacidsynthase [AHAS]) allows rapid detection of substances that inhibit the biosynthesis of leucin, isoleucin, and valin, e.g., sulfonylureas[30] and imidazolinones.[31] Another enzyme involved in the biosynthesis of these three amino acids is acetolactate reductoisomerase, which also may be used as a target enzyme for testing herbicides.[32] Glutamine synthetase is an enzyme for testing compounds that inhibit the synthesis of the amino acid glutamine. The above enzyme is the target for the microbial herbicide phosphinothricin (glufosinate).[33] EPSP-synthase in the shikimic pathway is the target enzyme of glyphosate,[34] but up to now no other useful herbicides have been found to inhibit this enzyme.

Using enzyme preparations as test systems, one must be aware that only compounds that have these special enzymes as targets can be found. This risk must be taken into account when only enzymic preparations are used as bioassays for finding new herbicides.

4. Cell and Tissue Culture

Plant cell and tissue cultures have been tested for their ability to identify and screen herbicides.[35-39] When the effects on whole plants (seedlings) were compared to those on white calli of the same species, some positive correlations were obtained. Better correlations were found between the herbicidal/phytotoxic effects on seedlings and on green calli.[36]

Comparing the results for plant seedlings and suspension cultures showed that photosynthesis inhibiting herbicides affected seedlings much more than the nonphotosynthetic suspension cultures. On the other hand, compound interferences with cell division or growth were more pronounced in cell suspensions than on whole seedlings.[36] These examples indicate that cell and tissue cultures have their limitations in screening herbicides. Cell and tissue cultures cannot be considered substitutes for bioassays with whole plants, but they may be used as supplementary tests. Cell cultures, however, are useful tools in metabolic studies.

B. GREENHOUSE TRIALS

The most common and successful way to screen chemicals for herbicidal activity is bioassay in greenhouse. This method is employed by the majority of agrochemical companies.

In most cases, the test compounds are not soluble in water and must therefore be prepared as suitable formulations. This can be done by first dissolving the chemical in an organic solvent (e.g., acetone, ethanol, etc.), adding an emulsifier, and diluting with water. According to the physico-chemical properties of the chemical solutions, emulsions or dispersions are obtained.

1. Soil Application

Compounds killing the seeds, inhibiting germination, or taken up by the seedlings via the root or the shoot, can be found with a soil application. The most common way to apply the compounds is to spray or to pour the compound onto the soil surface.

Sorptive and nonsorptive substrates can be used in these tests. Artificial soils, which are mixtures of peat, clay, sand, and fertilizers are often used as sorptive substrates. They offer the opportunity to maintain and use a standard soil type. Natural soils, however, are favored because they allow new compounds to be tested under situations that do not differ too much from field conditions. However, a constant supply of a uniform natural soil is not always readily available. Preferred soils are loamy sands, sandy loams, loams, or clays with about 1 to 3% organic matter content.

Quartz sand is often used as a nonsorptive substrate to test new compounds for their intrinsic herbicidal properties. The plants are grown in nutrient solutions (e.g., Hoagland[18]) containing the test compound. Instead of quartz sand it is also possible to use vermiculite as a test substrate (hydroponic test).[40]

a. Pre-planting Incorporation

If a chemical is volatile, the compound may evaporate before acting on the weed seeds or seedlings. This is known to occur with thiolcarbamates such as diallate, triallate, and EPTC.

To reduce losses due to volatilization the test preparations are incorporated into the soil, e.g., by mixing by hand or with a stirrer. After mixing, the prepared soil is filled into containers (plastic boxes or pots), and the seeds of the test plants are sown into the soil. Water is added to give adequate moisture; 70 to 80% of maximum water holding capacity is desirable to give optimum growing conditions and should be maintained throughout the test period by additional watering. Furthermore, to guarantee optimum growing conditions, the test containers should be kept in a greenhouse at approximately 18 to 21°C air temperature and 70 to 80% relative humidity with additional lighting if necessary. The growth period should be at least 3 weeks.

b. Pre-emergence Application

With this test procedure the chemical preparations are applied on the soil surface after the seeds have been sown. Often there is a period of 1 day between sowing and application where the soil is kept wet and the seeds soaked to initiate germination. All other procedures are identical to those mentioned in the preceding section.

2. Post-emergence Application

a. Foliar Application

Normally the application is done when the plants are in the 1 to 2 true leaf stage, but spraying at earlier (cotyledon) or later (more than 3 leaves) stages is also possible.

Most plants have different patterns of germination and development. Therefore, it is sometimes necessary to grow plants under different conditions and time schedules to make sure that they are in the appropriate stage of development at the time of application. Germinating and growing test plants from different origins and climates requires a good knowledge of their biology and ecology.

With foliar application, crop and weed plants are treated at the same time. The herbicide can act by contact through leaves and stems and/or via soil by root uptake. Some examples of herbicides that act only by foliar uptake are bentazon, glyphosate, paraquat, and phenmedipham. Examples of post-emergence herbicides that act both via foliage and root uptake are atrazine, diuron, and metribuzin.

Morphology of crop and weed plants has a great influence on herbicide absorption and can be a factor in selectivity.[41]

b. Directed Spray

This method is used when an over-the-top application is phytotoxic to the crop.[42] This is true especially for young crop plants. If crop plants are older, an application of herbicides between the crop rows is sometimes possible. This application is used, for example, for controlling late germinating weeds in established sugar beets with simazine. Special application methods, e.g., granules, have been developped for this treatment.

In this context the application of the nonselective herbicide glyphosate by roller wipers, ropewicks, or recirculating sprayers also should be mentioned, where weeds growing above the crop can be controlled without damaging the crop.

Special tests are necessary to screen herbicides for paddy rice. In this case the soil is flooded with water (1 to 2 cm) after sowing or transplanting of the rice, and the herbicide is applied into the water. Seeds of typical annual and/or perennial weeds (monocots, dicots) are also sown into the soil, as are tubers and rhizomes of respective weeds. The test is run under optimum growing conditions for rice, i.e., >25°C air temperature. It is very common to apply paddy rice herbicides as granules.

3. Test Plants

In Tables 1 and 2 the most important arable crops and weeds are listed with their codes[43] and scientific names. The lists are not complete and only give an overview on test plants usually used for screening herbicides in the greenhouse.

4. Evaluation Methods

The aim of screening tests is to separate phytotoxic compounds from those without activity. In most cases thousands of compounds are tested, and therefore a quick way of screening is essential.

The simplest way to evaluate the activity of a substance is to make visual observations of the treated plants compared to untreated. The following rating systems are currently used by various companies, as can be seen in patent applications, e.g., 0–9, 0–10, 1–5, 1–9, etc., indicating the damage/activity on a linear or logarithmic scale. The lower index means no activity, the higher index complete kill. A preferential rating system is the evaluation of percentage damage/activity (0 = 0% damage, 100 = 100% kill). Additionally, a letter code can be used to indicate special symptoms like burn, chlorosis, and defoliation.

Fresh weight and dry matter determinations, height or length measurements, and countings of weeds and/or crop plants are further methods to evaluate herbicides. They are, however, time-consuming and therefore not appropiate for routine tests.

From the above mentioned evaluations further results can be derived, like ED_{50} values (dosage giving 50% effect/activity) or the selectivity index S by calculating

$$S = \frac{ED_{50}\ \text{crop}}{ED_{50}\ \text{weed}} \tag{2}$$

For those compounds showing promise for further development, statistical analysis of the test results is recommended.

The regression analysis is a common method in the statistical analysis of bioassays.[44] Bioassays are usually summarized by ED_{50} and its confidence interval or standard deviation.[45] A model was developed to quantify herbicidal effects of applying nonlinear regressions to herbicide bioassay.[46]

C. GROWTH CHAMBER TESTS

After having selected candidate compounds with greenhouse and laboratory tests, it is appropriate to study the influence of environmental factors on the performance.

The behavior of herbicides may depend on a variety of factors such as: light, temperature, air humidity, soil moisture, irrigation, water capacity of soil, soil structure, adsorption, organic matter content, and plant availability in the soil.

Controlled environmental chambers or phytoboxes can be used to determine whether the above-mentioned conditions can influence the plant response either before, at the time of, or after herbicide application.[47] Techniques for investigating environmental effects have been published.[48] Such tests are useful to study particular environmental parameters. It is, however, sometimes difficult to simulate "natural" parameters in growth chambers, because plants

TABLE 1
List of Important Monocotyledonous Crops and Weeds and their Computer Codes[43] Used for Screening of Herbicides

Code	Genus
Crops	
AVESA	*Avena*
HORVX	*Hordeum*
ORYSS	*Oryza*
SECCS	*Secale*
SORSS	*Sorghum*
TTLSS	*Triticale*
TRZAX	*Triticum*
ZEAMX	*Zea*
Weeds	
AEGSS	*Aegilops*
AGRSS	*Agropyron*
AGSSS	*Agrostis*
ALOSS	*Alopecurus*
APESS	*Apera*
AVESS	*Avena*
BRASS	*Brachiaria*
BROSS	*Bromus*
COMSS	*Commelina*
CYNSS	*Cynodon*
CYPSS	*Cyperus*
DTTSS	*Dactyloctenium*
DIGSS	*Digitaria*
ECHSS	*Echinochloa*
ELUSS	*Eleusine*
ELOSS	*Eleocharis*
ERASS	*Eragrostis*
FESSS	*Festuca*
FIMSS	*Fimbristylis*
HETSS	*Heteranthera*
IMPSS	*Imperata*
ISCSS	*Ischaemum*
LEFSS	*Leptochloa*
LOLSS	*Lolium*
MOOSS	*Monochoria*
PANSS	*Panicum*
PASSS	*Paspalum*
PHLSS	*Phleum*
POASS	*Poa*
ROOSS	*Rottboellia*
SAGSS	*Sagittaria*
SCPSS	*Scirpus*
SETSS	*Setaria*
SORSS	*Sorghum*
SPDSS	*Sphenoclea*

grown under these conditions may be morphologically and physiologically different from those grown in the field.[47,49]

D. TESTS IN VEGETATION HALL

To overcome the problem of studying the performance of herbicides under nearly "natural" conditions, tests in a vegetation hall are recommended. A vegetation hall consists of a glasshouse and an open air area.

TABLE 2
List of Important Dicotyledonous Crops and Weeds and their Computer Codes[43] Used for Screening of Herbicides

Code	Genus
Crops	
ARHHY	*Arachis*
BEAVX	*Beta*
BRSSS	*Brassica*
GLXMA	*Glycine*
GOSHI	*Gossypium*
HELAN	*Helianthus*
LIUSS	*Linum*
SOLSS	*Solanum*
Weeds	
ABUSS	*Abutilon*
AMASS	*Amaranthus*
AMBSS	*Ambrosia*
ANVSS	*Anoda*
APHSS	*Aphanes*
BIDSS	*Bidens*
CAPSS	*Capsella*
CASSS	*Cassia*
CCHSS	*Cenchrus*
CHESS	*Chenopodium*
CIRSS	*Cirsium*
CONSS	*Convolvulus*
DATSS	*Datura*
DEDSS	*Desmodium*
ERYSS	*Erysimum*
EPHSS	*Euphorbia*
GAESS	*Galeopsis*
GASSS	*Galinsoga*
GALSS	*Galium*
HIBSS	*Hibiscus*
IPOSS	*Ipomoea*
KCHSS	*Kochia*
LAMSS	*Lamium*
LEPSS	*Lepidium*
LIDSS	*Lindernia*
MATSS	*Matricaria*
MENSS	*Mentha*
MERSS	*Mercurialis*
MOLSS	*Mollugo*
MYOSS	*Myosotis*
PHBSS	*Pharbitis*
PLASS	*Plantago*
POLSS	*Polygonum*
PORSS	*Portulaca*
RAPSS	*Raphanus*
ROTSS	*Rotala*
SASSS	*Salsola*
SEBSS	*Sesbania*
SIDSS	*Sida*
SINSS	*Sinapis*
SOLSS	*Solanum*
SONAS	*Sonchus*
STESS	*Stellaria*
THLSS	*Thlaspi*
VERSS	*Veronica*
VIOSS	*Viola*
XANSS	*Xanthium*

Plants are grown in pots (7 to 10 l volume) assembled on a revolving frame structure that can be moved from the open air area to the glasshouse and vice versa.[50] By this procedure plants can be grown under nearly field conditions, but they can be protected from adverse weather conditions (frost, wind, heavy rainfall, etc.) by easily moving them into the unheated glasshouse.

Selected herbicides can be tested by this way under semifield conditions. It is also possible to run long term trials over the whole vegetation period of a special crop, and in some cases, obtain yield data.[50]

E. FIELD TESTS

After glasshouse screening with bioassays, the aim of field tests is to select compounds suitable for practical use. The performance of a compound in the field will be influenced by environmental factors (e.g., light, moisture, wind, soil type), cultivation practices, application methods, plant morphology and physiology, and varietal differences. These factors can influence the performance dramatically. They can interact, and therefore it may take several years to investigate all aspects.

1. Field Screening

Screening tests in the greenhouse should be confirmed under field conditions. Furthermore, there should be a selection of the most interesting herbicidal compounds in respect to their weed spectrum and crop safety and their effective and selective doses. In the field, formulated test compounds are preferred. Such field screening can be done with small scale trials (plot sizes of 2 to 10 m^2), and the experimental design and test procedure should be simple and flexible.[51]

The formulated compounds can, in most cases, be applied with a knapsack sprayer[42] or with a lightweight bicycle sprayer.[51] Logarithmic sprayers have been used for such screening trials in the past, but spraying with constant dose rates is now preferred. To obtain information on the minimum rate necessary for weed control and the maximum tolerated dose by the crop, test series are designed with variable rates. In the field screening experiments, the speed of the manually operated sprayers is about 3.6 km h^{-1}, which is about half the speed of a tractor sprayer.[51]

Application of test compounds in the form of granules is the preferred method in paddy rice. Careful and experienced operators are indispensable for such applications. This is also true for applying herbicides in aquatic systems.

The field screening can be done with the relevant crop or with so called multicrop trials. In the latter case, several crops, e.g., cereals, rape, sugarbeets or soybeans, cotton, and maize, are sown side by side in rows and treated with different doses of the herbicide.

In addition to the native weeds, other dicot and monocot weeds can be sown in rows or distributed over the test field. Irrigation is recommended if needed to ensure adequate plant development and uptake of herbicides by plants. Several evaluations should be carried out during the test period. These are usually visual observations or ratings (on a decimal scale 0 to 100%) that compare the damage/weed control in treated plots to that in the untreated control. After having screened compounds in the described manner a few herbicides are selected for further tests.

2. Field Experiments

These tests are conducted over a period of about 2 to 4 years to confirm the performance of a herbicide under different cropping, soil, and climatic conditions. The experiments are normally run on medium sized plots (10 to 25 m^2) with a randomized block design. Statistical analysis of the obtained results is desirable. The crops should be grown under optimum farming conditions.[51] Special attention should be given to varietal or cultivar reactions to the applied herbicide in respect to phytotoxicity and yield. The number of such designed field experiments can exceed more than one hundred.

3. Large Scale Trials

These trials are carried out in the final stage of field evaluation. With such tests the performance of the product should be checked under normal agricultural conditions, i.e., with the equipment and method of the farmer.[51] The tested herbicide should be a part of the so-called "Integrated Pest Management" (IPM), which uses all available techniques and methods to maintain weed and pest populations below levels causing economic injury. New and effective herbicides can gain an important role in economic weed control systems as a part of IPM systems,[52] and the large scale trials offer the possibility to test new herbicides for the concept of weed control with economic threshold levels.[53]

Special trials are necessary to test herbicides for their use in perennial crops like citrus, orchards, vineyards, or plantation crops. In this context, cover crops must also be taken into consideration.

Field experiments in paddy rice are difficult to carry out because of cross-contamination between adjacent plots. This can be prevented only by building dikes between the plots or separating them with plastic walls.

4. Registration Trials

To obtain registration or approval for a new herbicide, special trials, and often official tests, have to be carried out. Most of the official trials are run at universities and/or government research stations. The requirements for such tests differ from country to country. In most cases trials from two growing seasons are necessary to get registration by the authorities. Timing of application, test requirements, size and number of plots, etc. are mostly defined in guidelines of national authorities and/or international organizations like the European Plant Protection Organization (EPPO). Special attention is paid to the yield and quality of the crop.

5. Application

Herbicides usually are formulated as wettable powders (WP), suspension concentrates (SC), dry flowables (DF), emulsifiable concentrates (EC), or granules (GR). In most cases the herbicidal formulations are applied with water as carrier. This is done with portable hand-operated or powered Knapsack sprayers, with tractor-powered sprayers, or with aircraft-mounted sprayers.[51,54] Special equipment such as controlled droplet applicators (CDA) or electrodyn sprayers are available, but currently they are not widely used to apply herbicides. Application of granules is performed by hand or machine, and it is very common in paddy rice.

III. BIOASSAYS OF FORMULATIONS AND MIXTURES

A. BIOASSAYS FOR TESTING FORMULATIONS

Under practical conditions herbicides are usually applied as formulated products rather than as active ingredient. Formulations should provide optimal conditions for the application, distribution, retention, and/or penetration of the herbicide. To compare the phytotoxicity and activity of a formulation with the pure chemical the "parallel-line-assay" is suitable (see Chapter 6).[46]

B. BIOASSAYS FOR HERBICIDE MIXTURES

In many cases herbicides are applied in mixture with other herbicides to improve the effectiveness or broaden the spectrum of weed control. Furthermore, herbicides can be mixed with safeners (antidotes, protectants), synergists, extenders, or adjuvants to improve herbicidal activity and/or selectivity.[55] Under practical conditions, mixtures with other agrochemicals such as insecticides, fungicides, fertilizers, etc. are quite common.[55] In all these cases the performance of such combinations must be checked with bioassays.

The joint action of such mixtures can be analyzed again with the method of parallel line assay.[56] Synergistic/antagonistic effects of combinations can be calculated (see Chapter 7).[57]

IV. TESTS TO STUDY THE ENVIRONMENTAL BEHAVIOR OF HERBICIDES

A. DISSIPATION OF HERBICIDES

After applying a herbicide its dissipation has to be checked.[58] Dissipation of a herbicide can be studied in laboratory, greenhouse, and field experiments, respectively.

Dissipation can be biotic and abiotic. Biotic dissipation is possible by microbial degradation[59] and uptake by plants.[60,61] Abiotic dissipation is governed by chemical degradation,[59] photodecomposition,[59,62] volatilization,[63,64] and adsorption to soil particles.[65] Most of these processes can be studied with instrumental methods, but bioassays are also a valuable tool[66-68] when instrumental methods are unavailable or too expensive.

Bioassays have advantages over instrumental techniques when information on the biological significance of the remaining herbicide, and/or its degradation product(s), is desired. This is especially true for studying herbicides persistence in the soil and the potential for carry-over on following rotational crops. Comparative studies of instrumental and bioassay methods for the analysis of herbicide residues in soils have shown a high correlation for most tested compounds.[69]

In case of differences between the two methods, the limited availability of a herbicide or the susceptibility of the test plant may be the reason. Half-life ($t_{1/2}$) of a herbicide is a common parameter in dissipation studies.[70]

B. LEACHING OF HERBICIDES IN SOILS

Nowadays there is much public concern over groundwater pollution with pesticides.[71-74] Registration authorities are therefore requiring intensive studies to clarify the behavior of pesticides in the soil. As a result of such investigations, strategies should be recommended to avoid groundwater contamination or carry-over effects on follow-up crops.

In this respect it should be mentioned that the limit of 0.1 μl^{-1} for any individual pesticide in drinking water set by the Council of European Communities[75] is an arbitrary and political one; it is not based on any toxicological considerations.

Mobility is a key process for the behavior of a herbicide in the soil. Therefore mobility studies are required by registration authorities.[76] In most cases model trials with standardized soil and irrigation are sufficient.

In the following sections, tests are described for studying the behavior of herbicides in the soil.

1. Soil Thick/Thin Layer Chromatography

In the case of thick layer chromatography, the mobility of a herbicide is checked by application on soil-filled leaching plates. After a measured amount of water has leached through the soil, seeds of susceptible test plants are placed on the soil. At those sites where the herbicide and/or phytotoxic metabolite(s) are located, germination or growth of the test plants is inhibited.[77]

Mobility studies can also be carried out with soil thin layer plates and the use of radioactive-labeled herbicides. In this case the radioactivity distribution on the soil plates is checked[78,79] and the mobility is classified from "none" to "high".

2. Soil Columns and Soil Cores

Glass cylinders filled with sieved soil are used for column studies.[80] The herbicide is applied on the top of the soil, and for 48 h the soil is irrigated with high amounts of water, which are not comparable with natural situations. Afterwards the herbicide eluted into the percolation water and remaining in different layers of the soil is analyzed. This can be done with analytical methods, with radioactive-labeled compounds, and last but not least, with bioassays.

Tests for studying the leaching of herbicides under more natural conditions can be carried out with undisturbed soil cores.[81]

3. Lysimeters

The most advanced system to study the behavior of a herbicide under nearly practical situations is the lysimeter test.[81,82] The distribution of an applied herbicide in the soil and in the leachate, and also the uptake by the crop and following crops, can be determined with lysimeter trials carried out under open air conditions. Using radioactive-labeled herbicides, a balance sheet after two vegetation periods under field conditions could be given.[83] Such lysimeter tests are very time consuming and expensive, therefore, their use should be limited to special cases.

An interesting attempt to clarify whether a pesticide has a potential for leaching or not is to calculate an index formed by its half-life in soil ($t_{1/2}$) and its partition coefficient between soil organic carbon and water (K_{oc}).[84]

Several models to assess the risk of groundwater or surface water contamination or the fate of herbicides in the soil have been proposed.[85-87] Some computer models, e.g., "Chemical Movement in Soils" (CMIS), "Pesticide Root Zone Model" (PRZM), and "Groundwater Loading Effects of Agricultural Management Systems" (GLEAMS), have been compared for leaching prediction.[88] Such simulations are a valuable tool in calculating possible risks, but they should not replace experiments or be the only basis for registration.

C. METABOLISM OF HERBICIDES

Knowledge of herbicide metabolism in soil, in/on crop plants, and in animals is required by registration authorities.[59,89] Common sense should be applied to limit the extent of required studies to levels that provide necessary and sufficient answers for risk assessment.[90]

1. Soil

Information on the pathway of transformation and the rapidity of degradation of the herbicide in several soil types, under aerobic and anaerobic conditions, is required.[59,91] In many cases the DT_{90} (time for 90% degradation of the parent compound) should be determined. If the DT_{90} is >100 days, further field studies are necessary, especially with respect to carry-over to following crops. Bioassays might be helpful in these studies.

2. Plants

Degradation and transformation studies in/on target plants are also essential.[89,92,93] The investigations are to be seen in context with the residue studies mentioned in the following section.

Different metabolic behavior in crop and weed plants, which may be a factor accounting for selectivity[94] or differential tolerance in cultivars,[95] are studied mainly with instrumental methods. Bioassays are involved only in determining the possible phytotoxic or herbicidal activity of main metabolites.

3. Animals

Metabolism and biokinetic studies in mammalian animals, especially rats and mice, are required for registration. Such studies must be seen in context with toxicological, ecotoxicological, and residue data, and they should help predict the potential effects of the proposed herbicide use on man and his environment.[96] Additionally, the toxicity of the main metabolite(s) in soil, plants, and animals must be checked.

D. RESIDUES OF HERBICIDES

1. Soil

After its application, a herbicide should have a certain length of biological activity to control weeds in certain crops. However, the herbicide should have disappeared at harvest to avoid phytotoxicity or carry-over problems to the following crop(s). Herbicides remaining in the soil after harvest are residues. They can be detected with analytical methods and/or with bioassays.

2. Persistence

Bioassays have proven invaluable in estimating whether a particular rotational crop can be grown on areas previously treated with a herbicide, e.g., a sulfonylurea herbicide.[97,98]

Persistence is not an inherent property of a particular herbicide, but it is dependent on many factors, such as pH, soil texture, water capacity, degradation, leaching, adsorption, etc. Persistence is the residence of a chemical in a specifically defined compartment of the environment.[99] Several attempts have been made to simulate persistence and forecast herbicide residues in soil.[100-102]

In this context "bound residues" should be mentioned, although they would not be expected to affect the natural processes of the soil because they are considered to be essentially biologically inert.[103]

3. Plants

The amounts of a herbicide that can be found at harvest in/on the crop plants or in/on parts of crops provided for consumption are called "residues". Residues can be the parent compound and/or its metabolites.

The "tolerated limit" is the maximum amount of a pesticide allowed in/on food. The tolerated limit is based on data from chronic toxicity studies. In chronic feeding studies, the amount of a pesticide, which induces no symptoms in rats after continuous admixture to the food, is determined. This amount called "no observable effect level"(NOEL) is expressed in mg kg^{-1} body weight. Dividing the NOEL by a security factor for man (usually 100) gives the "acceptable daily intake" (ADI) value. To get the "permissible level" the ADI value is multiplicated with 60 kg (medium body weight of man) and divided by the assumed daily consumption (in kg) of the respective food.

In many countries a further security factor ensures that the "tolerance" or tolerated limit of a pesticide is below the toxicologically permissible level. A tolerance is the legal maximum residue of a pesticide allowed in food and feed. The regulatory assessment of bound residues in plants is under consideration.[104]

For some compounds "waiting intervals" are laid down. They give the time between application of a pesticide and harvesting/consumption of the treated crop or crop part.

V. ECOTOXICOLOGICAL TESTS

The aim of ecotoxicological tests is to study the influence of a herbicide on aquatic and terrestrial biocoenoses, including nontarget organisms, and on soil fertility. First of all, the effects of a herbicide (pesticide) on representative organisms are tested in basic laboratory tests. These are followed by appropriate special tests that depend on the individual properties of the compound. Occurring effects have to be considered in view of their ecological and general acceptability.[105,106]

Standardized ecotoxicological tests are carried out in accordance with GLP (Good Laboratory Practice) regulations and special guidelines.

The tests are based on maximum bioavailability of the herbicide and intensive exposure of the test organism in order to obtain maximum toxicity of the substance. The potential hazard is calculated by relating the concentrations or doses of a pesticide to the toxicological effect data of the exposed organisms. If intolerable adverse effects cannot be excluded with sufficient certainty, further tests closer to practical conditions may be appropriate. In such tests the ecological significance of effects may also be assessed. The final step in the evaluation process is a risk/benefit analysis.

A. AQUATIC TESTS

It is important that algae, as primary producers of energy, not be seriously damaged by herbicide applications. Consequently, green algae are selected for ecotoxicological tests.

Generally, the unicellular algae *Scenedesmus subspicatus* or *Selenastrum* spp. are used as test species. Usually the EC_{50} (effective concentration giving 50% inhibition of growth or biomass), the NOEC (no observed effect concentration), and LOEC (lowest observed effect concentration) are determined using bioassays.

Daphnias, as primary planktonic consumers, are chosen for further tests. EC_{50}, NOEC, and LOEC are calculated with respect to mortality and, if appropriate, to reproduction.

Fish, as secondary consumers, can be affected by herbicides indirectly through the food chain (algae $\rightarrow$ daphnia $\rightarrow$ fish) or by direct uptake or sorption of the herbicide from contaminated water. Trout (*Salmo* spp.), carp (*Cyprinus* spp.), and golden orfe (*Leuciscus* spp.) are among the preferred test species. Biotests with fish, and aquatic model ecosystems or experimental ponds have been published.[107-109]

B. TERRESTRIAL TESTS

After having reached the soil, a herbicide may affect nontarget organisms and activities of the soil. Such effects may be inhibitory or stimulatory. The following nontarget organisms and/or their activities can be influenced: bacteria, fungi, algae, protozoa, nematodes, earthworms, collemboles, N-cycle, C-cycle, respiration, N-fixation, and dehydrogenase-activity.

Several tests are described or proposed to study the side effects of pesticides on soil organisms and activities.[110-115]

C. TESTS WITH OTHER NONTARGET ORGANISMS

For herbicides, the determination of the acute oral toxicity (LD_{50} = lethal dose for 50% of test organisms) for birds, e.g., quails (*Coturnix* spp.), is generally sufficient.

Honeybees may be affected from a post-emergence application of a herbicide to flowering weeds. Exposure can be by contact with spray droplets or by contact with spray deposit on plants. Appropriate laboratory and field methods for assessing the hazard of pesticides to bees are recommended.[116]

Guidelines for the assessment of the effects of pesticides on predatory and parasitic arthropods are available or under development.[117]

VI. TOXICOLOGICAL TESTS

Herbicides are, and will continue to be, an important part of world agriculture. The benefits of their correct use are enormous. Nevertheless, there are also risks by using herbicides and other pesticides. Therefore, much attention is paid to minimize the risks associated with the use of these agricultural chemicals.[118]

Regarding human health, laboratory studies with animals are the initial step for predicting potential hazards of pesticides, and the results form the basis for calculating the safety of agrochemicals to humans. Toxicity studies are therefore essential in registering a pesticide.[119] The requirements differ somewhat from country to country, but data on acute, subchronic and chronic toxicity are required, as well as mutagenicity, carcinogenicity, and teratogenicity studies.

Like ecotoxicological tests, all toxicological studies have to be carried out under GLP regulations in accordance with respective guidelines published by national and international organizations and authorities.

A. ACUTE TOXICITY

The purpose of acute toxicity tests is to define the median lethal dose (LD_{50}) or concentration (LC_{50}) and compare them in relation to known substances. Furthermore, initial information on the mode of toxic action can be obtained. Preferential test animals are rats, rabbits, and mice. Acute toxicity is based on the following tests: oral application, dermal application, inhalation, eye and dermal irritation, and dermal sensitization.

B. SUBCHRONIC TOXICITY

After determining its acute toxicity, the herbicide is then tested in subchronic studies. The test substance is administered to the test animals (rats or dogs) in their diet over a period of 90 days. Signs of toxicity, changes in body weight, organ weights, and blood composition, diet consumption, mortality, enzyme activities, and histopathology are parameters that are evaluated.[119] Neurotoxicity tests are carried out with hens.

C. CHRONIC TOXICITY

Chronic feeding studies are performed to evaluate the effects from long term exposure (1 to 2 years) to a herbicide. The test animals are rats and/or mice and a nonrodent species, usually dogs. With these tests the NOEL should be determined. As pointed out in Section IV.D.3, the NOEL is the basis for calculating the ADI value.

Furthermore the effects of a herbicide on carcinogenicity, embryotoxicity, teratogenicity, and on animal reproduction are required.

D. OTHER TOXICITY TESTS

Tests needed for getting registration also include mutagenicity studies with bacterial and nonbacterial systems. Several *in vitro* and *in vivo* methods are described.[120] The most meaningful method is the Salmonella/microsome test (Ames test).[121]

In addition to the described toxicological tests that are performed with the active ingredient (a.i.), several acute toxicity parameters must be evaluated with the formulated product.

VII. SUMMARY AND CONCLUSIONS

Weeds are competing with crop plants for space, water, light, and nutrients and thus causing tremendous yield reductions[122] or other disadvantages. The interference of weeds with the crop can be minimized by controlling the weeds with different methods. Chemical weed control with herbicides has proven to be the most successful and most economic one.[122]

Even after several decades of chemical weed control there is still a need for new herbicides, particularly in order to further increase the environmental and toxicological safety. Consequently, the evaluation of possible detrimental effects plays a key role in the development of a new herbicide. In 1988, the toxicological and ecobiological evaluations averaged about 32.6% of the total research and development costs for pesticides.[6] These high costs are justified in order to guarantee the safety of producers, applicators, consumers, and the environment; however, standardization and harmonization of test requirements are necessary to keep expenditures at an affordable level.

The biological performance of a new herbicide must also be evaluated thoroughly. Biological testing accounts for about 30.9% of the total costs.[6] Appropriate biotests are necessary for testing herbicides under laboratory, greenhouse, and/or field conditions.

Bioassays must be designed in such a way that they guarantee identification of herbicides with different properties. With the discovery of new classes of herbicides or new modes of action, the search for new bioassay methods is indispensable.

ACKNOWLEDGMENT

I thank Dr. R. H. Strang for his critical reading of the manuscript.

REFERENCES

1. **Giles, D. P.,** Principles in the design of screens in the process of agrochemical discovery. *Aspects of Applied Biology 21, "Comparing Laboratory and Field Pesticide Performance"*, 39, 1989.
2. **Vorpagel, E. R.,** Molecular modelling: a tool for designing crop protection chemicals, in *Pesticides, Minimizing the Risks*, Ragsdale N. N. and Kuhr, R. J., Ed., ACS Symp., Ser. 336, 115, 1987.
3. **Hurle, K.,** How to handle weeds? Biological and economic aspects, *Ecol. Bull.*, 39, 63, 1988.
4. **LeBaron, H. M. and Gressel, J., (Eds.),** *Herbicide Resistance in Plants*, John Wiley & Sons, New York, 1982.
5. **Martin, T.,** Broad versus narrow-spectrum herbicides and the future of mixtures, *Pest. Sci.*, 20, 289, 1987.
6. **Kraus, P.,** Global pest management in the future, in Pesticides: food and environmental implications, *Intern. Atomic Energy Agency IAEA — SM 297/38*, 1, 1988.
7. **Anon.,** *Agriculture, Towards 2000*, Food and Agriculture Organization, Rome, 2nd ed., 1987.
8. **Hurle, K. and Kemmer, A.,** Bioassays in weed science, *Ber. Fachg. Herbologie, Univ. Hohenheim*, 24, 153, 1983.
9. **Banki, L.,** *Bioassays of Pesticides in the Laboratory — Research and Quality Control*, Akademiai Kiado, Budapest, 1978.
10. **Horowitz, M.,** Application of bioassay techniques to herbicide investigation, *Weed Res.*, 16, 209, 1976.
11. **Gerber, H. R.,** Biotests for herbicide development, in *Crop Protection Agents and Their Biological Evaluation*, McFarlane, N. R., Ed., Academic Press, New York, 1977, 307.
12. **Arndt, W., Nobel, W., and Schweizer, B.,** *Bioindikatoren — Möglichkeiten, Grenzen und neue Erkenntnisse*, E. Ulmer Verl., Stuttgart, 1987.
13. **Weigel, H. J. and Jäger, H. J.,** Physiologische und biochemische Verfahren zum Nachweis von Schadstoffwirkungen, in Staub-Reinhalt, *Luft* 45, 269,1985.
14. **Anon.,** LI-6200 Portable Photosynthesis System; LI-1600 Steady State Porometer. LI-COR Inc., Lincoln, NE.
15. **Lichtenthaler, H. L.,** *Applications of Chlorophyll Fluorescence*, Kluwer Academic Publ., Dordrecht, 1988.
16. **Blaich, R., Bachmann, O., and Baumberger, J.,** Die Luminoskopie als zerstörungsfreie Methode zur Frühdiagnose von Herbizidschäden, *Ber. Fachg. Herbologie, Univ. Hohenheim*, 24, 119, 1983.
17. **Fedtke, C.,** Modes of herbicide action as determined with *Chlamydomonas reinhardii* and coulter counting, in *ACS Symp. Series 181, Biochemical responses induced by herbicides*, 1982, 231.
18. **Hoagland, D. R. and Arnon, D. I.,** The water culture method for growing plants without soil, *Circ. Calif. Agric. Station*, 347, 39, 1938.
19. **Fedtke, C.,** *Biochemistry and Physiology of Herbicide Action*, Springer-Verlag., 1982.
20. **Ashton, F. M. and Crafts, A. S.,** *Mode of Action of Herbicides*, John Wiley & Sons, New York, 1981.
21. **Moreland, D. E.,** Mechanism of action of herbicides, *Annu. Rev. Plant Physiol.*, 31, 597, 1980.
22. **Moreland, D. E.,** Measurements of reactions mediated by isolated chloroplasts, in *Research Methods*, Truelove, Ed., South. Weed. Sci. Soc., 1977.
23. **Truelove, B., Davis, D. E., and Jones, L. R.,** A new method for detecting photosynthesis inhibitors, *Weed Sci.*, 22(1), 15, 1974.
24. **Da Silva, J. F., Fadayomi, R. O., and Warren, G. F.,** Cotyledon disc bioassay for certain herbicides, *Weed Sci.*, 24, 250, 1976.
25. **Decleire, M., de Cat, W., and Bastin, R.,** Possibilite de detection d'herbicides dans l'eau par l'inhibition du verdissement des cotyledones de concombres etioles et exises. *Parasitica*, 29(4), 185, 1973.

26. **Bramley, P. M.,** The *in vitro* biosynthesis of carotenoids. *Adv. Lipid Res.,* 21, 243, 1985.
27. **Sauter, H.,** Pflanzliche Stoffwechselprofile: eine neue Testmethode für PF-Wirkstoffe, paper presented on a meeting in Bayreuth, West Germany, March 2–3, 1988.
28. **Vanstone, D. E. and Stobbe, E. H.,** Electrolytic conductivity — a rapid measure of herbicide injury, *Weed Sci.,* 25, 352, 1977.
29. **Fedtke, C.,** Physiological activity spectra of existing graminicides and the new herbicide 2-((2-benzothiazolyl-oxy)-*N*-methyl-*N*-phenylacetamide (mefenacet), *Weed Res.,* 27, 221, 1987.
30. **Chaleff, R. S. and Mauvais, C. J.,** Acetolactate synthase is the site of action of two sulfonylurea herbicides in higher plants, *Science,* 224, 1443, 1984.
31. **Shanner, D. L., Anderson, P. C., and Stidham, M. A.,** Imidazolinones (potent inhibitors of acetohydroxyacid synthase), *Plant Physiol.,* 76, 545, 1984.
32. **Schulz, A., Schönemann, P., Köcher, H., and Wengenmayer, F.,** The herbicidally active experimental compound HOE 704 is a potent inhibitor of the enzyme acetolactate reductoisomerase, *FEBS Lett.,* 228(2), 375, 1988.
33. **Ridley, S. M. and Mc Nally, S. F.,** Effects of phosphinothricin on the isoenzymes of glutamine synthetase isolated from plant species which exhibit varying degrees of susceptibility to the herbicide, *Plant Sci.,* 39, 31, 1985.
34. **Amrhein, N., Schab, J., and Steinrücken, H. C.,** The mode of action of the herbicide glyphosate, *Naturwissenschaften,* 67(7), 356, 1980.
35. **Owen, W. J.,** Use of plant cell and tissue cultures in studies of herbicide mode of action and metabolism, *Brighton Crop Prot. Conf.—Weeds,* 1989, 1197.
36. **Zilkah, S. and Gressel, J.,** Cell cultures vs. whole plants for measuring phytotoxicity, *Plant Cell Physiol.,* 18 (Pt. I); 641(Pt. II); 657(Pt.III), 815, 1977.
37. **Gressel, J., Zilkah, S., and Ezra, G.,** Herbicide action, resistance and screening in cultures vs. plants, in *Frontiers of Plant Tissue Culture,* Proc. 4th Int. Plant Tissue Cult. Congr., Calgary, 1978, 427.
38. **Camper, N. D. and McDonald, S. K.,** Tissue and cell cultures as model systems in herbicide research, *Rev. Weed Sci.,* 4, 169, 1989.
39. **Thiemann, J., Nieswandt, A., and Barz,W.,** A microtest system for the serial assay of phytotoxic compounds using photoautotrophic cell suspension cultures of *Chenopodium rubrum, Plant Cell Rep.,* 8, 399, 1989.
40. **Schmidt, R. R.,** Bioteste zur Erfassung sortenspezifischer Reaktionen von Sojabohnen auf Metribuzin, *Ber. Fachg. Herbologie, Univ. Hohenheim,* 24, 143, 1983.
41. **Hess, F. D.,** Relationship of plant morphology to herbicide application and absorption, in *Methods of Applying Herbicides,* McWhorter, C. G. and Gebhardt, M. R., Eds., Monograph 4 of the Weed Science Soc. of America, 19, 1987.
42. **Bode, L. E.,** Spray application technology, in *Methods of Applying Herbicides,* McWhorter, C. G. and Gebhardt, M. R., Eds., Monograph 4 of the Weed Science Soc. of America, 85, 1987.
43. **Anon.,** *Important Crops of the World and their Weeds (Scientific and Common Names, Synonyms and WSSA/WSSJ Approved Computer Codes),* Bayer, A.G., West Germany, 1986.
44. **Finney, D. J.,** Bioassay and the practice of statistical interference, *Int. Stat. Rev.,* 47, 1, 1979.
45. **Streibig, J. C.,** Herbicide bioassay, *Weed Res.,* 28, 479, 1988.
46. **Streibig, J. C.,** Measurement of phytotoxicity of commercial and unformulated soil-applied herbicide, *Weed Res.,* 24, 327, 1984.
47. **Devine, M. D.,** Environmental influences on herbicide performance: a critical evaluation of experimental techniques, *Proc. EWRS Symp. Factors Affecting Herbicidal Activity and Selectivity,* 1988, 219.
48. **Caseley, J. C.,** Techniques for investigation effects of weather on herbicidal performance, *Proc. EWRS Symp. The Influence of Different Factors on the Development and Control of Weeds,* 1979, 105.
49. **Garrod, J. F.,** Comparative responses of laboratory and field grown test plants to herbicides, *Aspects Appl. Biology,* 21, 51, 1989.
50. **Grosse-Brauckmann, E.,** Can a pot experiment be used for testing the yield capacity of cereals cultivars?, *Z. Acker Pflanzenbau,* 140, 214, 1974.
51. **Fryer, J. D. and Makepeace, R. J.,** *Weed Control Handbook,* Vol. I, 6th ed., Blackwell Scientific, Oxford, 1977.
52. **Auld, B. A., Menz, K. M., and Tisdell, C. A.,** *Weed Control Economics,* Academic Press, New York, 1987.
53. **Heitefuß, R., Gerowitt, B., and Wahmhoff, W.,** Development and implementation of weed economic thresholds in the F.R. Germany, *Brit. Crop. Prot. Conf.— Weeds,* 1987, 1025.
54. **Gangstad, E. O.,** *Weed Control Methods for Right-of Way Management,* CRC Press, Boca Raton, FL, 1982.
55. **Green, J. M. and Bailey, S. P.,** Herbicide interactions with herbicides and other agricultural chemicals, in *Methods of Applying Herbicides,* McWhorter, C. G. and Gebhardt, M. R., Eds., Monograph 4 of the Weed Science Soc. of America, 37, 1987.
56. **Streibig, J. C.,** Fitting equations to herbicide bioassays: using the method of parallel-line-assay for measuring joint action of herbicide mixtures, *Ber. Fachg. Herbologie. Univ. Hohenheim,* 24, 183, 1983.

57. **Colby, S. R.,** Calculating synergistic and antagonistic responses of herbicide combinations, *Weeds,* 15, 20, 1967.
58. **Nash, R. G.,** Dissipation from soil, in *Environmental Chemistry of Herbicides*, Vol.1, Grover, R., Ed., CRC Press, Boca Raton, FL, 132, 1989.
59. **Guth, J. A.,** The study of transformations, in *Interactions between Herbicides and the Soil,* Hance, R. J., Ed., Academic Press, New York, 123, 1980.
60. **Schmidt, R. R. and Pestemer, W.,** Plant availability and uptake of herbicides from soils, in *Interactions between Herbicides and the Soil,* Hance, R. J., Ed., Academic Press, New York, 179, 1980.
61. **Hance, R. J.,** Adsorption and bioavailability, in *Environmental Chemistry of Herbicides*, Vol.1, Grover, R., Ed., CRC Press, Boca Raton, FL, 1, 1989.
62. **Torstensson, N. T. L.,** Microbial decomposition of herbicides in soil, in *Progress in Pesticide Biochemistry and Toxicology,* Vol. 6, *Herbicides,* Hutson, D. J. and Roberts, T. R., Eds., John Wiley & Sons, New York, 249, 1987.
63. **Hance, R. J.,** Transport in the vapour phase, in *Interactions between Herbicides and the Soil,* Hance, R. J., Ed., Academic Press, New York, 59, 1980.
64. **Taylor, A. W. and Glotfelty, D. E.,** Evaporation from soils and crops, in *Environmental Chemistry of Herbicides*, Vol.1, Grover, R., Ed., CRC Press, Boca Raton, FL, 90, 1989.
65. **Calvet, R.,** Adsorption-desorption phenomena, in *Interactions between Herbicides and the Soil*, Hance, R. J., Ed., Academic Press, New York, 1, 1980.
66. **Kohn, G. K.,** Bioassay as a monitoring tool, *Residue Rev.,* 76, 99, 1980.
67. **Iwanzik, W. and Egli, H.,** Comparison of bioassay and chemical analysis for triasulfuron quantification in soil samples, *Brighton Crop Prot. Conf.— Weeds,* 1989, 1145.
68. **Hsiao, A. J. and Smith, A. E.,** A root bioassay procedure for the determination of chlorsulfuron, diclofop acid and sethoxydim residues in soils, *Weed Res.,* 23, 231, 1983.
69. **Eberle, D. O. and Gerber, H. R.,** Comparative studies of instrumental and bioassay methods for the analysis of herbicide residues, *Arch. Environ. Contam. Toxicol.,* 4, 101, 1976.
70. **Gallandt, E. R., Fay, P. K., and Inskeep, W. P.,** Clomazone disspation in two Montana soils, *Weed Technol.,* 3, 146, 1989.
71. **Boesten, J. J. T. I.,** Leaching of herbicides to ground water, a review of important factors and available measurements, *Proc. Brit. Crop Prot. Council Conf.— Weeds,* 1987, 559.
72. **Hance, R. J.,** Herbicide behaviour in the soil; with particular reference to the potential for groundwater contamination, in *Progress in Pesticide Biochemistry and Toxicology,* Vol. 6, *Herbicides,* Hutson, D. H. and Roberts, T. R., Eds., John Wiley & Sons, New York, 223, 1987.
73. **Milde, G. and Friesel, P.,** *Grundwasserbeeinflussung durch Pflanzenschutzmittel,* G.Fischer Verl., 1987.
74. **Tooby, T. E.,** Pesticide contamination of water — regulatory considerations, *Brighton Crop Prot. Conf.— Weeds,* 1989, 1165.
75. **Council of European Communities,** Council directive of 15 July 1980 relating to the quality of water intended for human consumption, *Off. J. Eur. Commun.,* No. L 229/11-29, 1980.
76. **Leistra, M.,** Transport in solution, in *Interactions between Herbicides and the Soil,* Hance, R. J., Ed., Academic Press, New York, 31, 1980.
77. **Gerber, H. R., Ziegler, P., and Dubach, P.,** Leaching as a tool in the evaluation of herbicides, *Proc. 10th Brit. Weed Contr. Conf.,* 1970, 118.
78. **Helling, C. S.,** Pesticide mobility in soils. III.Influence of soil properties, *Proc. Soil Sci. Soc. Am.,* 35, 743, 1981.
79. **Jotcham, J. R., Smith, D. W., and Stephenson, G.R.,** Comparative persistence and mobility of pyridine and phenoxy herbicides in soil, *Weed Technol.,* 3, 155, 1989.
80. **Anon.,** Versickerungsverhalten von Pflanzenschutzmitteln, *Biol, Bundesanstalt für Land- und Forstwirtschaft. Richtlinien für die amtl. Prüfung von Pflanzenschutzmitteln,* Teil IV, 4–2, 1986.
81. **Jarczyk, H. J.,** Behaviour of pesticides in soil as determined by undisturbed soil cores and lysimeter systems, *4th Intern. Congr. Pesticides Chemistry (IUPAC),* 1978, V-35.
82. **Jarczyk, H. J.,** Untersuchungen zum Sickerverhalten von Pflanzenschutzmitteln in einer Monolith-Lysimeteranlage, *Pflanzenschutz Nachr. Bayer*, 40, 49, 1987.
83. **Mittelstaedt, W. and Führ, F.,** [3-C^{14}] Metamitron-Vorauflaufspritzung zu Zuckerrüben im Freiland-Lysimeterversuch. Radioaktivitätsbilanz in Zuckerrüben, Folgekulturen und Boden, *Landwirtsch. Forsch.,* SH 37, 666, 1980.
84. **Gustafson, D. I.,** Groundwater ubiquity score: a simple method for assessing pesticide leachability, *Environ. Toxicol. Chem.,* 8, 339, 1989.
85. **Carter, A. D.,** The use of soil survey information to assess the risk of surface and groundwater pollution from pesticides, *Brighton Crop Prot. Conf. — Weeds*, 1989, 1157.
86. **Hutson, J. L. and Wagent, R. J.,** Predicting the fate of herbicides in the soil environment, *Brighton Crop Prot. Conf. — Weeds,* 1111, 1989.

87. **Leonhard, R. A.**, Herbicides in surface waters, in *Environmental Chemistry of Herbicides*, Vol.1, Grover, R., Ed., CRC Press, Boca Raton, FL, 45, 1989.
88. **Shaffer,R. D. and Penner, D.**, Evaluation of herbicide leaching prediction models, *Proc. North Cent. Weed Control Conf.*, 1988, 43, 59.
89. **Wolf, E.**, Die Prüfung und Zulassung von Pflanzenschutzmitteln (The testing and clearance of crop treatment agents), *Mitt. Biol. Bundesanst.*, 216, 1983.
90. **Sharp, D. B.**, Metabolism of pesticides — an industry view, in *Pesticide Science and Biotechnology*, Proc. 6th IUPAC Congr. of Pesticide Chemistry, Greenhalgh, R. and Roberts, T. R., Eds., Blackwell Scientific, Oxford, 483, 1987.
91. **Guth, J. A.**, Experimental approaches to studying the fate of pesticides in soil, in *Progress in Pesticide Biochemistry*. Vol. I, Hutson, D. H. and Roberts, T. R., Eds., John Wiley & Sons, New York, 85, 1981.
92. **Möllhoff, E.**, Experimental approaches to plant metabolism studies, in *Progress in Pesticide Biochemistry and Toxicology*, Hutson, D. H. and Roberts, T. R., Eds., John Wiley & Sons, New York, 29, 1985.
93. **Smith, A. E.**, Transformations in soil, in *Environmental Chemistry of Herbicides*, Vol.1, Grover, R., Ed., CRC Press, Boca Raton, FL, 172, 1989.
94. **Cole, D. J., Edwards, R., and Owen, W. J.**, The role of metabolism in herbicide selectivity, in *Progress in Pesticide Biochemistry and Toxicology*, Vol.6, *Herbicides*, Hutson, D. H. and Roberts, T. R., Eds., John Wiley & Sons, New York, 57, 1980.
95. **Fedtke, C. and Schmidt, R. R.**, Selective action of the new herbicide 4-amino-6-(1,1-dimethylethyl)-3-(ethylthio)-1,2,4-triazin-5(4H)-one in different wheat, Triticum aestivum, cultivars, *Weed Sci.*, 36, 541, 1988.
96. **Thomas, B.**, Pesticides and soil: a regulatory view, *BCPC Monograph 27, Symp. on Soils and Crop Protection Chemicals*, 1984, 15.
97. **Beyer, E. M., Duffy, M. J., Hay, J. V., and Schlueter, D. D.**, Sulfonylureas, in *Herbicides, Chemistry, Degradation and Mode of Action*, Vol.3, Kearney, P. C. and Kaufmann, D. D., Eds., Marcel Dekker, Inc., New York, 117, 1988.
98. **Walker, A. and Welch, S. J.**, The relative movement and persistence in soil of chlorsulfuron, metsulfuron-methyl and triasulfuron, *Weed Res.*, 29, 375, 1989.
99. **Greenhalgh, R., Baron, R. L., Desmoras, J., Ernst, R., Esser, H. O., and Klein, W.**, Definition of persistence in pesticide chemistry, *Pure Appl. Chem.*, 52, 2563, 1980.
100. **Hurle, K. and Walker, A.**, Persistence and its prediction, in *Interactions between Herbicides and the Soil*, Hance, R. J., Ed., Academic Press, New York, 83, 1980.
101. **Walker, A. and Barnes, A.**, Simulation of herbicide persistence in soil; a revised computer model, *Pestic. Sci.*,12, 123, 1981.
102. **Walker, A.**, Evaluation of a simulation model for prediction of herbicide movement and persistence in soil, *Weed Res.*, 27, 143, 1987.
103. **Calderbank, A.**, The occurrence and significance of bound residues in soil, *Rev. Environ. Contam. Toxicol.*, 108, 71, 1989.
104. **Kovacs, M. F., Jr.**, Regulatory aspects of bound residues (chemistry), *Residue Rev.*, 97, 1, 1986.
105. **Pflüger, W.**, Prüfung und Bewertung ökobiologischer Risiken im Pflanzenschutz, in *Toxikologische und klinisch-pharmakologische Prüfungen. Anforderungen, Methoden, Erfahrungen, Perspektiven*, Grosdanoff, P., Kraupp, O., Schütz, W., and Schulte-Hermann, R., Eds., Walter de Gruyter, Berlin, 343, 1990.
106. **Rothert, H., Brasse, D., and Bode, E.**, Process of weighing and decision making in the course of registration procedure of plant protection substances with special respect to side-effects on terrestrial fauna, *Gesunde Pflanze*, 42, 1, 29, 1990.
107. **Hansen, P. D.**, Bioteste — Stand und Entwicklung. Fischteste - wirkungsbezogene Biotestverfahren — ökologische Teste, in *Gewässerschutz, Schriftenreihe Verein Wasser- Boden- Lufthygiene*, Kühn, R. and Leschber, R., Eds., 57, 109, 1984.
108. **Gunkel, G.**, Untersuchungen zur ökotoxikologischen Wirkung eines Herbizids in einem aquatischen Modellökosystem. I. Subletale und letale Effekte, *Arch. Hydrobiol.* Suppl.65 (Monogr. Beiträge) 2/3, 235, 1983.
109. **Crossland, N. O. and Wolff, C. J. M.**, Outdoor ponds: their construction, management, and use in environmental toxicology, in *The Handbook of Environmental Chemistry*, Vol.2 (Part D), Hutzinger, O., Ed., Springer-Verlag, 1988.
110. **Greaves, M. P. and Malkomes, H. P.**, Effects on soil microflora, in *Interactions between Herbicides and the Soil*, Hance, R. J., Ed. Academic Press, New York, 223, 1980.
111. **Anderson, J. R.**, Some methods for assessing pesticide effects on non-target soil microorganisms and their activites, in *Pesticide Microbiology*, Hill, I. R. and Wright, S. J. L., Eds., Academic Press, New York, 247, 1978.
112. **Gerber, H. R., Anderson, J. P. E., Bügel-Mogensen, B., Castle, D., Domsch, K. H., Malkomes, H.-P., Somerville, L., Arnold, D. J., van de Werf, H., Verbeken, R., and Vonk, J. W.**, Revision of recommended laboratory tests for assessing side-effects of pesticides on the soil microflora.Bundesforschungsanstalt für Landwirtschaft, Braunschweig, 1989.

113. **Anderson, J. P. E., Castle, D., Ehle, H., Eichler, D., Laermann, H.-T., Maas, G., and Malkomes, H.-P.,** Richtlinie für die amtliche Prüfung von Pflanzenschutzmitteln, Teil VI, 1-1: Auswirkungen auf die Aktivität der Bodenmikroflora, Biol. Bundesanstalt, Braunschweig, 2. Aufl. 1990.
114. **Eijsackers, H. and van de Bund, C. F.,** Effects on Soil Fauna, in *Interactions between Herbicides and the Soil,* Hance, R. J., Ed., Academic Press, New York, 255, 1980.
115. **Huge, P. L.,** Chemische, physikalische und mikrobiologische Untersuchungen an drei Bodenarten nach achtjähriger Behandlung mit Tribunil, *Pflanz schutz Nachr. Bayer*, 34(2), 97, 1981.
116. **Felton, J. C., Oomen, P. A., and Stevenson, J. H.,** Toxiticy and hazard of pesticides to honey bees: harmonization of test methods, *Bee World,* 67, 114, 1986.
117. **Hassan, S. A.,** Standard methods to test the side-effects of pesticides on natural enemies of insects and mites developed by the IOBC/WPRS Working Group "Pesticides and Beneficial Organisms", *EPPO Bull.,* 15, 214, 1985.
118. **Young, A. L.,** Minimizing the risk associated with pesticide use: an overview, in *Pesticides, Minimizing the Risks,* Ragsdale, N. N. and Kuhr, R. J., Eds., ACS Sympos. Series, 1987, 336, 1.
119. **Cardona, R. A.,** Current toxicology requirements for registrations, in *Pesticides, Minimizing the Risk,* Ragsdale, N. N. and Kuhr, R. J., Eds., ACS Sympos. Series, 1987, 336, 14.
120. **Herbold, B. A. and Machemer, L. H.,** Mutagenitätsprüfung chemischer Agenzien. Grundlagen-Methoden-Folgerungen, *Pflanzen schutz Nachr. Bayer,* 34, 2, 153, 1981.
121. **Ames, B. N., Durston, W. E., Yamasaki, E.l, and Lee, F. D.,** Carcinogens or mutagens: a simple test system combining liver homogenates for activation and bacteria for detection, *Proc. Nat. Acad. Sci. U.S.A.,* 70, 2281, 1973.
122. **Eue, L.,** World Challenges in Weed Science, *Weed Sci.,* 34, 155, 1985.

Chapter 3

DOSE-RESPONSE CURVES AND STATISTICAL MODELS

Jens C. Streibig, Mats Rudemo, and Jens Erik Jensen

TABLE OF CONTENTS

0-8493-6603-8/93/$0.00+$.50

I. INTRODUCTION

The purpose of a biological assay, or bioassay, is to measure and compare the activities of physical, chemical, or biological stimuli by the response produced on living matter.[1-6] Its history goes back to early psychophysical measurements,[7] but the foundation of the modern theory of bioassay by statistical methods was laid in the '30s and '40s.[8-14]

A classical bioassay compares the potency of a test preparation and a standard preparation by application of dilution series to the experimental units. A basic technique is transformation of the dose and response variables, such that the transformed response is linearly related to the transformed dose with parallel lines for the test and standard preparations (Section II.D).

Although standardization is also a matter of concern in herbicide research, the object in herbicide bioassay is often more general: to estimate and compare the dose-response curves for a group of herbicides or mixtures of herbicides and relate the findings to the basic mechanisms of phytotoxicity.[15-17]

The importance of dose was already stated by Theophrastus Bombastus von Hohenheim (Paracelsus) more than 400 years ago.[18] The dose-response curve in Figure 1A below describes the effect of different herbicide doses on plant growth; sufficiently high doses stop growth or may kill the plants, and sufficiently low doses have no effect on growth. The range of plant growth in Figure 1A has an upper limit, which is usually defined by plant growth without herbicides, also called the untreated control, and a lower limit, which could be different from zero. The dose corresponding to plant growth half way between the upper and lower limit is defined as ED_{50}.

Sometimes plant responses are converted to percentage effect relative to growth of plants without herbicides. This is a common way to present results in the weed science literature, but it has some drawbacks. Firstly, the growth of plants without herbicides is subject to error like any other response describing the curve, and secondly, percentage growth or effect in relation to growth of plants without herbicides may hide large differences in plant growth, which renders comparison of interassay variation difficult.[19]

A common way to conduct bioassay experiments is to use a design with a constant dilution factor so that we get equidistant doses on a logarithmic scale (Section III.A). A log transformation of dose rates in Figure 1A usually gives close to a symmetric curve as shown in Figure 1B. This curve is symmetric around its ED_{50}. In our laboratory we have not yet found asymmetrically S-shaped dose-response curves for herbicides with whole plants in monoculture (see also Section I.C). However, in a field experiment with mixed stands of barley and rape, asymmetric response curves turned out to describe the effect of dose on plant growth better than did symmetrical curves.[20]

The response curve in Figures 1A and 1B is the logistic growth curve implying exact symmetry around ED_{50} when log dose is used on the horizontal axis. Another growth curve, that is sometimes used in bioassay, is the Gompertz curve, which is asymmetric around ED_{50}.[21]

The dose-response curve in Figures 1A and 1B describe the effect of herbicide doses at a given time, but another factor that influences the effect of a dose is the duration of exposure. Thus a complete description of the herbicide dose-response relationship should also include the effect on the growth curve. A hypothetical combined dose-response and growth surface is shown in Figure 2.[22]

In Figure 2, a dose-response curve is obtained by considering a plane parallel to the dose and response axes and perpendicular to the time axis. The growth curve at any one dose is sectioned parallel to the time and response axes and perpendicular to the dose axis. However, in the present treatise we will only consider dose-response models. To our knowledge generalized response surfaces as shown in Figure 2 have not been published for herbicides, probably because of the rather large experimental design needed.

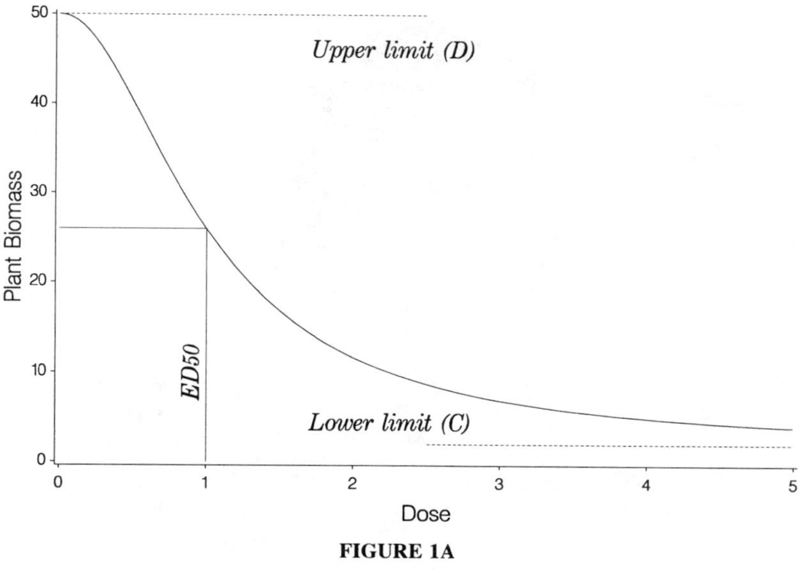

FIGURE 1A

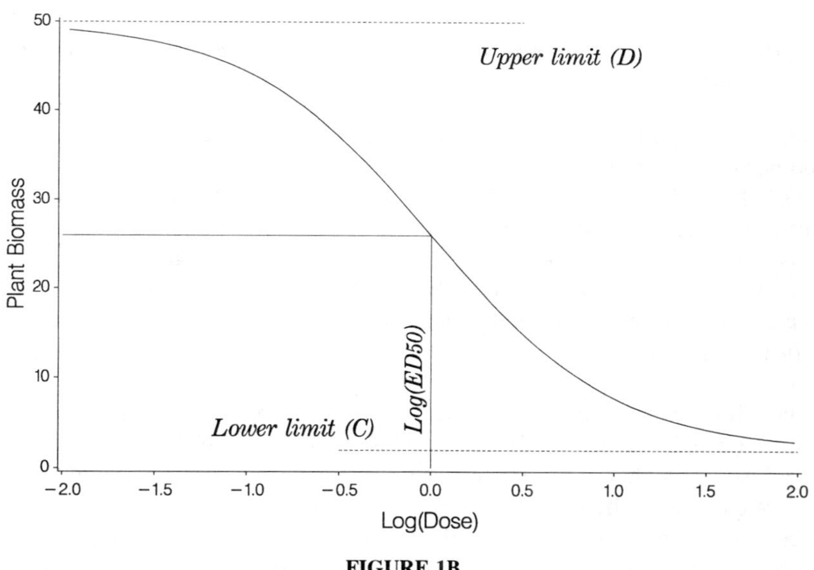

FIGURE 1B

FIGURE 1. The logistic dose-response curve. The figure shows (A) plant biomass plotted against dose (untransformed) and (B) plant biomass plotted against z = log(dose). The four parameter values used are the upper limit, $D = 50$, the lower limit, $C = 2$; $ED_{50} = 1$ and $b = 2$.

Experiments aimed at understanding the relationship and possible interactions between dose rate and duration of exposure are important if we wish to find the "optimal" time for spraying with the lowest possible dose.

Bioassays may be classified according to the type of response. The response of a plant to a herbicide dose is said to be quantal when the only recorded consequence of applying the herbicide is the presence or absence of a certain reaction, e.g., death, damage, infection etc. A quantal response may be expressed as an "all or none" variable taking the values 0 and 1, where 0 could denote undamaged and 1 damaged.

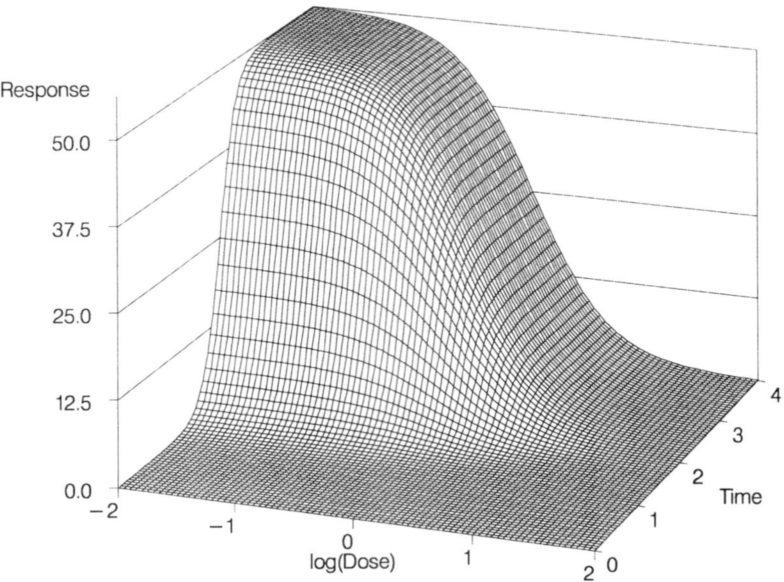

FIGURE 2. Hypothetical logistic dose-response growth surface. Log dose on x-axis, increasing to the right, time on y-axis, increasing inwards, and response on vertical axis.

If instead the response is graded, such as the biomass or height of plants, the contents of metabolites or the photosynthetic capture of CO_2, it is quantitative, which is the main type of assay used for herbicides. In a third type, called direct assay,[6] where the effect is basically all or none, it is possible to increase the amount applied to an experimental unit until a specific response is established. In this type of assay, for example when digitalis is applied to guinea pigs or cats until death occurs, it is usually found that the logarithm of the dose necessary to produce a given response is normally distributed.[23] Possible usage may be in some rapidly reacting *in vitro* systems, such as determination of I_{50} of photosynthesis using an oxygen electrode.

Let us now briefly describe the statistical methods used in quantal bioassay;[5,24] quantitative bioassays are treated in more detail in the following sections of this chapter. Let $z = \log x$ denote log dose (throughout this chapter we use the natural logarithm and not the logarithm of base 10 often used in the bioassay literature. This has, of course, no consequence except for difference in scale). Let $p(z)$ denote the probability of the "all" response, e.g., death of a plant, when the log dose z is applied. The probit model (probit suggested by Bliss[9] as a contraction of "probability unit") assumes that

$$p(z) = \Phi\left(\frac{z - z_0}{\sigma}\right) \tag{1}$$

where Φ denotes the cumulative distribution function of a standardized normal variant, z_0 denotes the median log dose, i.e., the dose for which half of the experimental units are expected to respond (often denoted ED_{50}, GR_{50}, I_{50}, or LD_{50}), and σ is inversely related to the steepness of the response curve at median log dose.

An alternative to the probit model is the logit model, for which it is assumed that

$$p(z) = \frac{\exp\left[b\left(z - z_0\right)\right]}{1 + \exp\left[b\left(z - z_0\right)\right]} \tag{2}$$

where z_0 denotes the median log dose as in Equation 1 and b measures the steepness of the response curve at median log dose. As will be further discussed in Section II.C, the probit and logit response curves are almost indistinguishable in practice. Therefore, the choice between the two models may be based on other considerations, such as ease of interpretation and mathematical or statistical convenience.[25]

From a statistical point of view, the logit model has some advantages as it gives an exponential family.[26,27] This simplifies computations considerably and allows the use of generalized linear models with true maximum likelihood.[28]

II. DOSE-RESPONSE CURVES

A. DOSE-RESPONSE CURVES IN GENERAL AND IN BIOASSAY

The relationship between a dose variable x, and a quantitative response variable y is often described by the model

$$y = f(x,\beta) + \varepsilon \tag{3}$$

where f is a function assumed to be known except for a p-dimensional parameter vector $\beta=(\beta_1,\ldots,\beta_p)$, and ε is a random error term. Thus, for the familiar straight line model we may put $\beta=(\beta_1,\beta_2)$, where β_1 is the intercept and β_2 is the slope. In the present section only the systematic part f of the dose-response model is considered.

The response function f may in some applications be deduced from theoretical considerations. This frequently holds true in physical and technical applications, but sometimes also in biology. For instance, in physiology and pharmacokinetics the differential equations for compartmental models may be solved. This gives the response as a function of time and dose with coefficients for exchange rates between compartments as unknown parameters.[29]

Most biological systems are so complex that current knowledge does not allow any deduction of theoretical response functions. This is the case for growth curves, where several different empirically motivated curve types have been studied.[30-32] A comprehensive discussion may be found elsewhere.[21]

The situation for herbicide dose-response curves is similar to that of growth curves; no theory seems available at present to deduce the form of the function f in Equation 3. This does not mean that the choice of f is unimportant. On the contrary, a carefully selected class of parametric functions can make the statistical analysis of bioassay data much more efficient than a casual choice. Some possible assumptions pertinent for response functions in herbicide bioassay will be discussed.

Let y denote weight or size of plants in response to x, an applied dose of a single herbicide. (Response curves for mixtures are considered in Chapters 6 and 7.) One natural requirement is that the expected weight or size of the test plants should decrease as dose increases,

$$f(x_2,\beta) < f(x_1,\beta) \quad \text{if} \quad x_2 > x_1 \tag{4}$$

that is $f(x,\beta)$ is a strictly decreasing function of x. For small nontoxic doses of many herbicides it turns out that plant growth may be stimulated,[33,34] and models have been suggested for response curves that include such a growth stimulation.[35] If the study of nontoxic effects or the so-called no-observable-effect level (NOEL) is not a major objective of an assay (see Chapter 8), it seems natural to choose a class of models satisfying Equation 4. Accordingly, we may choose the first nonzero dose level above the initial stimulating nontoxic levels if this stimulation is of appreciable size.

To compare the potency of a test preparation relative to a standard preparation, it is necessary to solve the equation

$$f(x,\beta) = y \tag{5}$$

for x given y and β. Then Equation 4 guarantees that the solution is unique if it exists. In such studies, Equation 4 is usually assumed to hold (see Section II.D).

The expected response should tend to a limit at very high doses. Thus it is natural to assume that a limit

$$\lim_{x \to \infty} f(x,\beta) = C(\beta) \tag{6}$$

exists and that $C(\beta)$ is either zero or a positive number. Typical examples with a positive limit in Equation 6 are assays where the test plants have already produced some dry matter before the herbicide has acted or because the herbicide, applied pre-emergence and taken up by the seed, first commences its action when the plant becomes autotrophic.[33]

B. POLYNOMIAL AND INVERSE POLYNOMIAL MODELS

In many applications where empirical response curves are needed polynomial response functions are employed, i.e., $f(x,\beta)$ in Equation 3 is a polynomial in x with coefficients forming the parameter vector β. Polynomials induce statistical models that are linear in the parameters, allowing use of easily available computer programs. By appropriately choosing the degree of the polynomial we get a class of response functions that flexibly adapt to observed values within the dose range studied. Outside the applied range of doses, however, the behavior of $f(x,\beta)$ is essentially out of control. In particular, the assumption that the limit in Equation 6 exists will not be satisfied except for a polynomial of degree zero.

The use of inverse polynomials, where the transformed response x/y is a polynomial in x, has been suggested.[36] For first and second order polynomials, the corresponding response curves are

$$y = \frac{x}{a + bx} \tag{7}$$

and

$$y = \frac{x}{a + b_1 x + b_2 x^2} \tag{8}$$

Equation 7 gives a response function that has been used in weed-crop competition studies,[37-38] and it is also well known in enzyme theory to describe Michaelis-Menten kinetics.[39] The inverse polynomial models give better control of the asymptotic behavior of y as x tends to infinity compared to polynomial models, but they do not seem to be much used in the analysis of bioassay data.

Section II.D describes another type of transformations that combined with a first order polynomial, have been widely used in the analysis of bioassays.

C. LOGISTIC AND RELATED MODELS

The four parameter logistic model

$$y = C + \frac{D - C}{1 + \left(x/x_0\right)^b} \tag{9}$$

is often used in herbicide bioassay and radioimmunoassay.[6,33-35,40,41] This response function may be written in several different ways. With z=log x we can write Equation 9 on the form

$$y = C + \frac{D - C}{1 + \exp\left[b\left(z - z_0\right)\right]} \tag{10}$$

where z_0=log x_0, i.e., the z value required (ED_{50}) to obtain a response half way between the upper limit, D, and the lower limit, C. The response functions Equation 9 and Equation 10 are shown in Figures 1A and 1B. On a log dose scale, the logistic curve is symmetric around the point $z=z_0$, where the slope is maximal.

If we look at the effect of time on dose-response curves (Figure 2), it is obvious that ED_{50} may depend on the duration of exposure. If the objective of a series of assays is to compare ED_{50}, we must do so at comparable stages of development of the test plant. Further, if the lower limit is greater than zero, the common practice of converting plant growth data to percent of growth in the untreated control may introduce bias. The advantage of using model Equation 10 is that ED_{50} for a series of independent experiments can be assessed on the basis of how well the test plants have grown in the untreated control, which is described by the upper limit, D.

The logistic curve meets the requirements discussed in Section II.A: it decreases monotonically and it has a limit as the dose tends to infinity. Furthermore, the parameters have simple interpretations: D is the initial response level for zero dose, C is the limit corresponding to infinitely large doses. The parameter z_0 is the log dose value corresponding to the point of inflexion, where a small increment in dose has a maximal effect on growth (see Equation 10 and Figure 1B). This dose is therefore often denoted ED_{50}, I_{50}, Gr_{50}, or LD_{50}. Accordingly, z_0 may be interpreted as the logarithm of the dose corresponding to 50% effect in Equation 9 (Figure 1A). With due caution ED_{50} may be used to compare toxicity between assays.[34] The fourth parameter, b, measures the steepness of the logistic curve around its point of inflexion.

The quantitative probit response function

$$y = C + (D - C)\left[1 - \Phi\left(\frac{z - z_0}{\sigma}\right)\right] \tag{11}$$

has a similar form as the probit response function for quantal data (Equation 1). In Figure 3 we have further drawn the two curves corresponding to Equation 10 and 11 with $\sigma=(2\pi)^{-1/2}b$, which implies that the curves have the same slopes at z_0. It is evident that the two curves are very close, and in practice we will almost never have enough observations to distinguish between the two types of response functions. The choice between Equation 10 and 11 is therefore essentially a matter of mathematical convenience, and the logistic curve has some advantages — we may, for instance, get an explicit solution of Equation 5. Further, it is possible to obtain the logistic function as a solution of a differential equation, which is useful to describe competition and growth,[42] and possibly it is useful in bioassay as well.

Bioassay data may sometimes deviate significantly from the hypothesis of symmetry in Equation 10. For such data, a five parameter logistic model

$$y = C + \frac{D - C}{\left[1 + \left(x/x_0\right)^b\right]^c} \tag{12}$$

has been suggested.[43]

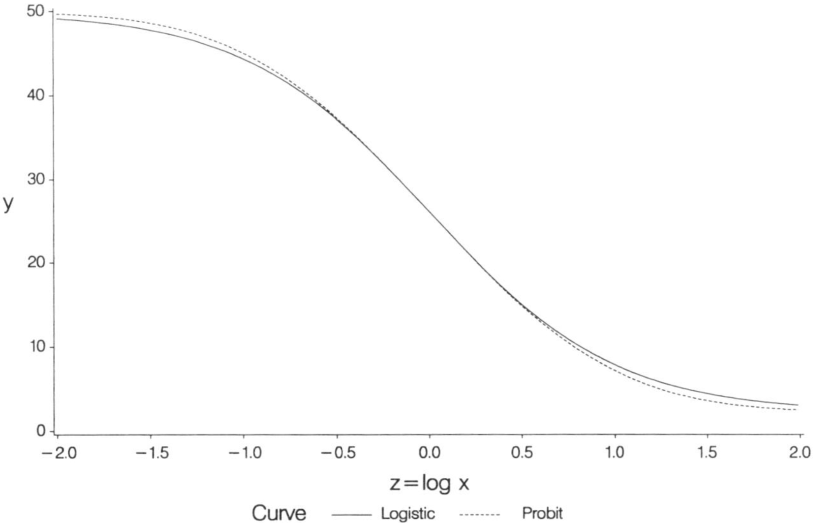

FIGURE 3. The logistic (Equation 10) and probit (Equation 11) curves shown in the same coordinate system. The response y is plotted against z = log(dose).

D. RELATIVE POTENCY — TRANSFORMATIONS TO STRAIGHT LINES

A single dose-response curve such as the one shown in Figure 2 does not in itself tell much about the efficacy and/or selectivity of a herbicide, because the assay to assay variation often is great. In a way bioassay can be compared with quantitative chemical analysis which implies the use of a standard to which an "unknown" can be related. The most straight forward use of bioassays is that of comparing biological activity of test compounds relative to some more or less arbitrarily chosen standard.

The relative potency is the exchange rate between herbicides and answers the question: "How much more or less of herbicide B must I apply to obtain the same effect as I get with x amount of herbicide A?" Conceptually, the relative potency resembles the exchange rate between currencies, e.g., 7 Dkr equals 1 $U.S. and thus the relative exchange rate is 7. At least in theory the exchange rate reflects the value of purchase of different currencies.

When the objective of an assay is to compare a different herbicides, the systematic part of the dose-response relationships may be written

$$y = f_i(x), \quad i = 1, \ldots, a \tag{13}$$

where f_i is the response function for the ith herbicide.

If the form of the a response curves are identical, except for individual scale factors, this means that there exist constants x_{0i}, i=1,...,a, and a function f that does not depend on i, such that

$$f_i(x) = f\left(\frac{x}{x_{0i}}\right), \quad x \geq 0, i = 1, \ldots, a \tag{14}$$

Response curves that satisfy Equation 14 are called similar. Typically x_{0i} may be chosen as the ED_{50} value of herbicide i.

For herbicides with similar response curves, the relative potency of herbicide i with respect to herbicide j is defined as

$$\rho(i|j) = \frac{x_{0j}}{x_{0i}} \tag{15}$$

With similarity we get, for all x, the same effect with $\rho(i/j)x$ units of herbicide j as with x units of herbicide i. If herbicide A is the reference, potencies of the other herbicides may be defined relative to herbicide A,

$$\rho_i = \frac{x_{01}}{x_{0i}} \tag{16}$$

The test of similarity hypotheses is an important part in all bioassays where we compare herbicides.[5,24]

When considering a parametric class of possible dose response curves, the similarity assumption (Equation 14) may be written

$$f_i(x) = f\left(\frac{x}{x_{0i}}, \gamma\right), \quad i = 1, \ldots, a \tag{17}$$

where γ is a vector of parameters that are common for all herbicides, while x_{0i} is the dose needed of herbicide i to achieve the effect $f(1, \gamma)$. This means that Equation 14 through 17 hold if the assayed herbicides contain the same active ingredient(s) and all other formulation constituents or adjuvants are biologically inactive. The relative potency is constant at any one response level. Conceptually, the relative potency is an exchange rate of herbicides having the same biological effect.

Before computer programs made it practicable to use maximum likelihood and nonlinear least squares analysis, transformations of variables to achieve linear dose-response curves were common in the analysis of bioassays. The linearizing transformation may be illustrated with the four parameter logistic response curve. The response curves for a herbicides are assumed to be

$$f_i(x) = C + \frac{D - C}{1 + \left(x/x_{0i}\right)^{b_i}}, \quad i = 1, \ldots, a \tag{18}$$

Note that the parameters C and D are the same for all herbicides, while the other two parameters, x_{0i} and b_i may vary between herbicides. To linearize the response curves of Equation 18, we transform the doses to

$$z = \log x \tag{19}$$

and the responses to

$$v = \log \frac{D - y}{y - C} \tag{20}$$

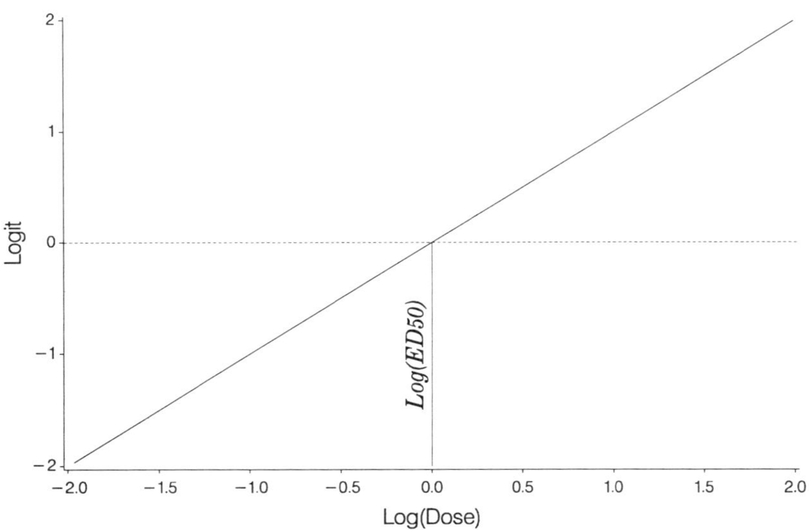

FIGURE 4. Logit-transformed responses based on Figure 1. See also Equation 20 and Equation 21 in the text.

Then the relations $y=f_i(x)$, $i=1,\ldots,a$, take the form

$$v = b_i\left(z - z_{0i}\right), \quad i = 1,\ldots, a \tag{21}$$

with $z_{0i}=\log x_{0i}$. Thus Equation 21 describes a family of straight line response curves in the transformed variables (Figure 4).

The similarity condition (Equation 14) now takes the form

$$b_1 = \cdots = b_a \tag{22}$$

which simply means that the lines in Equation 21 are parallel.

Assuming that the condition (Equation 22) holds,

$$f_i(x) = C + \frac{D-C}{1+\left(x/x_{0i}\right)^b}, \quad i = 1,\ldots,a \tag{23}$$

or, using Equation 16,

$$f_i(x) = C + \frac{D-C}{1+\left(\rho_i x/x_{0_1}\right)^b}, \quad i = 1,\ldots,a \tag{24}$$

This reparametrization is used in the statistical analysis of bioassay with similar response curves, sometimes called parallel line assays. For intrinsic nonlinear curves, the condition of similarity is sometimes called generalized parallelism.[43]

The discussion suggests the following simple procedure for analysis of quantitative bioassay data. Transform the dose and response according to Equations 19 and 20. Plot the data, test for parallel lines with well known methods for linear regression, and check with subjective methods the corresponding graphs. If the similarity hypothesis, here equivalent to the hypothesis of mutually parallel lines, is accepted, the horizontal distances between lines gives the logarithm of the corresponding relative potencies.

The method outlined above is widely used but suffers from one important drawback: the transformation Equation 20 depends on the parameters C and D, which typically are unknown and thus must be estimated from the data. This estimation has to be done in a rather peculiar way to ensure that Equation 20 is well defined; C must be chosen smaller and D must be chosen larger than all responses included in the data.

With the computer program packages available today, there are more satisfactory methods than the linearization method above, namely maximum likelihood and nonlinear least squares, described in Sections III.D and III.E.

For quantal data the linearizing transformation is in one way easier to handle, in that C=0 and D=1 are given and need not be estimated from the data. Then Equation 20 becomes the well known logit transformation. This is applied to relative frequencies of observations corresponding to the same dose. The maximal and minimal possible relative frequencies, y=1 and y=0, have to be treated with some care, as Fisher already observed in his early note on bioassay.[11] Today the best method of analysis for quantal bioassay data is the direct use of maximum likelihood.

The similarity of curves and hence constant relative potency between standard and test herbicides is a necessary but not sufficient condition for assuming biologically similar mechanisms of action. Consequently, assessing selectivity of herbicides with different modes of action in terms of relative potencies is more complicated. The relative potency now depends on the response level considered (see Chapter 4). Cornfield[44] advocated using nonparallel response curves and the corresponding functional relationship between the relative potency and dose. In herbicide research nonparallel curves are common, and in the future more interest should presumably be directed to estimation of dose-dependent relative potencies.

E. NONPARAMETRIC AND SEMIPARAMETRIC RESPONSE CURVES

In other branches of statistics and biometry, there has recently been an increasing interest in nonparametric methods that do not presuppose a particular parametric form for response curves. A notable example is nonparametric regression analysis of growth curves, and more generally, the nonparametric analysis of longitudinal data.[45]

It seems likely that these methods and, in particular, the closely related methods for self-modeling and shape invariant modeling could also be exploited in bioassay.[46-48] The general assumption of similarity such as in Equation 17 still applies, but the functional form of f is unspecified. The function f would then be estimated from the data, subject only to monotonicity assumptions such as in Equation 4. Methods based on isotonic regression models[49] may also be useful in bioassay.

As mentioned earlier, estimation of dose-dependent relative potencies is an important problem. It seems likely that one could devise nonparametric methods similar to those that have been developed for nonconstant shifts and hazard rates.[50,51]

For exploration and estimation of a response surface with several dose variables and for judging possible synergism, a nonparametric method based on level curve fitting has been suggested.[52] The recently developed "bootstrap" technique[66] is further proposed for assessing the fit.

III. EXPERIMENTAL DESIGNS AND STATISTICAL ANALYSIS

A. EXPERIMENTAL DESIGNS

All effort put into the design of experiments is to answer the question of how to allocate limited experimental resources optimally with respects to the topics being investigated.

Although most real problems require a series of experiments,[53] we will only discuss the design of single experiments.

Strategy in experimental design is based on knowledge of the specific application field and statistical methods. The effect of experimental error can be greatly reduced by appropriate design, and the experimental error can be estimated by a proper statistical analysis. This gives reliable measures of the precision of observed plant responses, which makes it possible to judge the validity of conclusions.

Let us consider an experiment that compares *a* different herbicides applied at different doses to plants. An experimental unit may consist of a group of plants in a pot. A common type of design is to use a minor modification of a randomized block experiment with *d* nonzero doses for each of the herbicides. Each block contains

$$n = n_0 + ad \tag{25}$$

units, where n_0 is the number of untreated control units in each block. If $n_0=1$ we have a conventional randomized block experiment with as many units in each block as there are treatments. The untreated control is, however, a reference for all the herbicides; therefore, it usually pays to have relatively more control units than units with other treatments. This is generally the case in experimental designs with a common control.[54] The total number of experimental units is

$$n_{tot} = rn = r\left(n_0 + ad\right) \tag{26}$$

where r is number of blocks.

One way to choose doses is first to specify a suitable maximal dose employed in the experiment for each herbicide, say a multiple of an estimated ED_{50}, and then to dilute the doses with appropriate dilution factors.

How should we choose the most appropriate dose ranges for different herbicides? Let x_{ij} denote the levels for herbicide i. The dilution method described above gives dose levels of the form

$$x_{ij} = t_j x_{0i}^*, \quad j = 1,\ldots,d \tag{27}$$

where x_{0i}^*, $i=1,\ldots,a$ are chosen so that x_{0j}^*/x_{0i}^*, is a preliminary estimate of the potency of herbicide i relative to herbicide j. We may for instance choose x_{0i}^* as a preliminary estimate of ED_{50} of the ith herbicide. Thus the doses (Equation 27) are scaled relative to the potencies of the different herbicides, and $t_1,\ldots,t_d$ are the normalized doses. A common type of design is to have equidistant doses on the log scale, that is

$$t_j = \Delta^{j-1} t_1, \quad j = 2,\ldots,d \tag{28}$$

where Δ denotes the dilution factor, and t_1 is the minimum normalized dose in relation to ED_{50}. In order to specify the design we then have to choose

$$n_0, r, d, t_1, \ldots, t_d \tag{29}$$

or

$$n_0, r, d, t_1, \Delta \tag{30}$$

if we use the log scale equidistant design (Equation 28).

The literature on optimal experimental designs for estimation of response curves and response surfaces is comprehensive.[55] Furthermore, special methods exist for optimal estimation in nonlinear regression models.[56] These methods should also be applicable for the response curves used in bioassay, in particular for the logistic curve.

For various reasons such as lack of space, manpower, etc. we may be forced to use block sizes smaller than the number of treatments, and in these situations the well developed theory of incomplete block designs could be exploited.[43,54] When comparing several herbicides with a standard and previous experience suggests similar response curves, designs with fewer doses for the tests than for the standard could be used.

B. ERROR STRUCTURE — TRANSFORMATIONS AND WEIGHTING

Equation 3 can be written on the form

$$y = f(x,\beta) + \sigma\varepsilon \tag{31}$$

where ε is assumed to have a standardized normal distribution, $N(0,1)$, and σ is the standard deviation.

If least squares is used instead of the maximum likelihood method, it is only necessary to assume that ε in Equation 31 has zero mean and unit variance. The response function $f(\cdot, \beta)$ is supposed to be given, e.g., as the logistic function Equation 9, except for the parameter vector β.

From a statistical point of view, two critical assumptions are constant variance, also called homoscedasticity, and normality of the residuals. The assumption of constant variance often turns out to be untenable. It may be detected by using plots of residuals discussed in Section III.E. If the variance is not constant we have heteroscedasticity, and the following weighting model may be employed,[57,58]

$$y = f(x,\beta) + \sigma g(x,\beta,\theta)\varepsilon \tag{32}$$

where $g(x,\beta,\theta)$ is supposed to be known except for the parameters β and θ, for instance

$$g(x,\beta,\theta) = \left[f(x,\beta)\right]^{\theta} \tag{33}$$

The weighting function Equation 33 has been studied for heteroscedastic assay data together with the four parameter logistic response curve.[59]

An alternative to the weighting model is the use of transformations, $h(\cdot,\lambda)$, e.g., the Box-Cox power transformation family[60]

$$h(y,\lambda) = \begin{cases} \left(y^{\lambda} - 1\right)/\lambda & \text{for } \lambda \neq 0 \\ \log y & \text{for } \lambda = 0 \end{cases} \tag{34}$$

where $\lambda=1$ is equivalent to untransformed. Actually, $\lambda=1$ corresponds to a linear displacement of y, but the results of, e.g., an analysis of variance, will be identical to those obtained without transformation.

Originally, Box-Cox transformation was only considered for the y variable, such as the transformed model (Equation 31)

$$h(y,\lambda) = f(x,\beta) + \sigma\varepsilon \tag{35}$$

This method has the drawback that in addition to adapting the residual distribution, it also changes the systematic part, i.e., the relation between y and x. If we want to keep the systematic part of the model unchanged, it is better to use the transform-both-sides method (TBS),[61-63] which gives the model

$$h(y,\lambda) = h\left[f(x,\beta),\lambda\right] + \sigma\varepsilon \tag{36}$$

Weighting and TBS methods also can be combined

$$h(y,\lambda) = h\left[f(x,\beta),\lambda\right] + \sigma g(x,\beta,\theta)\varepsilon \tag{37}$$

For Michaelis-Menten kinetics, where f is given by Equation 7, most of the methods of analysis suggested in the literature may be related to either weighting with

$$g(x,\beta,\theta) = x^{\theta} \tag{38}$$

or to TBS, or to combinations of the two methods.[39]

In all models discussed above, the random variation only applies to the response, while the dose has been assumed to be perfectly known. A modification of this assumption, interpreted either as a random effect in a controlled variable model or as a random coefficient effect, leads to models of the type in Equation 37[64] and will be briefly discussed in Section III.F.

C. STATISTICAL ANALYSIS

The purpose of developing mathematical dose-response curves is threefold: to describe assay data within acceptable limits, to test various hypotheses such as similar response curves of different herbicides, and to obtain estimates of the relative potencies with confidence intervals. Properly conducted assays with appropriately applied statistical methods may be used to summarize experimental data in a parsimonious way.

Let us consider a randomized block design with a herbicides and a logistic response curve for the ith herbicide,

$$f_i(x) = C_i + \frac{D - C_i}{1 + \left(x/x_{0i}\right)^{b_i}}, \quad i = 1,\ldots,a \tag{39}$$

This is slightly more general than Equation 18 because the lower limit, C_i, depends on i. Corresponding to zero dose, the upper limit, D, is of course the same for all herbicides. Based on experience discussed below, Equation 36 could be a tentative choice with $h(y, \lambda)$ given by Equation 34 and with either additive block effects or multiplicative block effects and residual errors as discussed below.

For an observation y corresponding to herbicide i, dose x, and a given block, the additive model is

$$h(y,\lambda) = h\left[f_i(x),\lambda\right] + \gamma_{\text{block}} + \sigma\varepsilon \tag{40}$$

with f_i from Equation 39. The residuals, ε, corresponding to different observations, are all supposed to be independent with a standardized normal distribution, a prerequisite for specifying the likelihood function.

Similarly, for multiplicative block effects and residual errors, the model is

$$h(y,\lambda) = h\left[f_i(x)\gamma_{\text{block}}\sigma\varepsilon,\lambda\right] \tag{41}$$

It may be argued that the multiplicative model (Equation 41) is more satisfactory than the additive model (Equation 40) from a biological point of view. The additive model might give negative estimates for the lower limit so that the model at high dose ranges predicts negative plant production. If block effects and random errors are relatively small, however, the differences between the two models are often negligible. The additive model gives a somewhat simpler statistical analysis.

Given the likelihood function and data, the maximum likelihood estimates of the parameters and the corresponding estimated residuals, $\hat{\varepsilon}$ are for the additive model

$$\hat{\varepsilon} = \frac{h\left(y,\hat{\lambda}\right) - h\left[\hat{f}_i(x),\hat{\lambda}\right] - \hat{\gamma}_{\text{block}}}{\hat{\sigma}} \tag{42}$$

where $\hat{\lambda}$ denotes the maximum likelihood estimate of a parameter λ, and $\hat{f}_i$ is is the response function (Equation 39) with all parameters replaced by their maximum likelihood estimates. A subjective test of the model could be performed by plotting residuals or studentized residuals in various ways.[63-65]

If the analysis of the residuals tentatively allows us to accept model Equation 40 or 41, the next step is to test if the hypothesis of similarity of the response curves is satisfied. For Equation 39 the similarity hypothesis takes the form

$$H_0:\ C_1 = \cdots = C_a,\quad b_1 = \cdots = b_a \tag{43}$$

This hypothesis may be tested by a likelihood ratio test. If the similarity hypothesis is accepted, we may proceed and compute confidence intervals for the relative potencies:

$$\rho_{\text{low}}(i|j) < \rho(i|j) < \rho_{\text{up}}(i|j) \tag{44}$$

Confidence intervals may be computed in several ways from estimated derivatives of the likelihood function.[63]

The different models considered above are nested in a natural way, and the corresponding hypotheses may be tested by using likelihood ratios, or equivalently, log likelihood differences. Under suitable assumptions, twice log likelihood differences, G^2, also called deviances, are asymptotically χ^2-distributed with df_{diff} degrees of freedom, where df_{diff} is the difference in the number of parameters between the alternative hypothesis and the null hypothesis.[28] An approximate analysis may be performed by nonlinear least squares, see for example Chapter 6 in Weisberg's book.[65]

D. COMPUTATIONAL CONSIDERATIONS

The models presented here are nonlinear, and the statistical analysis requires suitable optimization routines that use iterative methods. In order to complete the analysis, initial estimates of the parameters must be supplied. Good initial values of the parameters often allow an iterative technique to converge faster than would poor starting values. Poor starting values may result in convergence to an unwanted stationary point of the sum of squares or log likelihood surface, if there are local minima/maxima in addition to an absolute extremum. Furthermore, in practice, poor starting values often cause optimization routines to diverge.

With the logistic curves, starting values for β can rather easily be found by setting D equal to the maximum plant response of the untreated control and C equal to zero. If responses clearly indicate a lower limit greater than zero, then a value equal to or slightly smaller than the smallest observed plant response value may be used as a starting value for C.[33] A preliminary logit transformation of the plant response, according to Equation 20, plotted against z should give approximately straight line(s) and may be used to find good starting values of z_0=log(ED_{50}) defined at v=0 and the slope, b.

Most optimization routines will run faster and/or give more precise estimates and standard deviations if they are supplied with analytical first and maybe also second derivatives with respect to the model parameters. For nonlinear least squares, the derivatives of the response curve may be used. The partial derivatives of the response function (Equation 24) are given in the appendix (Section V).

In the maximum likelihood analysis, however, many routines use derivatives and second derivatives of the log likelihood, which can be difficult to calculate explicitly. Fortunately, most routines have an option permitting numerical determination of the derivatives. This makes optimization slower, but it minimizes programming work and allows optimization for a wide range of response functions.

Inference based on maximum likelihood or least squares nonlinear regression methods allows computation of approximate (asymptotic) confidence intervals and confidence regions for parameters. These methods may be too inaccurate or too complicated in some cases. Then resampling methods such as the bootstrap offer an alternative.[66,67] A systematic study comparing different methods, including the bootstrap method, for computing confidence intervals in nonlinear regression has recently been published.[68]

E. AN EXAMPLE WITH PHENOXY ACID HERBICIDES

To illustrate some of the statistical methods available an example is given in this section. The potencies of commercial formulations of MCPA (75% active ingredient), 2,4-D (50% active ingredient), mecoprop (50% active ingredient), and dichlorprop (66.7% active ingredient) were tested on white mustard (a=4).

The purpose of this assay was to test whether the response curves of four phenoxy acetic acids, having the same mode of action, were mutually similar, and if they were, to obtain the relative potencies for the compounds using MCPA as reference herbicide.

In the experiment, 25 seedlings in 6 l pots (surface area 133 cm^2) in greenhouse were sprayed (Hardi 4110-16 nozzle, 4 bar, and 275 l ha^{-1}) at their 2 to 2.5 true leaf stage and harvested 20 days after spraying.

The experimental layout was a randomized block design with r=3 blocks. The untreated control was replicated twice (n_0=2) within each block. There were d=6 nonzero doses based on active ingredient of each formulation. This gives 1+6 × 4=25 different treatments and a total of 78 observations (n_{tot}=3[2+4 × 6]=78).

The analysis was carried out in two steps using nonlinear least squares (procedure NLIN from the SAS® package) and maximum likelihood analyses (the MAXLIK module from the GAUSS™ mathematical and statistical language).

The maximum likelihood analysis has the advantage over nonlinear least squares in that it allows a wider range of transformation and/or weighting techniques with concomitant comparison of nested models. Therefore, the results presented here are based upon maximum likelihood analysis, but the data could also be analyzed by nonlinear least squares. The log

TABLE 1
Log Likelihood Values and the Number of Parameters for Various Models Analyzed

Model	Model number	Log likelihood value (no. of parameters)		
		no. block eff.	add. block eff.	mult. block eff.
Full model	—	352.50 (26)	348.29 (28)	349.20 (28)
Logistic response curves	(45)	359.61 (11)	356.13 (13)	356.69 (13)
Similar curves	(46)	360.91 (8)	357.55 (10)	358.24 (10)
Similar curves, TBS	(48)	356.33 (9)	352.50 (11)	353.22 (11)
Identical responses	(47)	377.27 (5)	375.10 (7)	375.73 (7)

Note: The so-called "full model" is the ordinary analysis-of-variance (ANOVA) model allowing all the 25 treatments (combinations of herbicides and doses) to have different responses. The improvement of fit after inclusion of block effects is fairly constant for the different models.

likelihood values and number of parameters for different models are summarized in Table 1.

In the following Equations 45, 46, and 47, y_{ijk} denotes the observed fresh weight affected by herbicide i=1,...,4, dose j=0,...,6 (j=0 corresponds to the untreated control), and block k=1,2,3. Note that not all 84 possible combinations of i, j, and k are represented. (We formally use i=1,2 and j=0 to denote the two controls in each block.)

To test the hypothesis that all herbicides had the same mechanism of action and thus similar response curves, the statistical analysis was performed in consecutive steps. In the first analysis the four response curves were assumed to have the same D and C but different b and x_0. The systematic part of the model is Equation 18. The block effects were included as additive effects corresponding to the statistical model

$$y_{ijk} = C + \frac{D - C}{1 + \left(x_j / x_{0i}\right)^{b_i}} + \gamma_k + \sigma\varepsilon_{ijk} \tag{45}$$

with 4 × 2+2+(3–1)+1=13 unknown parameters.

If Equation 45 describes the data just as well as an analysis of variance (ANOVA), then a lack of fit test should be nonsignificant. A full model (ANOVA) consists of 28 parameters, which is 15 more parameters than required by model Equation 45 (Table 1). Furthermore the parameters of Equation 45 can be interpreted biologically far better than the 28 parameters of the full model.

The likelihood ratio test for lack of fit compared to the full model with 28 parameters (25 treatments, 3 blocks, and variance) gave the test statistic G^2=2(356.13–348.29)=15.69, according to Table 3, which for the model is asymptotically χ^2-distributed with 28–13=15 degrees of freedom. The 95% percentile of the chi-square distribution is $\chi^2_{95,15}$=25.00. The P-value corresponding to the test statistic is P=40%, thus the reduced model showed no significant lack of fit. A constraint is laid on one of the blocks, in this case block number 3, due to overparameterization, which is illustrated for other models in Tables 2 and 3.

In this analysis the estimates of the four b_i (not shown) were rather close, as judged by their confidence limits, which indicates similar curves.

Even if the exact biochemical mechanism of phenoxy acetic acids is unknown, we tentatively expect that the herbicides tested act in the same way in the plant. This assumption corresponds to the reduced 10 parameter statistical model

TABLE 2
Summary of Parallel Line Regression (Equation 46) of White Mustard (*Sinapis alba* L.) Fresh Weight on MCPA (Reference), 2,4-D, Mecoprop, and Dichlorprop[a]

Parameter		Unit	Estimate	Standard error
D		g pot^{-1}	218.8	7.7
C		g pot^{-1}	24.7	13.1
b		—	1.60	0.25
MCPA	z_{01}	g ha^{-1}	59.8	9.4
2,4-D	ρ_2	—	1.58	0.27
Mecoprop	ρ_3	—	0.65	0.10
Dichlorprop	ρ_4	—	0.58	0.09
γ_1		g pot^{-1}	6.8	6.6
γ_2		g pot^{-1}	17.3	6.6
γ_3		g pot^{-1}	—	—
σ		—	23.7	1.9

Note: Additive block effects are included. The parameter estimates were obtained by maximum likelihood analysis. The most important parameters are the upper limit, D, z_{01} (ED_{50} for MCPA) and the relative potencies ρ_2, ρ_3, ρ_4 for 2,4-D, mecoprop, and dichlorprop.

[a] From Streibig 1989 (unpublished).

TABLE 3
Summary of Parallel Line Regression of White Mustard Fresh Weight on MCPA (Reference), 2,4-D, Mecoprop, and Dichlorprop[a]

Parameter		Unit	Estimate	Standard error
D		g pot^{-1}	214.8	9.7
C		g pot^{-1}	25.0	8.4
b		—	1.56	0.23
MCPA	z_{01}	g ha^{-1}	58.8	7.8
2,4-D	ρ_2	—	1.40	0.22
Mecoprop	ρ_3	—	0.65	0.09
Dichlorprop	ρ_4	—	0.58	0.08
γ_1		—	0.68	0.71
γ_2		—	1.5	1.3
γ_3		—	—	—
σ		—	1.9	1.6
λ		—	0.49	0.17

Note: Additive block effects are included. The transform-both-sides method was used corresponding to the model in Equation 48 (see text). The parameter estimates were obtained by maximum likelihood analysis.

[a] From Streibig 1989 (unpublished).

$$y_{ijk} = C + \frac{D - C}{1 + \left(x_j / x_{0i}\right)^b} + \gamma_k + \sigma\varepsilon_{ijk} \tag{46}$$

where all four curves have the same upper and lower limit, D and C, and the same steepness parameter, b.

Again, the test for the reduced model was nonsignificant, i.e., we can assume that the four curves are mutually similar. More precisely the likelihood ratio test yielded G^2=2.84, (df_{diff}=3), which is far from significant, $\chi^2_{.95,3}$=7.81, and P=42%. We conclude, in agreement with the above considerations, that the similar effect model gives a reasonable description of the data. Estimates and standard errors of the parameters are shown in Table 2.

The interesting parameters in Table 2 are the upper limit, D, the z_{01} (ED_{50}), and the relative potencies, ρ_2, ρ_3, ρ_4 of 2,4-D, mecoprop, and dichlorprop. The upper limit describes the plant growth without herbicides, which is rather important if we compare, e.g., ED_{50} and relative potencies obtained from different assays with the same test plants and herbicides. The relative potency of 2,4-D (ρ_2) was 1.58, which shows that 2,4-D was 1.58 times more potent than MCPA. An approximate 95% confidence interval for ρ_2 is obtained by adding plus/minus twice the standard error of 0.27, which renders the relative potency of 2,4-D significantly different from 1.00. Mecoprop and dichlorprop had almost identical relative potencies significantly smaller than unity, and they were thus less potent than MCPA.

The next question that arises is: Can the model be reduced any further? One straightforward and biologically meaningful reduction is

$$y_{ijk} = C + \frac{D-C}{1+\left(x_j/x_0\right)^b} + \gamma_k + \sigma\varepsilon_{ijk}, \quad i=1,\ldots,a, \quad j=1,\ldots,r \tag{47}$$

in which the response curves for the four herbicides are assumed to be identical. This model has seven parameters. The likelihood ratio test gave a G^2 value of 35.09 which, under the hypothesis that the model is correct, is χ^2-distributed with 3 degrees of freedom. Comparison with the fractile $\chi^2_{.999,3}$=16.27 shows that this is strongly significant (P=10^{-7}). Thus, the curves were not identical, and we conclude that the model (Equation 46) should be used for further analysis, which was also indicated by the confidence intervals for the relative potency parameters ρ_2, ρ_3, and ρ_4.

As previously mentioned, block effects may be included in at least two ways: additive block effects or multiplicative block effects. Additive blocks gave a slightly better fit than did multiplicative block effects. The differences, however, were small (Table 1).

It seems biologically more reasonable to include block effects as multiplicative. The predicted responses in such a model are not allowed to be negative. We have chosen, however, to use additive block effects because of computational advantages. Inclusion of block effects gave an improvement of fit in most models (significant at the 5% level). The P-values were between 3 and 7%.

A plot of residuals ($\hat{\varepsilon}$) vs. predicted values from Equation 46 indicated that variance was dependent upon the size of the response (Figure 5A), which seems quite likely when the plant production in the untreated control was more than 10 times greater than the production at the highest doses. To obtain variance homogeneity of the residuals and thus improve the test statistic and the confidence intervals of parameters, the TBS method (Equation 36) with $h(y, \lambda)$ given by Equation 34 was used. This model can be expressed as

$$h\left(y_{ijk},\lambda\right) = h\left[C + \frac{D-C}{1+\left(x_j/x_{0i}\right)^b}, \lambda\right] + \gamma_k + \sigma\varepsilon_{ijk} \tag{48}$$

The transformation significantly improved the fit (G^2=10.10, df_{diff}=1, and P=0.15%). The maximum likelihood estimate and standard errors of λ were $\hat{\lambda}$=0.49 and $s_{\hat{\lambda}}$=0.17, respectively (Table 3); hence, λ was significantly different from 1(no transformation) and 0 (log trans-

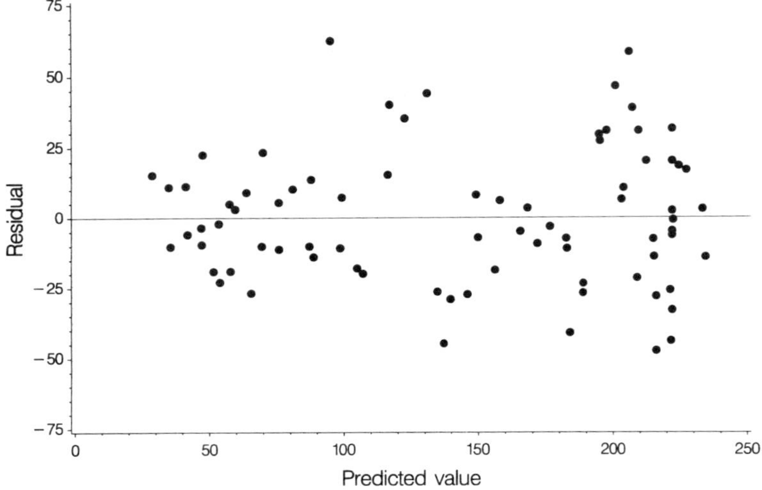

FIGURE 5A

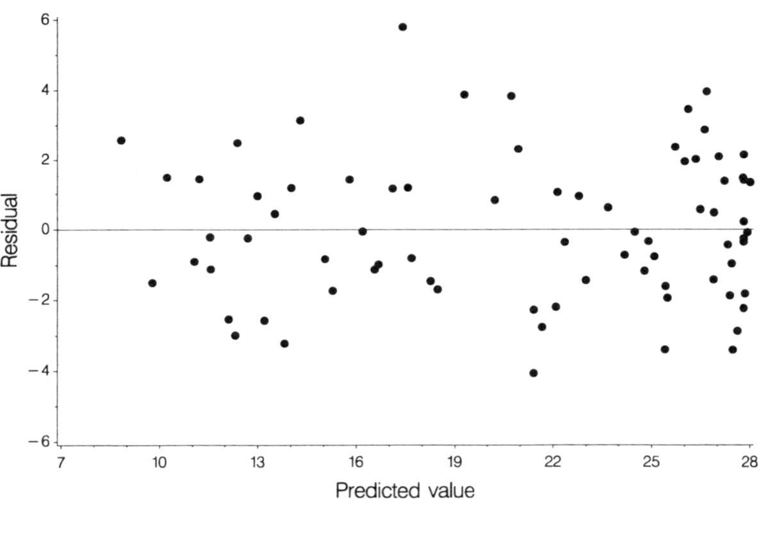

FIGURE 5B

FIGURE 5. Residual plots from A) the untransformed model (46), showing a tendency towards increasing variance for increasing predicted values, and B) model (48), where the transform both sides method (TBS) was used to remove the pattern seen in A.

formation). The size of λ indicated that a square root transformation would stabilize the variance. The residual plot for Equation 48 showed no further systematic deviations from variance homogeneity (Figure 5B).

The parameter estimates corresponding to the systematic part of the model were only slightly affected by the TBS method, but they tended to give lower standard errors of the parameters after transformation (Table 3).

The fitted curves for the reference herbicide (MCPA) and mecoprop are shown in Figure 6.

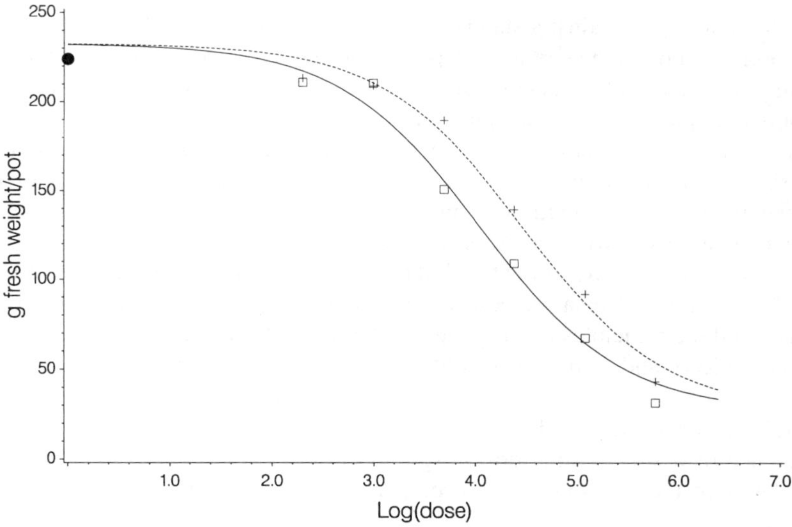

FIGURE 6. Regression curves for MCPA (Herbatox M750™: □), and mecoprop (Herbatox MP500™: +) (untreated control: ●) (From Streibig 1989, unpublished.)

The TBS technique in Table 3 did in fact change our conclusion about the relative potency of 2,4-D. In Table 2, it appeared to be significantly different from 1.00, but in Table 3 it is $1.40 \pm 2 \times 0.22$, barely significant from 1.00. The relative potencies of mecoprop and dichlorprop were unaffected by the transformation.

A summary of this assay would be that the four herbicides had similar response curves, which indicate that they might have similar a mechanism of action in the plant. Furthermore, the potencies of the herbicides could be ranked: 2,4-D ≥ MCPA > mecoprop = dichlorprop.

F. MODELS WITH RANDOM DOSE EFFECTS

The nonlinear regression models Equations 31 through 37 all have the error structure assembled in one random variable, ε. For some data, however, it is more realistic to consider a regression model with several variance components. In the analysis of response curves we may, for instance, include random errors in the dose. Two closely related random-effect models corresponding to additive and multiplicative effects of variation in dose have been suggested.[64] When the dose-response curve is nonlinear in dose, the variance components typically turn out to be identifiable even without replicate measurements of dose. In a bioassay example with four herbicides applied either as technical grades or as commercial formulations, a vast improvement in fit was obtained by comparing the TBS method to analyses with no weighting or transformation. For the data set studied, a further significant improvement in fit[64] was obtained by use of the variance component models.

If the error in dose is caused by a mixture error in the dilution process, it will be accumulated in a dilution series, which can be modeled by a random walk process.[69]

IV. DISCUSSION

A. FUNDAMENTAL AND STATISTICAL VALIDITY

Being rooted in agricultural traditions, herbicide research has not seriously dealt with fundamental and statistical validity of assay work.

Behind the experimental design of an assay and behind the final statements about the result such as alleged parallel lines, relative potency, etc. lie several assumptions. Some are of chemical and biochemical nature, others are about the nature of a test preparation whose potency is to be assayed against a standard. Assumptions about how herbicides generate the plant response, let alone the chemical properties of the herbicide and its biochemical action in the plant, are important aspects of bioassay work. Growth conditions of the plants and application techniques of herbicides play an important role in an array of events leading to the final conclusions for a bioassay. Mathematical and statistical assumptions are also essential. Jerne and Wood[14] pointed out that even philosophical assumptions about the relation between the theoretical abstraction of pure mathematics and the reality of bioassay are important for getting the whole picture of assay work.

The insistence of similarity of the test and the standard preparation were put forth by Jerne and Wood[14] and Finney[4,6] as a prerequisite for a fundamentally valid assay. Let us describe the approach and the principles as suggested by Jerne and Wood[14]. They distinguish between fundamental validity and statistical validity. A fundamentally valid assay requires

1. Validity of experimental design
2. Existence of a single value dose-response curve
3. Similarity of the test and standard preparations in a parallel line assay

The most important difference between herbicide bioassay and the general bioassay *sensu stricto* is that dose-response curves for herbicides also describe the selectivity of the compounds. Consequently, the similarity of test and standard preparation is only met in some instances for herbicides.

One object of the corresponding experimental designs is that the difference between several dose-response curves within an assay are wholly caused either by differences in dose or by random sampling. This quite natural assumption may of course be modified if certain factors, affecting the response, are known and recorded in such a way that their influence can be taken into account. For bioassays with whole plants, the uniformity and size of the plant material at the time of herbicide spraying is important. Also the number of plants per unit area is important. However, if the number of plants per experimental unit differ slightly, it is often of no consequence, because the plasticity of plants makes it possible for them to fill the space available. Plants can therefore compensate for small differences in the environment; however, assays, particularly with aerial application of herbicides, must, be quite accurately characterized as to stage of development and number of plants per unit area if we want to compare results from different experiments. The same applies to assays with soil acting herbicides, but here it is a difficult task to quantitatively assess the roots penetration of the soil volume. The stage of development at the time of transfer of seedlings to the nutrient solution must be properly described.

The existence of a single valued dose-response curve has been commented on in Section II.A. Many herbicides may stimulate growth at subtoxic doses. This growth can be modeled[35] but also easily masked if the objective is not the study of herbicides at subtoxic doses.

The similarity of herbicides assayed is a prerequisite if the assay should be classified as an analytical assay (see Section VI.B) in that the active constituents of the standard and test preparation are assumed to be similar. Thus, if the test is less potent than the standard, the relative potency is merely a measure of the dilution of the test compound.

The estimate of potency obtained in a fundamentally valid assay is independent of the experimental conditions or environment, the response measurement chosen, and the test species or variety of the test species used. This is a quite important statement that must be explicitly outlined whenever an estimate of relative potency is to be used in a generalized manner. Similarity as in Equation 14 must be true for any kind of test plant and any measurement chosen as response; $f(x)$ may well change with test plant or measurement, but the relative potency must be invariant. As Jerne and Wood[14] concluded: "Whenever it is possible to carry out assays using different test plants, responses or techniques, or to perform parallel analysis by chemical or physical methods, the additional evidence thus provided is most valuable."

The statistical validity of assays according to Jerne and Wood depends upon

1. The normal distribution of responses
2. Homoscedasticity of responses
3. The fact that the response model is "correct"

Today non-normal distributions and heteroscedastic responses can be effectively analyzed with high speed computers and good statistical programs, (Sections II and III).[63]

The "correctness" of a model is still based on the experimenter's knowledge of the test system. Some researchers distinguish between empirical and mechanistic models. An empirical model is described by a family of functions that is sufficiently flexible to describe the data reasonably well. A mechanistic model is described by a family of functions deduced from the kinetics of the mechanism of action of a herbicide producing the response. The distinction between the two types of models is blurred in complex systems, such as plant growth and herbicide action on whole plants. Models originally thought of as mechanistic are often based on such oversimplified assumptions that they are little more than empirical, yet, they may be given more credence because of their alleged mechanistic origin.

On the basis of the assumptions of fundamental validity and statistical validity, it can be concluded that if any of the assumptions of fundamental validity are untrue for an assay a "correct" answer cannot be obtained, whatever mathematical manipulation we apply. Conversely, if the statistical validity is untrue the assay data may be satisfactory, but the method of summarizing the findings of a potency estimate and its confidence interval is inappropriate and may lead to incorrect interpretation.

B. ANALYTICAL AND COMPARATIVE ASSAYS

With the concept of fundamental validity *sensu stricto*[6,14] as a starting point, one can distinguish between two types of assays: analytical assays, i.e., invariant relative potencies in different assay systems, and comparative assays, i.e., the relative potencies may change under differing experimental conditions.

If an assay should be classified analytical, all the assumptions in Section IV.A must be met. These quite rigorous assumptions are very seldom found in assays with herbicides and have not been explored in the weed science literature. An analytical assay often is used as a substitute for chemical analysis of compounds.

The only area where analytical assays are important in herbicide research is probably in determination of soil residues and in the newly developed immunological assays, which all compare dilution series of unknown samples to standard curves. With soil residues we often can assume that degradation products of a compound are biologically inert.[70,71] In this kind of work untreated soil used for preparation of the standard dose-response curve must be sampled at the same time as the soil with herbicide residues. Otherwise, plant growth in soils sampled at different times may vary and invalidate the common upper limit of responses for standard and test curves. If only one kind of test plant is used, we can never be sure whether the comparison of the standard with the test preparation is based upon one effective constituent only or on two or more in different proportions.

The purpose of conducting assays with herbicides is seldom to quantitatively analyze the active constituent per se, but to determine the selectivity of herbicidal compounds, be it in different test plants or of different herbicides. Herbicides are designed to kill plants and their selectivity is of great interest for the end user and the producer (Chapter 4). The best way to accumulate evidence about selectivity in early stages of herbicide development is to conduct well planned bioassay experiments. Consequently, most work on selectivity of herbicides can be classified as comparative assays. When working with herbicides with similar mode of action, but in different formulations and with different adjuvants, we may entertain the idea of similarity, although we know it need not be true (see Chapter 6). The diluents may be inert for some species but not for others. This is purposely used to change the selectivity of many herbicides (see Chapter 12). In such cases, the distinction between effective constituents and diluents tends to disappear.

C. STATISTICS AND BIOASSAY — POTENTIAL RESEARCH AREAS

The statistical models in bioassay are basically nonlinear, which caused a lot of problems before well developed programs for maximum likelihood and nonlinear least squares were widely available. This has now changed, although one can still find (June 1991) some large statistical packages without a general purpose maximum likelihood procedure.

Many other fields use intrinsically nonlinear models where a parallel development has taken place. The study of phenology and developmental stages for cereal crops is another area in biometry with heavily nonlinear models and a long history of computational tricks.[72]

Dose-response models and the corresponding statistical analysis seem to be potentially useful in many branches of herbicide research. They summarize the findings in a parsimonious way. They also allow checking the quality of data. The well established techniques from other branches of immunological assays could be used for the newly developed immunological assays for some herbicides.[40,73]

Some areas where the statistical methods for bioassay still seem to be insufficient and where future research is required are

1. Designs, including choice of dose levels, for optimal estimation of relative potencies
2. Optimal designs to test similarity
3. Tests of the similarity hypothesis without parametric assumptions
4. Estimation methods for nonparametric response functions under similarity hypothesis
5. Optimal designs for mixtures of herbicides (see Chapter 7)

V. APPENDIX

If the response function (Equation 24) corresponding to the parallel line assay is used, the response function can be formulated as

$$f(z) = C + \frac{D - C}{1 + e^{b\left[z + \log(\rho_i) - \log(x_{01})\right]}} \tag{49}$$

where $z=\log(x)$ and the partial first derivatives are given by

$$\frac{\partial f(x)}{\partial C} = 1 - \frac{1}{1 + e^{b\left[z + \log(\rho_i) - \log(x_{01})\right]}} \tag{50}$$

$$\frac{\partial f(x)}{\partial D} = \frac{1}{1 + e^{b\left[z + \log(\rho_i) - \log(x_{01})\right]}} \tag{51}$$

$$\frac{\partial f(x)}{\partial b} = -\frac{(D-C)\left[z+\log(\rho_i)-\log(x_{01})\right]e^{b\left[z+\log(\rho_i)-\log(x_{01})\right]}}{\left(1+e^{b\left[z+\log(\rho_i)-\log(x_{01})\right]}\right)^2} \tag{52}$$

$$\frac{\partial f(x)}{\partial \rho_i} = -\frac{b(D-C)e^{b\left[z+\log(\rho_i)-\log(x_{01})\right]}}{\rho_i\left(1+e^{b\left[z+\log(\rho_i)-\log(x_{01})\right]}\right)^2} \tag{53}$$

$$\frac{\partial f(x)}{\partial x_{01}} = \frac{b(D-C)e^{b\left[z+\log(\rho_i)-\log(x_{01})\right]}}{x_{01}\left(1+e^{b\left[z+\log(\rho_i)-\log(x_{01})\right]}\right)^2} \tag{54}$$

REFERENCES

1. **Bliss, C. I. and Cattell, McK.,** Biological assay, *Annu. Rev. Physiol.*, 5, 479, 1943.
2. **Gaddum, J. H.,** Bioassays and mathematics, *Pharmacol. Rev.*, 5, 87, 1953.
3. **Bliss, C. I.,** The principles of bioassay, *Am. Sci.*, 45, 449, 1957.
4. **Finney, J.,** The meaning of bioassay, *Biometrics,* 21, 785, 1965.
5. **Finney, J.,** *Probit Analysis,* 3rd ed., Griffin, London, 1971.
6. **Finney, J.,** *Statistical Methods in Biological Assay,* 3rd ed., Griffin, London, 1978.
7. **Fechner, G. T.,** *Elemente der Psychophysik,* Breitkopf and Härtel, Leipzig, 1860.
8. **Gaddum, J. H.,** Reports on biological standards. III. Methods of biological assay depending on a quantal response, *Med. Res. Counc., Spec. Rep. Ser.,* no. 183, 1933.
9. **Bliss, C. I.,** The method of probits, *Science,* 79, 38, 1934.
10. **Bliss, C. I.,** The method of probits — a correction, *Science,* 79, 409, 1934.
11. **Fisher, R. A.,** The case of zero survivors in probit analysis. Appendix to Bliss: the calculation of the dosage-mortality curve. *Ann. Appl. Biol.,* 22, 164, 1935.
12. **Finney, J.,** The principles of biological assay., *J. R. Stat. Soc.,* Suppl. 9, 46, 1947.
13. **Plackett, R. L. and Hewlett, P. S.,** Statistical aspects of the independent joint action of poisons, particularly insecticides. I. The toxicity of a mixture of poisons, *Ann. Appl. Biol.,* 35, 347, 1948.
14. **Jerne, N. K. and Wood, E. C.,** The validity and meaning of the results of biological assays, *Biometrics,* 5, 273, 1949.
15. **Blackman, G. E., Templeman, W. G., and Halliday, D. J.,** Herbicides and selective phytotoxicity, *Annu. Rev. Plant. Physiol.,* 2, 199, 1951.
16. **Blackman, G. E.,** Studies in the principles of phytotoxicity. I. The assessment of relative toxicity, *J. Exp. Bot.,* 3, 1, 1952.
17. **Sampford, M. R.,** Studies in the principles of phytotoxicity. II. Experimental designs and techniques of statistical analysis for the assessment of toxicity, *J. Exp. Bot.,* 3, 28, 1952.
18. **Paracelsus,** *Drey Bücher,* Heirs of Arnold Byrkmann, Cologne, 1564.
19. **Morse, P. M.,** Some comments on the assessment of joint action in herbicide mixtures, *Weed Sci.,* 26, 58, 1978.
20. **Christensen, S., Streibig, J. C., and Haas, H.,** Interaction between herbicide activity and weed suppression by spring barley varieties, in *Proc. EWRS Symp. 1990, Integrated Weed Management in Cereals,* 1990, 367.
21. **Seber, G. A. F. and Wild, C.,** *Nonlinear Regression,* Wiley, New York, 1989.
22. **Hartung, R.,** Dose-response relationships, in *Toxic Substances and Human Risk,* Tardiff, R. G. and Rodricks, J. V., Eds., Plenum Press, New York, 1987, 29.

23. **Bliss, C. I.,** The U. S. P. collaborative cat assay for digitalis, *J. Am. Pharm. Assoc.,* 33, 225, 1944.
24. **Hewlett, P. S. and Plackett, R. L.,** *The Interpretation of Quantal Responses in Biology,* Edward Arnold, London, 1979.
25. **Berkson, J.,** Why I prefer logits to probits, *Biometrics,* 7, 327, 1951.
26. **Haberman, S. J.,** Maximum likelihood estimates in exponential response models, *Ann. Stat.,* 5, 815, 1977.
27. **Cox, D. R. and Snell, E. J.,** *Analysis of Binary Data,* 2nd ed., Chapman and Hall, London, 1989.
28. **McCullagh, P. and Nelder, J. A.,** *Generalized Linear Models,* Chapman and Hall, London, 1983.
29. **Jacquez, J. A.,** *Compartmental Analysis in Biology and Medicine,* Elsevier, New York, 1972.
30. **Robertson, L. B.,** On the normal rate of growth of an individual and its biochemical significance, *Roux' Arch. Entwicklungsmech. Organismen,* 25, 581, 1908.
31. **von Bertalanffy, L.,** Quantitative laws in metabolism and growth, *Q. Rev. Biol.,* 32, 217, 1957.
32. **Richards, F. J.,** A flexible growth function for empirical use, *J. Exp. Botany,* 10, 290, 1959.
33. **Streibig, J. C.,** Models for curve-fitting herbicide dose response curves, *Acta Agric. Scandi.,* 30, 59, 1980.
34. **Streibig, J. C.,** Herbicide bioassay, *Weed Res.,* 28, 479, 1988.
35. **Brain, P. and Cousens R.,** An equation to describe dose responses where there is stimulation of the growth at low doses, *Weed Res.,* 29, 93, 1989.
36. **Nelder, J. A.,** Inverse polynomials, a useful group of multi-factor response functions, *Biometrics,* 22, 128, 1976.
37. **Vleeshouwers, L. M., Streibig, J. C., and Skovgaard, I.,** Assessment of competition between crops and weeds, *Weed Res.,* 29, 273, 1989.
38. **Fredshavn, J., Jørnsgaard, B., and Streibig, J. C.,** Assessment of crop weed competition under greenhouse and field conditions, in *Proc. EWRS Symp. 1990, Integrated Weed Management in Cereals,* 1990, 239.
39. **Ruppert, D., Cressie, N., and Carroll. R. J.,** A transformation/weighting model for estimating Michaelis-Menten parameters, *Biometrics,* 45, 637, 1989.
40. **Finney J.,** Radioligand assay, *Biometrics,* 32, 721, 1976.
41. **De Lean, A., Munson, P. J., and Rodbard, D.,** Simultaneous analysis of families of sigmoidal curves: applications to bioassay, radioassay, and physiological dose-response curves, *Am. J. Physiol.,* 235, E97, 1978.
42. **Skovgaard, I.,** Models for growth of competing plant populations (in Danish). Tech. Note, Royal Agricultural and Veteterinary University, Copenhagen, 1990.
43. **Finney J.,** Bioassay and the practice of statistical inference, *Int. Stat. Rev.,* 47, 1, 1979.
44. **Cornfield, J.,** Comparative bioassays and the role of parallelism, *J. Pharmacol. Exp. Ther.,* 144, 143, 1964.
45. **Müller, H. G.,** *Nonparametric Analysis of Longitudinal Data,* Springer, New York, 1988.
46. **Lawton, W. H., Sylvestre, E. A., and Maggio, M. S.,** Self-modeling nonlinear regression, *Technometrics,* 14, 513, 1972.
47. **Stuetzle, W., Gasser, Th., Molinari, L., Largo, R. H., Prader, A., and Huber, P. J.,** Shape-invariant modelling of human growth, *Ann. Human Biol.,* 7, 507, 1980.
48. **Kneip, A. and Gasser, T.,** Convergence and consistency results for self-modelling nonlinear regression, *Ann. Stat.,* 16, 82, 1988.
49. **Barlow, R. E., Bartholomew, D. J., Bremner, J. M., and Brunk, H. D.,** *Statistical Inference under Order Restrictions,* Wiley, New York, 1972.
50. **Doksum, K. A. and Sievers, G. L.,** Plotting with confidence: graphical comparison of two populations, *Biometrika,* 63, 421, 1976.
51. **Dabrowska, D. M., Doksum, K. A., and Song, J-K.,** Graphical comparison of cumulative hazards for two populations, *Biometrika,* 76, 763,1989.
52. **Tibshirani, R.,** Smoothing methods for the study of synergism, *Bull. Int. Stat. Inst.,* Proc. 47th Session, Book 3, 555, Paris, 1989.
53. **Box, G. E. P., Hunter, W. G., and Hunter, J. S.,** *Statistics for Experimenters,* Wiley, New York, 1978.
54. **Cochran, W. G. and Cox, G. M.,** *Experimental Designs,* 2nd ed., Wiley, New York, 1957.
55. **Mead, R. and Pike, D. J.,** A review of response surface methodology from a biometrics viewpoint, *Biometrics,* 31, 803, 1975.
56. **Box, G. E. P. and Lucas, H. L.,** Design of experiments in nonlinear situations, *Biometrika,* 4675, 77, 1959.
57. **Carroll, R. J. and Ruppert, D.,** A comparison between maximum likelihood and generalized least squares in a heteroscedastic linear model, *J. Am. Stat. Assoc.,* 77, 878, 1982.
58. **Davidian, M. and Carroll, R. J.,** Variance function estimation, *J. Am. Stat. Assoc.,* 82, 1079, 1987.
59. **Davidian, M., Carroll, R. J., and Smith, W.,** Variance functions and minimum detectable concentrations in assays, *Biometrika,* 75, 549, 1988.
60. **Box, G. E. P. and Cox, D. R.,** An analysis of transformations, *J. R. Stat. Soc.,* B26, 211, 1964.
61. **Carroll, R. J. and Ruppert, D.,** Power transformations when fitting theoretical models to data, *J. Am. Stat. Assoc.,* 79, 321, 1984.

62. **Snee, R. D.,** An alternative approach to fitting models when reexpression of the response is useful, *J. Qual. Technol.,* 18, 211, 1986.
63. **Carroll, R. J. and Ruppert, D.,** *Transformation and Weighting in Regression,* Chapman and Hall, London, 1988.
64. **Rudemo, M., Ruppert, D., and Streibig, J.,** Random-effect models in nonlinear regression with applications to bioassay, *Biometrics,* 45, 349, 1989.
65. **Weisberg, S.,** *Applied Linear Regression,* 2nd ed., Wiley, New York, 1985.
66. **Efron, B.,** The Jackknife, the Bootstrap and other Resampling Plans, CBMS-NSF, Monogr. 38, Soc. Ind. Appl. Math., Philadelphia, 1982.
67. **Wu, C. F. J.,** Jackknife, bootstrap and other resampling methods in regression analysis (with discussion), *Ann. Stat.,* 14, 1261, 1986.
68. **Huet, S., Jolivet, E., and Messéan, A.,** Some simulation results about confidence intervals and bootstrap methods in nonlinear regression, *Statistics,* 21, 369, 1990.
69. **Racine-Poon, A., Weihs, C., and Smith, A. F. M.,** Estimation of relative potency with sequential dilution errors in radio-immuno-assay, *Biometrics,* 47, 1235, 1991.
70. **Nyffeler, A., Gerber, H.-R., Hurle, K., Pestemer, W., and Schmidt, R. R.,** Collaborative studies of dose-response curves obtained with different bioassay methods for soil-applied herbicides, *Weed Res.,* 22, 213, 1982.
71. **Krauskopf, B., Wetcholowsky, I., Schmidt, R. R., Blair, A. M.,Anderson-Taylor, S. B., Eagle, D. J., Friedländer, H., Hacker, E., Iwanzik, W., Kudsk, P., Labhart, C., Luscombe, B. M., Madafiglio, G., Martin, T. D., Nel, P. C., Pestemer, W., Rahman, A., Retzlaff, G., Rola, J., Schmider, F., Stefanovic, L., Straathof, H. J. M., Streibig, J. C., Thies, E. P., Wakerly, S. B., and Walker, A.,** Collaborative bioassays to monitor the behavior of metsulfuron-methyl and metribuzin in the soil, in *BCPC Monogr. No. 47 Pesticides in Soils and Water,* 1991, 109.
72. **Jensen, J. E. and Rudemo, M.,** First passage time models for prediction of developmental stages, *Biometrical J.,* 33, 3, 1991.
73. **Vølund, A.,** Application of the four-parameter logistic model to bioassay: comparison with slope ratio and parallel line models, *Biometrics,* 34, 357, 1978.

Chapter 4

ASSESSMENT OF HERBICIDE SELECTIVITY

Michael R. Bartley

TABLE OF CONTENTS

0-8493-6603-8/93/$0.00+$.50

I. INTRODUCTION

Using herbicides that can be applied for weed control within a crop without causing undue damage to that crop has become increasingly important since 1941. In that year it was discovered that the salts of the chlorinated phenoxyacetic acids could provide excellent control of a range of dicotyledenous weeds occurring within cereal crops. Of the over 200 herbicides listed in Worthing and Hance's *The Pesticide Manual*,[1] approximately 85% are used in a crop-selective manner. It is evident that a very high proportion of herbicides have been developed, because of their utility for controlling troublesome weeds within a crop.

The definition of a selective herbicide used in this chapter is restricted to a substance that controls a range of weed species relatively susceptible to the substance in a crop or crops that are relatively tolerant. While differences in susceptibility to herbicides are very common between different weed species, crop species, and even different biotypes or cultivars of the same species, the information of most practical use is the range of weed species the farmer can selectively control with a particular herbicide in a specified crop.

A considerable body of literature addresses the biochemical and physiological processes underlying herbicide selectivity. The reader is encouraged to consult the many excellent reviews[2-11] for an insight into these mechanisms.

The purpose of this chapter is to describe how herbicide selectivity can be determined in a quantitative manner, so that the degree of selectivity of large numbers of herbicides, different formulations of one herbicide or different adjuvants in combination with one herbicide, can be compared using the relatively simple bioassays commonly employed in commercial agrochemical laboratories.

II. HERBICIDE SELECTIVITY — A SCREENING CASCADE

A. PRIMARY SCREENS

Commercial agrochemical companies commonly screen many thousands of substances every year for herbicidal properties. The first step in a typical screening cascade (Figure 1) is to determine whether a substance has any herbicidal effect whatsoever. This step may be carried out *in vitro* using cell cultures or isolated target enzymes, but compounds of interest will also be applied to whole plants, usually using a small number of indicator species. At this primary screening stage, single, relatively high rates of substances are used (2 to 5 kg ha^{-1}), and no useful information on crop selectivity is obtained.

If a substance is found to have sufficient herbicidal activity, it will be subjected to a further screen, where a wide range of different species encompassing the major crops and weeds of agronomic significance are included. At this stage, screens are likely to be single replicate tests, but they may include two or three rates of each herbicide treatment to facilitate judgements of relative potency (see Chapter 3).

Larger, replicated tests cannot usually be carried out because of the restricted amount of herbicide available. Some indications of crop selectivity can be determined, and these are used as the basis for selecting herbicides for further testing in crop specific secondary screens.

B. SECONDARY SCREENS

It is at the secondary screening stage that more detailed qualitative and quantitative crop selectivity responses of herbicides are examined. These screens are usually conducted on a crop-by-crop basis, although sometimes two or more crops (e.g., wheat and barley) may be included in a single screen if these crops are associated with a similar weed flora. Species should include major weeds of economic importance in the relevant crop. Tests are replicated and multiple herbicide doses are used. This type of screen forms the basis for determining

In vitro activity test
Purpose: To detect herbicide activity at enzyme sites-of-action, or in cell culture

First stage Primary Screen
Purpose: to detect herbicide activity at a single (high) rate on a small range of indicator species

Second stage Primary Screen
Purpose: To determine relative herbicide potency using 2-3 doses across major crops and weeds. Gives an indication of selectivity pattern

Secondary Screen
Purpose: To determine the relative herbicide potency (multiple rates) for a crop and for weeds associated with a crop. To determine the range of weeds controlled selectively, and to quantify the degree of crop selectivity

Field Screening
Purpose: To determine the relative herbicide potency (multiple rates) for a crop and for weeds associated with a crop, to test the selective weed spectrum and degree of crop selectivity under field conditions

Physiological and biochemical selectivity studies
Purpose: To determine the selectivity mechanism of herbicides

Advanced Selectivity Screen
Purpose: To investigate the influence of environmental factors on crop selectivity

Field Evaluation
Purpose: To establish the rate of herbicide required for control of a given weed spectrum, and to confirm crop selectivity, over a range of environmental conditions

Source : Bartley, unpublished.

FIGURE 1. A typical herbicide selectivity screening cascade. (From Bartley, unpublished.)

the relative potency of different herbicides on a range of weed species within a crop and for estimating the crop response. The range of species that can be selectively controlled within the crop can also be derived. The results of secondary screens can be used for selecting candidate herbicides for field trials.

When a novel herbicide has been shown to have useful *in vivo* crop tolerance, studies of intrinsic selectivity mechanisms are initiated using biochemical and physiological techniques. These studies investigate the interspecific differences in the penetration, uptake, translocation, metabolism, and target site activity of herbicides.[7]

The results of these experiments alongside *in vivo* screening and physico-chemical data may influence the direction of synthesis towards the production of chemical analogues with improved crop selectivity.

C. FIELD SCREENS

Field screens are used to investigate the activity and selectivity (and hence crop-selective weed spectrum) of new herbicide leads. Field screens are usually small plot, multiple rate, multiple species tests with similar fundamental objectives to laboratory secondary screens. However, these tests are carried out in field and environmental conditions similar to those that occur for crops and weeds in commercial crop production, and they are therefore more realistic indicators of practical crop selectivity than laboratory bioassays.

D. ADVANCED SELECTIVITY SCREENS

A series of field screens not only indicates the relative selectivity of different herbicides, but may also yield information on how different environmental factors influence selectivity. However, it is not usually possible to study the importance of any one factor (such as temperature, light, or soil type) in isolation of other factors in single field screens. Laboratory bioassays using controlled environment rooms can help significantly in distinguishing the major environmental influences on crop selectivity.

E. FIELD EVALUATION OF PRACTICAL CROP SELECTIVITY

Extensive field trials are required to confirm the crop selectivity of herbicides using normal cropping practices. These trials are an essential part of establishing the limits for crop selective herbicide use, and they may be required by government pesticide regulatory authorities. By correlating crop selectivity with the environmental conditions across a trials series, it is possible to test and refine hypotheses developed in controlled environment bioassays and convert these hypotheses to practical recommendations to the farmer for the correct selective use of herbicides.

III. THE QUANTITATIVE ASSESSMENT OF HERBICIDE SELECTIVITY

A. DIFFERENT WAYS OF EXPRESSING SELECTIVITY

Herbicide selectivity is commonly expressed in two different ways. The first of these descriptions may be termed "vertical assessment".[12,13] Using this method, selectivity is defined as the difference in response of the crop and weed species of interest at a single preset dose rate of herbicide. The selectivity of several herbicides may then be compared by examining the magnitude of the difference in response between the crop and weed species to each herbicide at the predetermined dose. Table 1 illustrates a hypothetical example of determining selectivity using this vertical assessment method.

From the data in Table 1, compounds 1 and 3 would appear to be the most selective, with differentials in herbicidal response between crop and weed of 79 and 77%, respectively. Compound 2 would appear to have poor selectivity to the crop. However, the dose chosen

TABLE 1
Comparison of the Crop Selectivity of Three Herbicides Applied at a Single Dose

	Weed response	Crop response	Selectivity (weed response – crop response)
Compound 1	94	15	79
Compound 2	99	66	33
Compound 3	82	5	77

Note: Data are expressed as percentage herbicidal response compared to an untreated control.

From Bartley (unpublished).

will greatly affect the herbicidal response obtained for all species, and therefore the ranking of selectivity for compounds tested. This is illustrated in Table 2, where the same herbicides are compared at several doses.

From this simple example, it is apparent that any quantitative analysis of selectivity must be based on a comparison of the dose-response relationships for crop and weed species to the herbicides of interest. Figure 2A illustrates the dose-response curves for crop and weed to the herbicides used in our example. From these curves it is readily seen that the horizontal displacement of dose-response curves for crop and weed is similar for all three herbicides. If the dose required to achieve a chosen response level is interpolated from these curves for both weed and crop, they will differ for each herbicide. However, the ratio of the dose required to elicit the chosen response in the weed to the dose required for the response in the crop is the same for all three herbicides. Although the potency of these herbicides differs, the degree of selectivity of weed control in the crop is the same. The comparison of the doses that produce similar responses in crop and weeds as a means to quantifying the degree of selectivity between species has been described as "horizontal assessment",[12,13] refering to the horizontal displacement of herbicide dose-response curves for the different species when curves are plotted on a logarithmic scale. If these dose-response curves are parallel, an expression of selectivity that is not influenced by dose or response level can be derived (see Chapter 3).

B. THE HERBICIDE DOSE-RESPONSE CURVE

In order to assess the relative potency of several herbicides on several species, and hence be able to quantitatively compare the selectivity of these herbicides to one or more crop, it is essential to carry out dose-response experiments for each herbicide on each of the species of interest.[14] In practice, most commercial agrochemical laboratories, carry out one experiment (a secondary screen) where the dose-responses of several herbicides are compared for a group of plants treated at the same time. Ideally, the selection of doses used in this type of experiment should be chosen such that one would expect to achieve a broad range of herbicidal response for each species, encompassing the response levels of interest. In glasshouse screening tests, the number of individuals treated by any one herbicide dose is usually low, and the assessment methods used are subjective, visual estimations of herbicide response relative to an untreated control. Sometimes more objective assessment methods may be used, such as measurements of dry or fresh weight or plant height. It is usual to express both subjective visual estimations and objective measurements of plant response as percentage of the untreated control; therefore the measured response is of a *quantitative* nature, as opposed to a *quantal* or all-or-nothing response.[14,15] The results from dose-response studies of this type are sometimes analyzed by multiple comparison techniques such as Duncan's Multiple Range test or the use of least significant differences. However, the use of these statistical methods obscures the inherent relationship between each dose of a dose-response experiment. As pointed out by Cousens,[16]

TABLE 2
Comparison of the Crop Selectivity of Three Herbicides Applied at a Range of Doses

	Compound 1			Compound 2			Compound 3		
Dose $g\ ha^{-1}$	Weed response	Crop response	Selectivity (weed response–crop response)	Weed response	Crop response	Selectivity (weed response–crop response)	Weed response	Crop response	Selectivity (weed response–crop response)
10	29	0	29	82	5	77	11	0	11
30	57	2	55	94	15	79	29	0	29
100	82	5	77	98	37	61	57	2	55
300	94	15	79	99	66	33	82	5	77
1000	98	37	61	100	87	13	94	15	79

Note: Data are expressed as percentage herbicidal response compared to an untreated control.

From Bartley (unpublished).

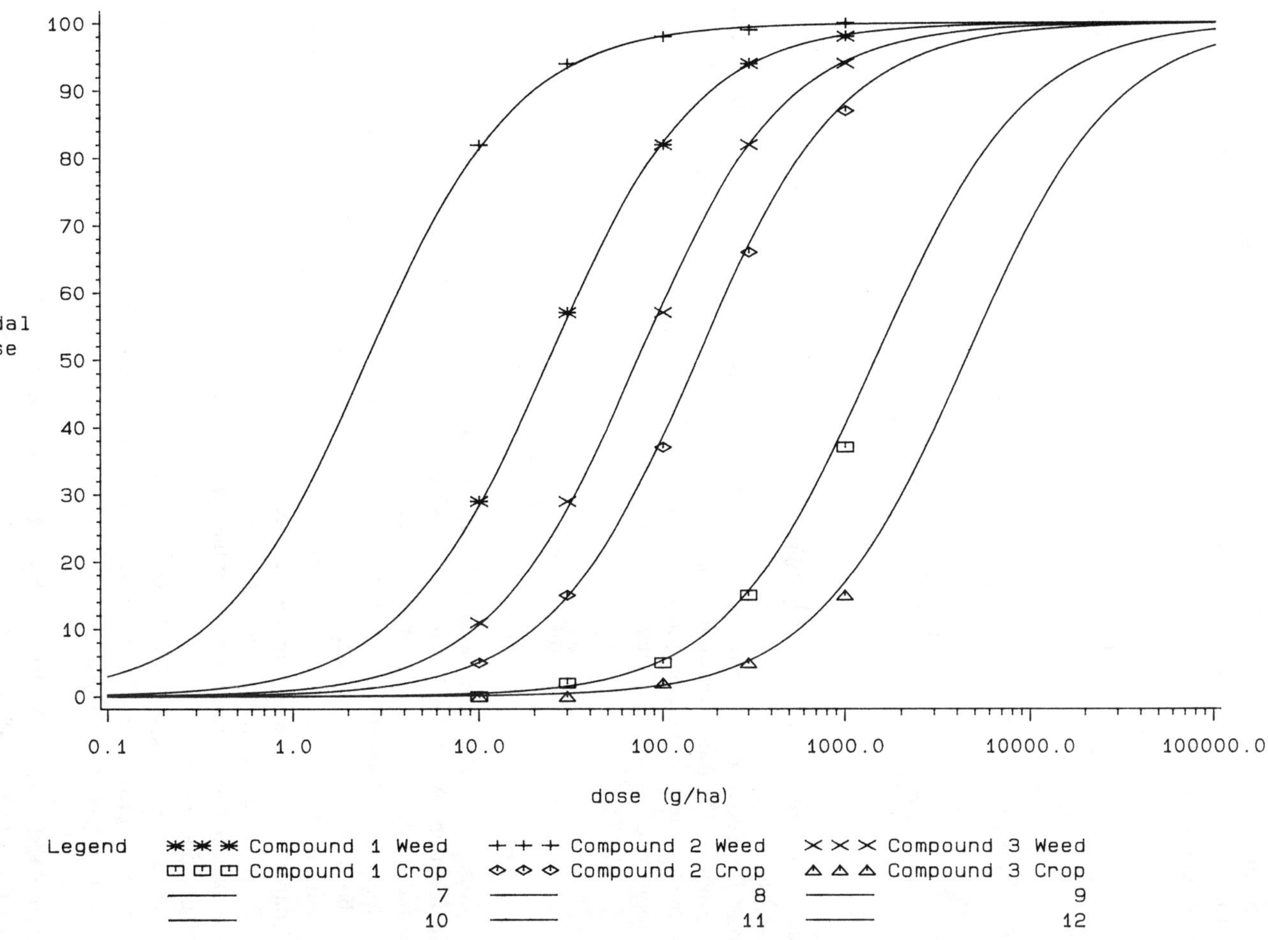

FIGURE 2A. Plot of percentage herbicidal response for a crop and a weed against the logarithm of dose for three herbicides.

the data derived from a dose-response experiment require analysis using regression techniques, and the use of multiple comparison techniques is not appropriate.[17,18]

If herbicidal response is plotted against the logarithm of dose, a sigmoidal curve will in general be observed,[12,14,19-21] as in Figure 2A. There are several sigmoidal curves that can be used to model the relationship between log (dose) and herbicidal response. Two of the most well known are the logistic curve,

$$P = \frac{100}{1 + \exp-\left[\alpha + \beta \log(x)\right]} \tag{1}$$

and the integrated normal curve,

$$P = \frac{100}{\sqrt{2\pi}} \int_{-\infty}^{\alpha+\beta x^{-5}} \exp\left(-\frac{1}{2} \times u^2\right) du \tag{2}$$

$$-\infty < x < \infty$$

Where P = percentage herbicidal response at dose x and α and β are parameters to be estimated (see Chapter 3).

In practice, a convenient way of fitting either one of these functions is to use a transformation of the response. The transformation is made so that when the transformed response is plotted against the log (dose x) there is a linear relationship. For the logistic curve the transformation is the logit transformation,[22] where

$$L = \text{logit}(P) = \log\left(\frac{P}{100 - P}\right) = \alpha + \beta \log(x) \tag{3}$$

The slope and intercept of the straight line, α and β , are the parameters of the original logistic function. A similar linearizing transformation (the probit transformation) can be carried out on the integrated normal curve.

The logit and probit transformations of herbicide response data will generally be very similar; however, the models will diverge at extreme response levels close to 0 or 100%, at which point neither model is more correct than the other (see Chapter 3). For the purpose of this paper the logit transformation has been used, and the response curves displayed in Figure 2A can be linearized using this transformation (Figure 2B). After the herbicidal response data are logit transformed, simple linear regression techniques can be used to fit the straight line to the data.

In cases where low doses of herbicide stimulate plant growth, plots of a response directly related to plant growth (e.g., fresh weight, dry weight, plant height) will deviate from the sigmoidal curve previously described.[12,23]

C. DOSE-RESPONSE EXPERIMENTS TO DETERMINE HERBICIDE SELECTIVITY — PRACTICAL CONSIDERATIONS

As described in Section II, crop selectivity of new herbicide treatments is frequently determined in a secondary screen, usually conducted in a glasshouse. Herbicides are applied at several doses using methods that mimic farmer practice. These applications are made, in the case of post-emergence treatments, to groups of plants of several different species of interest, which are usually grown in individual pots. Pre-emergence tests can be carried out using trays filled with a soil-based compost into which seed or vegetative propagules of the

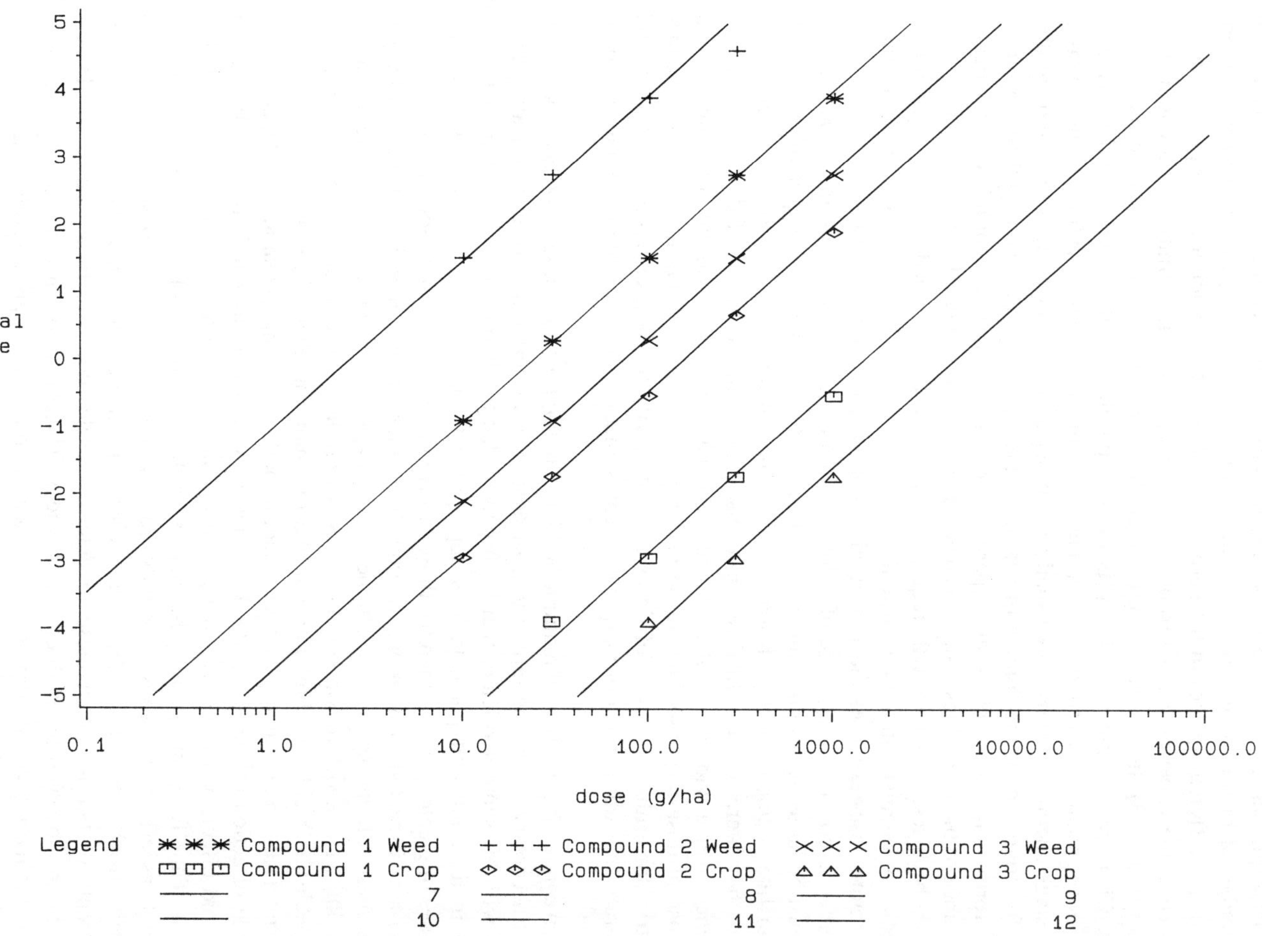

FIGURE 2B. Plot of percentage herbicidal response for the crop and weed where the response has been logit-transformed.

test species are sown — the test treatments are then applied to the soil surface shortly after sowing.

The experimental design of dose-response tests to determine crop selectivity is dependent on a number of factors including (1) the degree of prior knowledge of the relative susceptibility of the different species to the herbicides tested, (2) the amount of test plant material and space at the experimenter's disposal (both being commonly limiting in a commercial laboratory), and (3) the amount of active ingredient or formulation available to the experimenter. Chemical sample availability is often a problem to the weed scientist trying to determine the relative selectivity of several novel chemical analogues within a particular crop. At this stage chemical samples are usually synthesised in gram or even milligram quantities in an effort to maximize the number of analogues prepared within a given synthetic resource. Little information is available to the experimenter on the relative susceptibility of all species of relevance to the herbicides tested, because up to this stage only a limited number of doses will have been tested. These will frequently only have been applied to a small portion of the species of interest within one particular crop. In this case, the minimum number of rates to determine dose-responses must be used to conserve chemical supply.

In order to cover a wide range of responses in all species concerned, the steps between rates must be large. A design commonly used in our laboratories for this type of experiment is a four rate geometric series of doses, with a dilution factor of three between each dose. There is, however, no theoretical reason to use this geometric series of doses — although experience over a wide range of chemical series has shown that this rule-of-thumb design yields useful dose-response data in the majority of cases. As Cousens[16] suggests, neither geometric nor arithmetic series of doses are necessarily the best experimental design, and an arbitrary series based on prior experience can sometimes yield more useful dose-response curves.

When conducting this type of multiple herbicide, multiple species dose-response experiment, it is not necessary to use the same range of doses for all herbicides, or even for all species treated with the same herbicide. If it is suspected that the herbicides to be compared differ greatly in their inherent potency, it is appropriate to use a range of lower rates for the herbicides of greater potency than for those of lower potency. If certain species (usually crop species) are suspected to be more tolerant than other species, then it may be worthwhile treating these tolerant species with higher doses. In both these instances, the intelligent use of prior knowledge will maximize the success of producing useful dose-response curves for each species/herbicide combination with the minimum number of treatments.

The intervals and frequency of assessment of herbicidal response should be dependent on the mode of action of the herbicides being tested. Contact-acting herbicides usually have a more rapid effect than herbicides dependent on translocation to their site of action within the plant. Thus for contact-acting herbicides, early assessments of herbicidal response are important, as effects to crops may only be exhibited early in the experiment, the crop recovering from such damage by later assessments. In general terms, the most important assessments of herbicidal response to determine selectivity are those where the largest responses are observed for both crop and weeds. These may occur at different assessment times, as the more tolerant crop will often recover from early injury, whereas the herbicidal effect on more susceptible weeds can increase to a maximum at the end of the experiment.

The end point of each of these experimental designs is the production of dose-response curves for each species/herbicide combination tested, from which the comparative selectivity of each herbicide can be calculated.

D. SELECTIVITY INDICES

The dose of each herbicide required to give a chosen herbicide response for each species can be readily obtained from the linearized dose-response curves. It has been customary for many researchers to choose a 50% herbicide response level to compare the potencies of different herbicides.[12,14,23-26] This dose is variously referred to as the LD_{50} (lethal dose for 50% kill), ED_{50} (equivalent, or effective dose for 50% response) and GR_{50} (dose for 50% growth reduction). The term ED_{50} is preferred here, as it can be used to cover all types of herbicidal effect. The difference in response by two different species to a herbicide may be expressed as a selectivity index (*S*) , which is simply the ratio of the doses giving the chosen response level for the two species of interest. If the objective of deriving the selectivity index is to compare the physiological selectivity of different herbicide treatments, then the comparison of doses is based on the same response level, usually 50%, and the selectivity index is defined thus:

$$S = \mathrm{ED}_{50(\text{Species II})} \big/ \mathrm{ED}_{50(\text{Species I})}$$

and corresponds to the relative potency discussed in Chapter 2.

If species II is the crop of interest, then a large selectivity index indicates a high degree of selectivity between the weed (species I) and the crop.

If the logit transformed regression lines for the dose-responses of each species to the same herbicide treatment are parallel, the selectivity index will be identical irrespective of the response level chosen. If the lines significantly deviate from parallelism, then the response level chosen to calculate the selectivity index will influence the result.

While a selectivity index based on the 50% response for both crop and weed is useful for studying fundamental aspects of selectivity, a more appropriate value can be obtained to indicate the practical likelihood of achieving selective control of a weed or range of weeds in a particular crop. This is the ratio of the dose giving the maximum acceptable response in the crop to the dose giving the minimum acceptable response in the weed.[14] If a 10% response is defined as the maximum acceptable for the crop (i.e., if a response of greater than 10% is likely to cause yield depression or a cosmetic effect unacceptable to the farmer) and 90% is defined as the minimum acceptable level of weed control, the selectivity index can be defined as follows:

$$S = \mathrm{ED}_{10(\text{Crop})} \big/ \mathrm{ED}_{90(\text{Weed})}$$

In fact, any response levels for crop and weed can be chosen by the experimenter to derive this index, the choice being influenced by such factors as the degree of selectivity exhibited by standard herbicides, the demands of farmers in the particular countries, and the crops of interest. As this ratio is based on set response levels, a guide to the relative practical selectivity of a range of herbicide treatments can be ascertained even if the (linearized) dose-response curves are not parallel for the crop and weeds concerned. The ranking of several herbicides for selectivity will, however, be influenced by the choice of response levels.

In such cases, an element of the vertical assessment may be employed usefully for guidance. Agrochemical companies and advisory agencies are necessarily concerned about the effect of herbicidal injury on the crop if a herbicide overdose is mistakenly applied by the farmer. Such herbicide overdoses most commonly occur because of the mismatching of swaths when applications are made by broadcast spray. This results in strips of crop being treated with a double dose of herbicide. It is therefore of interest to estimate the level of crop injury incurred by twice the dose required for acceptable weed control.

An indication of the reliability of crop safety for a herbicide treatment can also be gained by examining the slope of the logit-transformed dose-response curve for the crop, defined as in Equation 3. A shallow slope indicates that minor overdoses will not result in major crop damage. A steep slope, by contrast, indicates the contrary, and also suggests that adverse environmental conditions influencing the uptake and translocation of a herbicide within the crop or affecting the ability of the crop to detoxify the herbicide may result in significant damage.

E. THE USE OF SELECTIVITY INDICES TO INDICATE THE CROP-SELECTIVE WEED SPECTRUM

If selectivity indices are calculated for each weed relative to the crop of interest, a good indication of the spectrum of weed species that can be selectively controlled in the crop can be determined. Weed species may be grouped into categories depending on their selectivity indices. If a species has an index ≥2, this weed can be controlled at a rate that will not significantly damage the crop even if misapplication results in double dosing. If a species has an index between 1 or 2, marginal selectivity is indicated; the weed will be selectively controlled if significant overdosing does not occur. A species with an index of <1 will not be selectively controlled. Table 3 illustrates such groupings for two commonly used pre-emergence maize herbicides, atrazine and metolachlor. This example clearly indicates the utility of atrazine for selective control of dicotyledonous weeds in maize and of metolachlor for control of graminaceous weeds.

IV. LABORATORY SELECTIVITY MEASUREMENTS AND FIELD RESPONSES

A. THE LIMITATIONS OF BIOASSAYS IN THE PREDICTION OF FIELD RESPONSES

Glasshouse and growth room experiments designed to determine relative selectivity are essential tools for the experimenter wishing to distinguish between large numbers of herbicides over a wide species range. It is not practical to conduct field trials for more than a few herbicides at one time due to the large resource demands of such field programs. However, the results of the dose-responses obtained from laboratory bioassays can be strongly influenced by factors related to the stage of growth and development of the test plants, the type of growing medium used, and the environment in which tests are conducted.[27,28] As selectivity indices are determined from these dose-response relationships, these indices will also be influenced by such environmental factors, particularly as herbicidal response for different species is often influenced by such environmental factors to different extents.

Even when conducting experiments to determine selectivity in controlled environment rooms, it is impossible to mimic the field environment in anything other than a crude fashion. Although temperature, lighting, and air humidity regimes can be controlled to match a particular field situation, it is not practical to mimic other conditions such as rainfall pattern, wind, soil type, soil structure, and moisture status. Most selectivity bioassays are, of necessity, carried out in glasshouses rather than controlled environment rooms, to accommodate large numbers of treatments. Such glasshouses rarely have very precise environmental controls, and thus can only approximately mimic any field environment. In this situation, numerous compromises must be made in order to carry out relatively simple, reproducible screens to examine large numbers of herbicides quickly.

The environmental factors (in which bioassays are carried out) may influence herbicide activity and selectivity in two ways. They may directly influence the uptake, movement, and action of herbicides in the plant, and they may also indirectly influence plant anatomy, morphology, and physiology prior to and following herbicide application. The principle factors that influence test results are listed below

1. **Temperature**. Small differences in temperature can influence plant development[29-31] and hence affect herbicide uptake, translocation, and action. The glasshouse temperature regime chosen must be as appropriate as possible for the crop and weed species under

TABLE 3
Species Controlled Selectively in Maize Using Common Pre-Emergence Herbicides

Metolachlor

Selectively controlled	S	Marginally selectively controlled	S	Not selectively controlled	S
Setaria faberii	13.5	*Solanum nigrum*	1.3	*Bidens pilosa*	0.4
Echinochloa crus-galli	12.9			*Chenopodium album*	<0.3
Panicum miliaceum	3.0			*Abutilon theophrasti*	<0.3
Brachiaria platyphylla	3.0			*Xanthium strumarium*	<0.3
Amaranthus retroflexus	2.9			*Ipomoea hederifolia*	<0.3
Sorghum halepense	2.6				

Atrazine

Selectively controlled	S	Marginally selectively controlled	S	Not selectively controlled	S
Chenopodium album	28.3	*Xanthium strumarium*	1.9	*Setaria faberii*	<0.3
Solanum nigrum	11.2	*Echinochloa crus-galli*	1.0	*Brachiaria platyphylla*	<0.3
Bidens pilosa	5.1			*Panicum miliaceum*	<0.3
Ipomoea	<0.3			*Sorghum halepense*	<0.3
hederifolia	2.7				
Amaranthus retroflexus	2.2				
Abutilon theophrasti	2.0				

Note: Species are ranked by their selectivity index, $S = ED_{10(maize)}/ED_{90(weed)}$.

From Bartley (unpublished).

test, but compromise is inevitable due to the range of climatic conditions in which different weed species associated with one crop grow.

2. **Daylength, light quality, and irradiance**. The effect of the light environment on plant morphogenic processes are well documented,[32] and variations in morphology between field and glasshouse or growth room grown plants will indirectly affect responses to herbicides. Both photoperiod and irradiance may affect the development of plant cuticles,[33,34] thus they may also affect the uptake of foliar-applied herbicides in a species-dependent manner, thereby influencing selectivity. Levels of irradiance affect the levels of activity observed with several chemical classes of herbicides,[35-39] and activity may vary with irradiance in a species-specific manner, thus influencing selectivity. Glasshouses used for agrochemical screening are usually equipped with supplementary lighting systems to compensate for the reduction in irradiance caused by the glasshouse structure.[40] Nevertheless, the results of selectivity tests conducted for crops grown in regions with naturally high irradiances during the growing season (such as rice, maize, and soybean) may be influenced by the lower irradiance levels encountered in more temperate areas, particularly during winter.
3. **Watering regime and air humidity**. The regular watering regimes used for bioassay tests are generally designed to avoid the occurrence of soil water deficits.[30] Air humidity can be regulated using controlled environment chambers, but it is not possible to mimic

the very wide range of conditions observed in the field, which will include dew formation on the leaves of crop and weeds, rainfall of different intensities and durations, and large fluctuations in soil moisture status.[40] All of these factors may cause differences between apparent herbicide selectivity determined in the glasshouse or growth room and selectivity observed in the field.

4. **Wind effects**. Experiments done in glasshouses or growth rooms are carried out in the absence of wind. In the field, the surface wax composition of leaves can be markedly altered by abrasion between leaves or by impact from soil and dust particles.[28] Such changes in leaf wax may vary from species to species. For example, cereal crops planted in rows may act as wind breaks, resulting in high amounts of wind abrasion. In contrast, a prostrate weed, such as *Stellaria media*, in a cereal crop may be protected from the wind. The efficacy of herbicide formulations and adjuvant systems is affected greatly by the amount and composition of leaf surface wax. Therefore, the masking of species-differential wind abrasion under experimental conditions may change the selectivity ranking of formulations obtained under laboratory conditions compared to that obtained in field trials.
5. **Plant growth stage**. The susceptibility of crops and weeds to herbicides is usually related to their size and developmental stage at treatment time. In the field, different species grow at different rates, thus the selectivity of herbicides will be affected greatly by the timing of application. Bioassays to determine selectivity for post-emergence herbicides should be set up to closely mimic the relative growth stages of crop and weeds common under field situations. Ideally, these bioassays should be repeated at different growth stages to determine the relative selectivity of different herbicide treatments over a wide range of timings.
6. **Soil conditions**. Soil type, nutrient status, seed and vegetative propagule planting depth, and the effect of constraining plant rooting systems within pots and trays will all have some influence on the effects of herbicides seen in laboratory experiments.[41] The behavior of pre-emergence herbicides is affected very substantially by the type of soil to which they are applied, particularly by the relative amounts of sand, clay, and loam present, the amount of organic matter, and soil pH. Several herbicides with rather poor physiological selectivity to crops can be used in the field because of the protection gained by relatively deep planting of crop seed; uptake of the surface-applied herbicide by the crop during germination is minimized. Examples of such "depth protection" include the use of trifluralin and terbutryne in wheat and chlorpropham in rice.[42] Thus, the apparent selectivity of pre-emergence herbicides determined in laboratory tests will be influenced by the planting depth chosen for each species included. Crop planting depths should be chosen to match those most commonly used by the farmer. Choosing weed planting depth is more problematic, as most species germinate at a range of depths within limits determined by physiological factors related to dormancy mechanisms. Initial selectivity bioassays should use standard planting depths based on a midpoint within the range of depths from which seedlings emerge in the field. These tests can be followed by comparisons of selectivity of small numbers of herbicide treatments, where the relative planting depths of crop and weed are varied.

The artificial environment used to conduct selectivity bioassays demands that the selectivity index so generated should be regarded as indications of relative, rather than absolute, selectivity, preferably compared to a well chosen standard tested in the same experiment. Ideally, the standard herbicide will have previously been tested under field conditions, and thus a measure of its selectivity will have been obtained in practical agronomic situations. The standard should also have a similar mode of action and (preferably) similar chemical structure to the novel herbicide treatments. The use of such a standard allows the experimenter to compare results for the standard herbicide used in glasshouse tests to those obtained in the field. If the glasshouse results are substantially at variance with those obtained in the field, then the validity of results for all herbicides included in the test may well be questionable. If, however, there is reasonable agreement between field and glasshouse on the absolute selectivity margin for the standard, then greater confidence can be given to the selectivity indices derived for the novel herbicide treatments. If a common standard is used in a number of selectivity experiments, it is also possible to determine the relative selectivity of herbicide treatments tested in different experiments by comparing the ratio of selectivity indices of the novel treatments relative to the selectivity indexes of the standard herbicide treatment derived from the same test. If this approach is required in order to compare the selectivity of a great number of chemical analogues or different formulations of the same herbicide, as much care as possible must be taken to minimize differences in the experimental conditions between tests. The following parameters can reasonably be controlled even for tests not conducted in controlled environment chambers:

1. Growth and developmental stage of the plant material (for post-emergence tests)
2. Depth of seeding
3. Soil or compost type
4. Watering regime
5. Plant and seed density

B. THE USE OF SELECTIVITY BIOASSAYS TO DETERMINE ENVIRONMENTAL INFLUENCES

The bioassay tests previously described are primarily used to select herbicide treatments for further investigation under field conditions. Selectivity in the field can be determined using similar dose-response experiments.

Many variables will not be under the trialists' control in the field, but general trends of how selectivity can be influenced by environmental factors, soil type, and crop/weed growth stage can be derived from a trial series if these variables are accurately recorded for each trial. For example, the selectivity index for a herbicide can be determined for trials carried out on different soil types, and correlations may be derived indicating whether selectivity is related to soil organic matter, sand content, clay content, pH, or some other variable. Similar correlations can be made with other factors, such as crop or weed growth stage, temperature, rainfall, daylength, total irradiance, and so on. This process will of course provide much useful information on herbicide selectivity over a wide range of conditions. However, it does not necessarily lead to the establishment of a causal relationship between crop selectivity and any single factor.

At this stage in the development of a new herbicide product, it may be prudent to conduct experiments using controlled environments that are designed to test the effect of varying environmental conditions, implicated in field trials as important to selectivity. This can be done by carrying out dose-response experiments where all factors except one are kept constant. The effect of the single varying environmental factor on the selectivity of a herbicide can then be determined by examining the relationship between the selectivity index for a herbicide and the environmental factor of interest. From these relationships, hypotheses can be developed and tested once more under field conditions, leading ultimately (for the commercial agrochemical company) to recommendations to the farmer indicating the optimum conditions and (perhaps more importantly!) precise limitations for the crop-selective use of the herbicide.

ACKNOWLEDGMENTS

The author is grateful to Mr. T. Kelly for statistical advice and to other colleagues for their constructive comments on the manuscript.

REFERENCES

1. **Worthing, C. R. and Hance, R. J.,** *The Pesticide Manual,* 9th ed., British Crop Protection Council, Farnham, U.K., 1991.
2. **Dodge, A. D.,** The mode of action and metabolism of herbicides, in *Weed Control Handbook: Principles,* 8th ed., Hance, R. J. and Holly, K., Eds., Blackwell, Oxford, 1990, chap. 7.
3. **Owen, W. J.,** Metabolism of herbicides — detoxification as a basis of selectivity, in *Herbicides and Plant Metabolism,* Dodge, A. D., Ed., Cambridge University Press, Cambridge, 1989, 171.
4. **Hathway, D. E.,** *Molecular Mechanisms of Herbicide Selectivity,* Oxford University Press, Oxford, 1989.
5. **Cole, D. J., Edwards, R., and Owen, W. J.,** The role of metabolism in herbicide selectivity, in *Progress in Pesticide Biochemistry and Toxicology,* John Wiley & Sons, Chichester, 1987, chap. 2, p. 6.
6. **Owen, W. J.,** Herbicide detoxification and selectivity, in *Proceedings of the British Crop Protection Conference — Weeds,* British Crop Protection Council, Farnham, U.K., 1987, 309.
7. **Hathway, D. E.,** Herbicide selectivity, *Biol. Rev.,* 61, 435, 1986.
8. **Hess, F. D.,** Herbicide absorption and translocation and their relationship to plant tolerances and susceptibility, in *Weed Physiology,* Vol. 2, Duke, S. O., Ed., CRC Press, Boca Raton, FL, 1985, chap. 8.
9. **Shimabukuro, R. H.,** Detoxification of herbicides, in *Weed Physiology,* Vol. 2, Duke, S. O., Ed., CRC Press, Boca Raton, FL, 1985, chap. 9.
10. **Hassall, K. A.,** *The Chemistry of Pesticides,* The MacMillan Press, London, 1982, chap. 11.
11. **Jensen, K. I. N.,** The roles of uptake, translocation, and metabolism in the differential intraspecific responses to herbicides, in *Herbicide Resistance in Plants,* LeBaron, H. M. and Gressel, J., Eds., John Wiley & Sons, New York, 1982, chap. 8.
12. **Streibig, J. C.,** Herbicide bioassay, *Weed Res.,* 28, 479, 1988.
13. **Streibig, J. C., Thonke, K. E., and Kudsk, P.,** The effect of formulations on the herbicidal activity of phenmediphan, in *Adjuvants and Agrochemicals,* Vol. I, *Mode of Action and Physiological Activity,* Chow, P. N. P., Grant, C. A., Hinshalwood, A. M., and Simundsson, E., Eds., CRC Press, Boca Raton, FL, 1986, chap. 9.
14. **Holroyd, J., Holly, K., Jordan, D., Robson, T. O., and Rosher, P. H.,** The evaluation of a new herbicide, in *Weed Control Handbook, Vol. 1, Principles,* 6th ed., Fryer, J. D. and Makepeace, R. J., Eds., Blackwell, Oxford, 1977, chap. 7.
15. **Finney, D. J.,** *Statistical Method in Biological Assay,* 3rd ed., Charles Griffin, London, 1978.
16. **Cousens, R.,** Underlying principles in the design and interpretation of experiments, *Aspects Appl. Biol.,* 10, 1, 1985.
17. **Dawkins, H. C.,** The misuse of t-tests, LSD and multiple range tests, *Bull. Br. Ecol. Soc.,* 12, 112, 1981.
18. **Dawkins, H. C.,** Multiple comparisons misused: why so frequently in response-curve studies?, *Biometrics,* 39, 789, 1983.
19. **Allot, D. J. and O'Neill, J. A.,** A parallel line assay method for the determination of herbicide soil residues, *Rec. Agric. Res.,* 18, 21, 1969.
20. **Streibig, J. C.,** Models for curve-fitting herbicide dose response data, *Acta Agr. Scand.,* 30, 59, 1980.
21. **Streibig, J. C.,** The herbicide dose-response curve and the economics of weed control, in *Proceedings of the Brighton Crop Protection Conference — Weeds,* British Crop Protection Council, Farnham, U.K., 1989, 927.
22. **Ashton, W. D.,** *The Logit Transformation with Special Reference to Bioassay,* Charles Griffin, London, 1972.
23. **Brain, P. and Cousens, R.,** An equation to describe dose responses where there is stimulation of growth at low doses, *Weed Res.,* 29, 93, 1989.
24. **Horowitz, M.,** Application of bioassay techniques to herbicide investigations, *Weed Res.,* 16, 209, 1976.
25. **Nyffeler, A., Gerber, H.-R., Hurle, K., Pestener, W., and Schmidt, R. R.,** Collaborative studies of dose-reponse curves obtained with different bioassay methods for soil-applied herbicides, *Weed Res.,* 22, 213, 1982.

26. **Schmidt, R. R.,** Development of herbicides, in *Herbicide Bioassays,* Streibig, J. C. and Kudsk, P., Eds., CRC Press, Boca Raton, FL, 1992, chap. 2.
27. **Pestemer, W.,** Biological determination of photosynthetic inhibitors in soils and water and application of bioassays to herbicide investigations. *Z. Naturforsch,* 34, 964, 1979.
28. **Garrod, J. F.,** Comparative responses of laboratory and field grown test plants to herbicides, *Aspects of Applied Biology 21, Comparing Laboratory and Field Pesticide Performance,* 51, 1989.
29. **Woodward, F. I.,** The climatic control of the altiduninal distribution of *Sedum rosea* (L) Scop and *S. telephium* L. II. The analysis of plant growth in controlled environments. *New Phytol.,* 75, 335, 1975.
30. **Davies, W. J. and Blackman, P. G.,** Growth and development of plants in controlled environments and in the field, *Aspects of Applied Biology 21, Comparing Laboratory and Field Pesticide Performance,* 1, 1989.
31. **Reed, D. W. and Tukey, H. B., Jr.,** Light intensity and temperature effects on epicuticular wax morphology and internal cuticle ultrastructure of carnation and Brussel's sprouts leaf cuticles. *J. Am. Soc. Hort. Sci.,* 109, 417, 1982.
32. **Smith, H.,** *Plants and the Daylight Spectrum,* Academic Press, London, 1981.
33. **Tribe, I. S., Gaunt, J. K., and Parry, D. W.,** Cuticular Lipids in the Graminae, *Biochem. J.,* 109, 8, 1968.
34. **Wilkinson, R. E.,** Sicklepod hydrocarbon response to photoperiod, *Phytochemistry,* 11, 1273, 1972.
35. **Dodge, A. D.,** The role of light and oxygen in the action of photosynthetic inhibitor herbicides, in *Biochemical Responses Induced by Herbicides,* Moreland, D. E., St. John, J. B., and Hess, F. D., Eds., American Chemical Society, Washington, D.C., 1982, chap. 4.
36. **Orr, G. L. and Hess, F. D.,** Mechanism of action of the diphenyl ether herbicide acifluorfen-methyl in excised cucumber (*Cucumis sativus*), light activation and the subsequent formation of lipophilic free radicals, *Plant Physiol.,* 69, 502, 1982.
37. **Pallett, K. E. and Dodge, A. D.,** Modifications of chloroplasts of flax cotyledons treated with monuron:myelinoid figures formed under low light conditions, *Plant Cell Environ.,* 3, 183, 1980.
38. **Potter, J. R. and Wergin, W. P.,** Role of light in bentazon toxicity to cocklebur: physiology and ultra structure, *Pest Biochem. Physiol.,* 5, 458, 1975.
39. **Sandmann, G. and Böger, P.,** Mode of action of herbicidal bleaching, in *Biochemical Responses Induced by Herbicides,* Moreland, D. E., St. John, J. B., and Hess, F. D., Eds., American Chemical Society, Washington, D.C., 1982, chap. 7.
40. **Legg, B. J.,** Micrometerological conditions in plant canopies under glasshouse and field conditions, *Aspects of Applied Biology 21, Comparing Laboratory and Field Pesticide Performance,* 13, 1989.
41. **Blair, A. M. and Martin, T. D.,** Effects of laboratory and field environments on plant root growth and consequences upon pesticide uptake by roots, *Aspects of Applied Biology, 21, Comparing Laboratory and Field Pesticide Performance,* 51, 1989.
42. **Holly, K.,** Selectivity in relation to formulation and application methods, in *Herbicides-Physiology, Biochemistry, Ecology,* Vol. II, 2nd ed., Audus, L. J., Ed., Academic Press, London, 1976, chap. 8.

Chapter 5

QUANTITATIVE STRUCTURE ACTIVITY RELATIONSHIPS: ADDING VALUE TO HERBICIDE BIOASSAYS

Daniel A. Kleier and Gary Gardner

TABLE OF CONTENTS

0-8493-6603-8/93/$0.00+$.50

I. INTRODUCTION

The results of herbicide bioassays are essential input for many herbicide optimization programs, and on a more fundamental level, for many studies of plant biomechanisms. If biological assay results are obtained for a series of molecules of sufficient diversity, quantitative relationships can be constructed between the measured biological activity and descriptors that characterize the molecular diversity (e.g., size, shape, lipophilicity, and electronic charge distribution). These quantitative structure-activity relationships (QSARs) can add significant value to the bioassay results from which they are constructed. The value added by the QSAR construction will depend upon the quality and suitability of the building materials (measures of activity and descriptors of molecular structure), the effectiveness of the analysis tools selected for relating them, and the skill of the QSAR architect and builder.

The suitability of a biological assay for subsequent QSAR analysis will depend upon the goal of a structure-activity study. If the goal is a fundamental understanding of a receptor-herbicide binding interaction, *in vitro* binding or functional inhibiton assays will be more appropriate than *in vivo* assays, since the latter are often confounded by metabolism and transport factors. If the goal is herbicide optimization, the selection of bioassays is less clear. QSAR analysis of *in vivo* greenhouse or even field data would seem to be closer to the goal of finding an optimal herbicide, but it may be so confounded by multiple and poorly controlled factors that no significant QSAR can be established. In this case, a multifaceted approach could prove to be more productive. For example, QSAR analysis of results from an *in vitro* assay related to the herbicidal lesion might be coupled with complementary analysis of transport and/or metabolism. These issues will be addressed in Section II.C.

The selection of molecular descriptors for consideration as possible correlates of biological activity is often guided by some hypothesized mode of action. For example, symptomology may prompt the hypothesis that photosystem I is the site of action of a class of herbicides. Since the critical lesion of photosystem I action often involves reduction of the parent herbicide, this hypothesis might be tested by considering molecular reduction potential for inclusion as a descriptor in the QSAR. The consistency of this hypothesis may then be evaluated by calculating various significance measures for the resultant QSAR. The possible relevance of a variety of descriptor types to various hypothesized modes of herbicide action will be described in Section II.B, and significance measures for hypothesis testing will be discussed briefly in Section III.A.

The selection of analysis tools for constructing a QSAR will depend on the goal of the QSAR architect as constrained by the quality of the biological data and the availability of suitable molecular descriptors. For example, if a detailed image of a herbicide receptor site, including the ability to predict binding energies, is the desired goal of a QSAR analysis, then an analysis tool such as distance-geometry (Section III.E.1) or comparative molecular field analysis (Section III.E.2) will be the tool of choice. These sophisticated forms of QSAR analysis, as well as many other forms that are based upon the extrathermodymanic postulate (Section III.A), are constrained by a requirement for dose-response data preferably from an *in vitro* binding assay. At the other extreme, if the goal of a QSAR analysis is simply to predict a molecule's activity category (e.g., active or inactive), then categorical data such as that obtained from many primary herbicide screens will suffice as building material, and methods of discriminant analysis (Section III.D) will be the analysis tools of choice.

II. MATERIALS FOR BUILDING QSARs

In order to build a sound QSAR, quality building materials as well as analysis tools are required. Building materials include bioassay results for a well designed series of related herbicides, relevant descriptors of molecular structure, and measures of biological activity whose quality is at least as good as the predictions desired. Typically the end result of a QSAR analysis is a functional relationship between potency and molecular descriptors, x_i:

$$\text{Potency} = f\left(x_1, x_2, \ldots\right) \tag{1}$$

The functional relationship may be explicit, as in multiple linear regression analysis, or implicit, as in some forms of graphical and artificial intelligence approaches.

A. ANALOGUE SERIES — DESIGN PRINCIPLES

QSAR analysis is facilitated by a well designed analogue series. The members of an analogue series are variations on a theme, the theme being defined by a basic molecular framework that is common to all molecules in the series. What is needed for a well designed analogue series is a representative sampling of all variations that can be built from this theme. The variations often take the form of functional groups (substituents) that replace hydrogens at various locations on the common framework.

The fractional factorial design approach[1,2] is one of several procedures for selecting a representative sampling of structurally diverse analogues for synthesis and testing. It can be performed either by hand[3] or with the assistance of a computer.[4] Once a set of structural descriptor types (e.g., size, lipophilicity, etc.) has been chosen, a full 2^n factorial design requires that all possible combinations of high and low values for these descriptors be realized in the set. A fractional factorial design requires a systematically chosen subset of compounds from the full factorial set. The set of analogues resulting from a factorial design are diverse in the sense that the variance of the structural descriptors is large while the correlation between descriptors is low. Alternative methods for designing an analogue series with sufficient diversity include cluster analysis[5] and difficulty information approaches.[6] The latter method addresses the criticism of synthetic impracticality by using synthesis difficulty ratings as an additional analogue selection criterion.[7] Highly fractional designs for analogue sets may require descriptor transformations and nontraditional analysis techniques that can mask the physicochemical meaning of the resultant QSAR relations.

B. DESCRIPTORS OF MOLECULAR STRUCTURE

The molecular descriptors (the x_i in Equation 1) used in the QSAR analysis of an analogue series may characterize each molecule as a whole, or they may characterize fragments or substituents located at equivalent positions within each molecule. They may be directly measured, empirically estimated, or theoretically calculated. Empirical estimations are usually based on a collection of transferable fragment properties. One advantage of using descriptors that are readily estimated or theoretically calculated is that a QSAR, once established in terms of these descriptors, can be readily used to estimate the activity of hypothetical molecules.

The process of choosing the best subset of descriptors from an infinite number of possible choices is usually trial and error. The initial selection of descriptor types may be based on an hypothesized mode of action and should usually include at least one descriptor related to lipophilicity. Various significance measures for the correlation of biological activity with the chosen descriptor types taken individually and in combination are then calculated. Improvement in these measures may be sought by testing additional descriptors not initially considered. They may also involve transformation among the descriptors. Of course, testing too many descriptor types for inclusion in a QSAR relationship increases the possibility of chance correlations,[8] but failing to test may mean missing a descriptor that is significantly correlated with activity.

1. Lipophilicity Descriptors

Due to the ubiquitous influence of lipophilicity on both specific and non-specific binding, as well as on uptake and transport, descriptors of lipophilicity often form the foundation of herbicide QSAR relationships.

a. Log K_{ow} Values

The most generally used measure of lipophilicity is the octanol-water partition coefficient, K_{ow}. The first published application of log K_{ow} values in a QSAR analysis was to a study of plant growth regulators.[9] The high correlation that is often observed between biological activity and log K_{ow} may be attributed to the fact that partitioning into a water saturated octanol phase is a good model for partitioning into biological membranes.[10] To neglect log K_{ow} as a molecular descriptor could result in a QSAR with a rather weak foundation, especially when correlates of *in vivo* activity are being sought. Log K_{ow} values can be measured by shake flask methods or estimated from HPLC retention times.[11] Algorithms and computer programs for estimating log K_{ow} values from either fragment constants[5,12] or atomic constants[13] are also available. An algorithm for estimating log K_{ow} from molecular surface, volume, weight, and charge densities has recently been reported.[14]

b. π Values

In some instances the distribution of lipophilicity within a molecule may be a more important determinant of activity than the overall value. This will be especially true if activity is determined in part by filling a hydrophobic pocket in some target receptor protein. Hence, it may be preferable (or simply more convenient) to use substituent lipophilicity rather than whole molecule log K_{ow} as a descriptor in QSAR analysis. A commonly used descriptor of substituent lipophilicity is the π value. The π value of a substituent X is defined as the contribution of X to the log K_{ow} of a substituted benzene (Ph-X) relative to the contribution of a hydrogen atom:

$$\pi_X = \log K_{ow}(\text{Ph - }X) - \log K_{ow}(\text{Ph - }H) \tag{2}$$

Thus, by convention $\pi_H = 0$ and lipophilic substituents have positive π values ($\pi_{Me} = 0.56$), while hydrophilic substituents have negative π values ($\pi_{OH} = -0.67$). Extensive tables of π values and fragment constants are available.[5]

c. Hydrophobic Moments

Hydrophobic moments are defined in a manner analogous to electronic dipole moments.[15] They should also serve as useful descriptors of the distribution of hydrophobicity within a molecule. Molecules that are hydrophobic at one end and hydrophilic at the other will have large values. Hydrophobic moments may prove useful in situations where activity requires interaction with an amphiphilic surface (e.g., detergents with a membrane).

2. Size and Shape Descriptors

After lipophilicity, size and shape descriptors are among the most commonly used to build QSARs. These descriptors relate to the ability of a molecule to fit into a receptor. They may also account for the accessibility by various reagents, including binding site residues, to reactive centers within the molecule. Descriptors of size and shape should be routinely considered for inclusion in a QSAR. A valuable review of size and shape descriptors for QSAR analysis has been published.[16]

a. Molecular Volume and Molar Refractivity

Molecular volume can be calculated theoretically based upon one of several assumptions concerning molecular boundaries.[17] Molar refractivity (MR) is really a "corrected" form of molar volume[18] and is one of the most extensively used descriptors of volume in QSAR analysis. MR can be included in QSAR analysis as a whole molecule descriptor or as a substituent descriptor. Extensive tables of fragment[5] and atomic[19] MR values are published from which both substituent and whole molecule MR values can be estimated by summation. Computer programs for performing these estimations are available.[12] MR values can also be experimentally determined from measured refractive indices.

b. Taft E_s Values

The extent to which bulky R groups reduce the rate of acid catalyzed hydrolysis of methyl esters RCOOMe[20] defines the steric parameter E_s:

$$E_s = \log\left(\frac{k_R}{k_{Me}}\right) \tag{3}$$

Here k_R and k_{Me} are rate constants for the acid catalyzed hydrolysis of RCOOMe and methyl acetate (MeCOOMe) respectively. Note that E_s values become more negative with increasing size so that a negative coefficient for E_s in a QSAR indicates that activity increases with increasing bulk. The original set of Taft E_s values has been extended to nonalkyl substituents[21] and corrected to factor out confounding electronic effects.[22] E_s values have been used extensively in the QSAR analysis of herbicidal activity, including, e.g., a successful application to *N*-benzylacylamides.[23] A number of extensions of Taft's original definition have been described.[16]

c. Sterimol Parameters

The sterimol parameter L is a measure of the length of a substituent, while the parameters B_1, B_2, B_3, and B_4 are measures of substituent widths in four different directions. Tables for common substituents are available and an algorithm for their computation has been described.[24] In further developments of the Sterimol concept B_2, B_3, and B_4 have been superceded by B_5, a new maximum width parameter.[25] They should be considered for use whenever activity appears to depend more on substituent shape rather than simple bulk. For example, they have been used in QSAR analysis of several herbicide classes to describe subtle shape requirements for activity.[26-30]

d. Steric Field

The comparative molecular field analysis (CoMFA) method[31] uses the steric field as a rich source of molecular descriptors (see Section III.E.2). The steric field is defined for each point on a three dimensional grid enmeshing a molecule as the van der Waals interaction energy between the molecule and a probe (e.g., a methyl group). The steric field is particularly useful for describing the effect of size and shape on binding data. The effect of steric field parameters on a QSAR can be presented graphically as the image of a receptor site. This feature should be of value in designing novel compounds that bind to the imaged site.

3. Geometric Descriptors

Geometric descriptors, such as interatomic distances, angles, and dihedral angles, describe the relative spatial arrangement of functional groups within a molecule. Their use in QSAR analysis is suggested when there are two or more functional groups (e.g., a pair of hydrogen bond acceptors) whose relative positions or orientations vary within a group of analogues. Of course, molecules within an analogue series may be rather flexible, so that a given geometric descriptor may reasonably assume many values within a bounded range. The bounds themselves may then find use in various forms of QSAR such as the distance geometry method.[32] A sample application of the distance geometry method to photosystem II herbicides is described in Section III.E.1.

When a specific geometric arrangement of functional groups within a molecule determines biological activity, the identity of the functional groups and their geometric arrangement defines what is often called a pharmacophore. The process of validating a putative pharmacophore may be viewed as another form of QSAR analysis in which the functional relationship between biological activity and geometric descriptors is formulated as a set of rules that place restrictions on the geometric descriptors. A successful application of this approach to the discovery of new plant growth regulators that inhibit auxin transport has recently been described.[33]

4. Topological Descriptors

Topological descriptors contain information about the number and types of atoms or functional groups in a molecule and the connectivity among them. For example, the presence or absence of key structural features can be indicated by using descriptors whose value is either one or zero respectively. Such indicator variables may be used alone as in Free-Wilson analysis,[34] but are often used in combination with continuous variables. Indicator variables are often considered for inclusion in a QSAR when a particular structural feature seems to impart unusually large or unexpected changes in activity that cannot be accounted for by other continuous decriptors already included in the QSAR.

A number of schemes exist for encoding the topology of a structure into a compact set of numerical descriptors,[35] and some have been used as correlates of herbicidal activity. For example, connectivity indices[36] have been used in the QSAR analysis of photosystem II herbicides[37] and are used in building QSAR models of pesticide toxicity.[38] Connectivity indices have the advantage of being readily calculated for most molecules, but a QSAR in terms of connectivity indices is difficult to interpret chemically, and this may limit their value for both understanding and optimizing herbicidal activity.

5. Electronic Structure Descriptors

Binding to or reactivity at a receptor usually has an electronic component; hence, electronic structure descriptors are often included in QSAR analysis. Descriptors of electronic structure convey information about where electrons are in a molecule and where they would like to be under perturbations such as electric fields and electromagnetic radiation. Their significance in herbicidal QSARs is often less than that of either lipophilicity or size and shape descriptors, but their inclusion may be quite important if electrostatic interactions determine the strength of receptor binding or if chemical reactions are involved in the mode of action, activation, or deactivation.

a. Substituent Constants

Several constants have been described that quantify the electron withdrawing or donating ability of a substituent. This ability may influence herbicide binding, metabolism, and transport. Generally, these constants are defined in such a manner that electron withdrawing substituents have positive values and electron donating substituents have negative values. Of the many descriptors of electronic effects, σ values are the most commonly used in QSAR analysis. Hammett σ values[39] are defined in terms of substituent effects on the dissociation constant K_a of benzoic acid:

$$\sigma_X = \log\left[K_a(X\text{-Ph-COOH})\right] - \log\left[K_a(\text{Ph-COOH})\right] \quad (4)$$

Hammett σ constants are roughly additive (i.e., the effect of several substituents on a phenyl ring is the sum of the σ values) and constitutive (i.e., the σ values depend upon whether the substituent is in a meta or para position). Taft σ^* constants[20] are defined in terms of the base catalyzed hydrolysis of acetate esters as a reference reaction:

$$\sigma_X^* = \frac{\log\left[k_b(\text{XCOOR})\right] - \log\left[k_b(\text{CH3COOR})\right] - E_s(X)}{2.48} \quad (5)$$

Here k_b is the rate constant for hydrolysis and $E_s(X)$ is the Taft steric constant for X. Taft σ^* values are best suited for describing the electronic effects of substituents at saturated carbons.

Following the pioneering work of Taft and Lewis,[40] Swain and Lupton separated the electronic effect of a substituent into two roughly independent terms, symbolized by F and R.[41] The F value for a substituent X was defined in terms of the acid dissociaton constant of the corresponding 4-X-substituted bicyclo [2.2.2] octane-1-carboxylic acid (Structure 1) in much the same way as Hammett σ constants are defined in terms of the dissociation constants of benzoic acids (Equation 4).

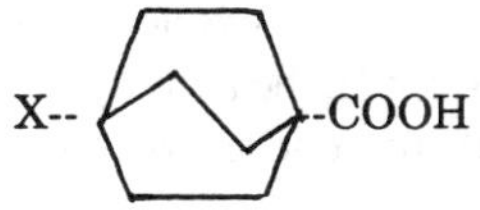

X-- [bicyclo[2.2.2]octane] -COOH

F values may be viewed as a measure of substituent ability to withdraw electrons via induction through single bonds and/or via a field effect through space.

The R value of a substituent X was determined from the relationship $R = (\sigma_p - 0.921{*}F)$, where σ_p is the Hammett σ value for the X substituent in a para position. The value, -0.921, was chosen so that the resonance parameter of trimethylammonium is zero. R values can be viewed as a measure of the ability of a substituent to withdraw electrons via a resonance effect (e.g., through a conjugated system of double and single bonds). F and R values are position independent, so that the same pair of values may be used to describe a substituent at any position on a phenyl ring. Of course, the relative weighting of F and R in a QSAR will be positionally dependent.

Collections of electronic substituent constants can be found in many standard sources.[5,42] As a rule, their significance in QSARs should be tested, but use is sometimes limited by incompleteness of the standard collections. If a desired substituent constant is missing from the standard collections, it can often be estimated by analogy to substituents of similar electronic structure whose constants are included in the tables.

b. Atomic Charges, Dipole Moments, Electrostatic Fields, and Polarizability

The distribution of charge within a herbicide will influence its electrostatic interaction with a binding site as well as its chemical reactivity. Electrostatic fields provide extremely detailed spatial information about charge distributions, while useful summaries are provided by atomic charges and multipole moments (e.g., dipole moments). Atomic charges,[43,44] dipole moments, and electrostatic fields may be calculated from molecular wavefunctions[45] as provided by a number of readily available electronic structure programs.[46] A number of more empirical methods for calculating atomic charges[47,48] and dipole moments[49] that do not require quantum mechanical calculations are also available.

Polarization of a herbicide charge distribution by a receptor may also contribute to binding via dispersion forces. MR values (see Section II.B.2.a) are measures of polarizability and hence the appearance of MR terms in a QSAR may be related to dispersion interactions.[50] Molecular polarizability may also be partitioned into atomic contributions,[51,52] and these have found value in some QSAR analyses.

Most descriptors of electronic charge distribution are readily computed, and hence their determination does not rely so heavily on databases as does that of substituent constants. The use of charge distribution descriptors will be particularly advantageous when structural variations within an analogue series are not simple substituent alterations.

c. Orbital Energies

The highest occupied molecular orbital (HOMO) and lowest unoccupied molecular orbital (LUMO) of a molecule have significant influence on chemical reactivity. Hence, HOMO and LUMO orbital energies are sometimes used directly as descriptors in QSAR analysis because of their expected influence on biological activity. Since HOMO and LUMO orbital energies are related to molecular redox properties, they are natural descriptors when nucleophilic or electrophilic attack or a redox step is suspected to be part of the mode of action of a herbicide (see Section II.B.6).

d. Properties of Electronic Excited States

Herbicides exert their effects in an environment that is generally bathed in sunlight. Some herbicides are photoactivated; others are photodegraded. Thus, herbicidal performance often depends upon the properties of electronic excited states, and these may serve as useful descriptors in QSAR analysis. For example, the activity of singlet oxygen generating polythienyl herbicides has been correlated with the energy of the lowest triplet state.[53]

6. Reaction Equilibrium Constants

Biological activity and chemical reactivity are often related. Reaction equilibrium constants are thus natural choices for inclusion as descriptors in QSAR analysis. Equilibrium constants for simple proton and electron transfer processes are among the most important to consider.

Many herbicides have acid dissociation constants (pK_a values) in the physiological pH range. pK_a values may determine the fraction of a compound in its active form or the ability of a compound to translocate from the site of application to the site of action.[54] Methods of empirically estimating[42] and theoretically calculating[55] pK_a values are available.

Reduction potentials measure the tendency of a molecule to participate in electron transfer reactions. For herbicides whose mode of action involves redox chemistry, reduction potentials are natural choices for inclusion as descriptors in a QSAR analysis. For example, the activity of bipyridinium herbicides[56] and the ability of quinones to displace DCMU (diuron) from thylakoid membranes[57] have both been correlated with reduction potential. Cyclic voltammetry provides a convenient method for measuring reduction potentials. Methods for calculating reduction potentials are available.[58] Methods for estimating reduction potentials include those based upon calculated LUMO energies (Section II.5). For example, Figure 1 illustrates a

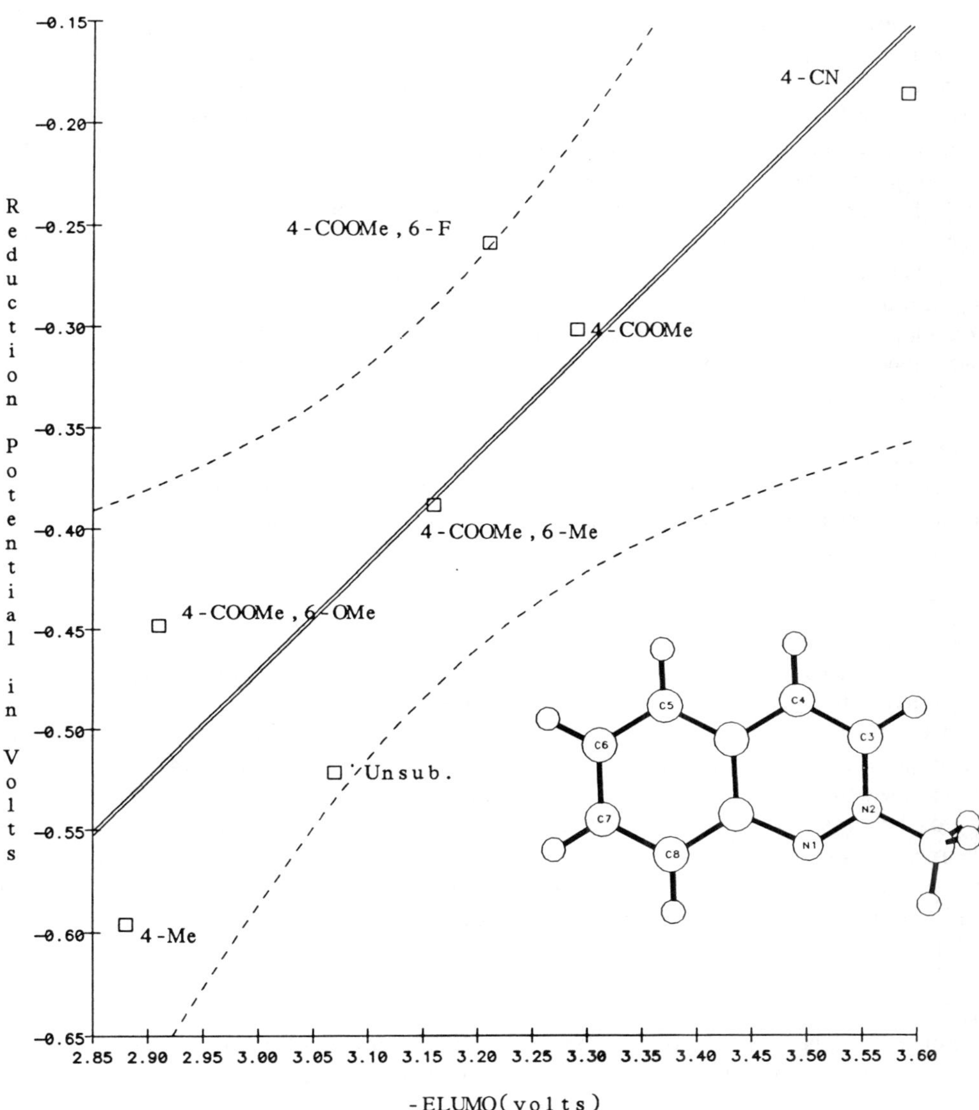

FIGURE 1. Linear correlation of reduction potential with LUMO orbital energy (ELUMO) for a series of substituted 2-methyl-cinnolinium herbicides.

significant correlation of measured reduction potentials with calculated LUMO energies that may be used to empirically estimate reduction potentials for the cinnolinium class of photosystem I herbicides.

Other equilibrium constants to consider for inclusion as descriptors are complexation constants with metal ions or hydrogen bond partners. Unfortunately, measured values of equilibrium constants for reactions related to the suspected mode of herbicidal action are seldom available. However, reaction equilibrium constants are often intimately related to electronic structure, so that inclusion of electronic structure descriptors such as σ values, LUMO energies, etc. (Section II.5) may indirectly account for important equilibrium effects.

C. DESCRIPTORS OF BIOLOGICAL ACTIVITY

As indicated above, a QSAR analysis usually describes a functional relationship between biological activity and molecular descriptors. Predictions from such an analysis cannot be any better than the biological data on which they are based and will also be compromised by poorly chosen or erroneously determined descriptors. In herbicide design programs attention must be paid not only to considerations about precision of the data, but also to the effect of environmental factors on both the plant and the herbicide molecule and to the level of analyses (*in vivo* vs. *in vitro*). QSAR analysis of *in vivo* data may be closer to the immediate goal of a herbicide optimization program, but it suffers from confounding factors such as metabolism and transport that may compromise the quality of the QSARs. On the other hand, analysis of *in vitro* data focuses largely on those factors important for intrinsic activity and neglects those confounding factors that may be essential for whole plant herbicidal activity. Together, *in vivo* and *in vitro* analyses have in a number of instances enabled a deeper understanding of how structure influences intrinsic activity as well as factors such as metabolism and transport.

1. Garbage In, Garbage Out

Martin and her colleagues[59,60] have provided thorough and thoughtful reviews on important considerations in the choice of biological data for QSAR studies in drug design. Since most of these considerations are directly applicable for herbicides, only a few key points will be summarized here.

a. Precision

The product of most QSAR analyses is, either explicitly or implicitly, a quantitative relationship (usually a regression equation) that can be used to predict the biological properties of a series of molecules. The merit of these predictions depends on the statistical significance of the relationship, which in turn, is a function of the precision of the potency values, the descriptors, and the range of both within the analogue series. Most herbicide design approaches are based on an attempt, either stated or implied, to understand the physical chemistry of the interaction between the herbicide and a specific constituent within the plant (the receptor). To gain understanding at the physicochemical level it is critical that the biological activity be quantified in physicochemical terms such as equilibrium or rate constants.

b. Potency

Potency can be defined as the reciprocal of the dose or concentration required to produce some predetermined biological response; it should be expressed on a logarithmic scale. The lower the dose required, the higher the potency. The dose most often used to evaluate potency is the median effective dose ED_{50}, the dose that is required to produce 50% of the maximal effect. EC_{50} is used where the molar concentration of the compound is considered rather than the dose. For biochemical responses to herbicides, such as inhibition of photosynthetic electron transport, the term I_{50} (the molar concentration at which 50% of the activity is inhibited) is often used. An analogous term for whole plant phytotoxicity is the GI_{50}, the concentration at which growth is inhibited by 50%.

c. Selection of Data

In general, QSAR calculations should be based on data generated from a test in which a continuous response has been measured at several doses. A dose-response analysis is essential to provide quantitative descriptions of the activity of a compound, such as threshold, ED_{50}, slope, and maximum effect, and to distinguish dose ranges at which different activities of the same compound occur. In seeking to predict properties from a series of compounds, one should have information that inactive compounds are truly inactive and that all compounds have the same mechanism of action. QSAR analysis of *in vivo* data for a herbicide class with multiple modes of action will be complicated, especially if GI_{50} or ED_{50} values for both modes of action are comparable.

As indicated above, precision of the ED_{50} data is critical for accurate QSAR predictions. The transformation of dose-response information that best linearizes the data is often used for calculating the ED_{50} or I_{50}, though nonlinear analysis techniques are becoming more common (Chapter 3). For responses which follow a Michaelis-Menten type equation, a plot of response vs. log dose is linear between 20 and 80% of the maximal response. Thus, optimization of the precision of potency requires that only points from 20 to 80% of the maximal response be used. This requirement can often pose a problem in herbicide bioassays, since the GI_{90} (concentration at which growth is 90% inhibited) is often preferred over the GI_{50} as a measure of economic weed control. However, lower precision will reduce the effectiveness of QSAR analyses if the the reciprocal of the GI_{90}, rather than the GI_{50}, is used as the measure of potency.

2. *In Vitro* vs. *In Vivo*

It seems reasonable that the precision of *in vitro* assays such as enzyme inhibition or photosynthetic electron transport would be greater than that for whole plant phytotoxicity. However, it is also logical that predictions of potency are only valid for the biological assay in which the potency was measured. How, then, do QSAR predictions from *in vitro* assays correlate with herbicidal activity?

Extensive QSAR studies have been reported on compounds that inhibit electron transport through photosystem II of higher plant chloroplasts (the Hill reaction), beginning with the pioneering work of Hansch and his colleagues (e.g., Hansch and Deutsch[61]). Electron transport inhibition has proven extremely useful as a basis for QSAR predictions of new active molecules (see, e.g., Kakkis, et al.[62]), but few studies have been reported that compare QSAR based on functional vs. herbicidal bioassays. One report that does make that comparison is that of Takemoto et al.[63] Multiple regression analysis related structural properties of phenyl-*N*-methoxy-*N*-methylureas to post-emergent herbicidal activity in radish plants and to inhibition of photosynthetic electron transport in isolated radish chloroplasts. Inhibition of electron transport was correlated with the hydrophobic constant, π, and the electronic constant, σ. Correlations of herbicidal activity with these or other constants were poor but could be improved if electron transport inhibition itself was used as an independent variable along with π. In this example, evidence was thus generated that herbicidal activity was related to functional inhibition at the chloroplast level. Another study that compared QSAR analyses of postemergent herbicidal activity with photosynthetic electron transport was that of Mitsutake et al.[64] Their conclusions were similar, in that inhibition of electron transport was enhanced by hydrophobicity of the compounds, and herbicidal activity was directly related to inhibition of electron flow.

The discovery by Tischer and Strotmann[65] that several classes of photosystem II inhibitors bind competitively at the same site in thylakoid membranes added a new level of analysis for these compounds. Now, one could assay the same compounds at three levels: at a specific receptor site, at an integrated and functional, sub-cellular electron transport chain, and at the whole plant level. Which of these bioassays would provide better data for QSAR studies? Gardner et al.[66] examined a large number of fluorophenyl urea herbicides for their ability to inhibit photosynthetic electron tranport and compete for atrazine binding sites in chloroplasts. Although QSAR analyses were not carried out in that study, binding activity showed a better correlation with whole plant phytotoxicity than did Hill reaction inhibition. This suggests that potency derived from binding data would be a better basis for QSAR studies than those from electron transport.

As indicated above (Section II.C.1.c), dose-response studies allow one to determine dose ranges at which different kinds of biological activity are exhibited in the same group of compounds. The ureas described by Takemoto et al.[63] cause bleaching, as well as inhibition of electron transport. These workers compared the QSAR for bleaching along with those for inhibition of electron transport and herbicidal activity (described above). Although a good QSAR was found for bleaching, it was quite different than that for Hill inhibition, and bleaching activity did not significantly correlate with post-emergent herbicidal activity. Furthermore, the bleaching activity occurred at a concentration range higher than that required for inhibition of photosynthesis. Thus, QSAR analyses helped to distinguish which of two biochemical effects of a series of compounds was responsible for herbicidal activity.

One of the most extensive comparisons of *in vitro* and *in vivo* herbicide QSAR has recently been reported[67] for the sulfonylurea class of herbicides. These compounds inhibit acetolactate synthase, the first enzyme in the biosynthetic pathway of the amino acids isoleucine and valine. Regression equations were developed for postemergent whole plant activity and enzyme inhibition. The equations were similar, and the parameters that determined *in vivo* activity were the same as for enzyme inhibition. However, the correlation coefficients for the whole plant equations were weaker and the standard deviations were broader.

3. Confounding Factors

We have seen that in some cases *in vivo* activity does correlate with *in vitro* activity, but that variability is usually greater for *in vivo* QSARs and thus the predictive ability of the QSAR is reduced. Some of this variability may be ascribed to a lower degree of precision in the potency measurement itself, but for *in vivo* tests there are additional biological and environmental variables that interfere with the ability of a compound to reach its active site and hence confound analysis of the results. A herbicide molecule must penetrate the waxy cuticle of the plant and enter into the vascular system. It must be transported to the site of action, and it must be resistant to inactivation by metabolism. Physical properties of the molecule that optimize uptake, transport, and metabolic stability may be quite different from those favored for high potency at the target site. In principle, QSAR techniques may be used to analyze each of these biological processes independently and thus optimize several properties in parallel.

Environmental factors also can alter herbicide potency both directly and indirectly via effects on the target plant. For example, soil type and the interaction of the molecule with the soil will determine its effectiveness as a pre-emergence herbicide. Microbial degradation in the soil may also lead to inactivation. Photochemical instability in sunlight may affect the available concentration on a leaf surface. The solubility of a molecule in water and its partition coefficient will determine whether or not it can survive rainfall shortly after application.

The developmental growth stage of the target plant may affect its response to a herbicide, as may the water status of the plant and the soil. If the target system requires light to function, as in photosynthesis, then light may be required for herbicidal activity. On the other hand, weed control may be enhanced by applying the compound at the end of the day or at night to allow thorough distribution of the compound in the plant before toxicity is manifested. Thus, time of day as well as light intensity may limit herbicidal activity. Finally, the ambient temperature may affect the plant response and the physical properties of the molecule itself.

Three key factors in herbicide design are activity, selectivity, and environmental stability. Although we have focused on activity, QSAR analyses have the potential to be equally useful in optimizing selectivity (at the level of uptake, metabolism, or the target site) and in choosing physical properties that minimize environmental insult and maximize economic utility.

III. TOOLS AND PRODUCTS OF QSAR ANALYSIS

The tools for building structure-activity relationships range from statistical analysis to molecular graphics analysis and methods based upon artificial intelligence. For statistical methods such as regression analysis, the functional relationship between biological activity and structural descriptors is a closed form equation. For some molecular graphical methods, the relationship may depend on the ability of functional groups of a molecule to contact binding sites or avoid excluded regions of a receptor image. For artificial intelligence methods the functional relationship may be encoded in a set of rules or in the strength of connections in a neural network. In fact, use of a combination of these tools should enhance the value of the resultant structure-activity picture.

A. MULTIPLE LINEAR REGRESSION (MLR)

The product of a multiple linear regression (MLR) analysis of structure-activity relations is a linear equation:

$$\text{Potency}_j = \Sigma_i C_i X_{ij} \tag{6}$$

Here Potency$_j$ is a measure of the biological activity of molecule j and X_{ij} is a structural descriptor of type i (e.g., π, MR, or σ value) for molecule j. The coefficients C_i are determined by a least square procedure that minimizes the variance between the experimentally measured values of potency and those given by Equation 6. As a rule, MLR requires at least 5 or 6 compounds per descriptor type considered for inclusion in Equation 6.[8] The partial least squares (PLS) method (Section III.B) promises to transcend this limitation on the number of compounds required.

The functional form of Equation 6 has its origins in physical organic chemistry, where the role of potency is played by the logarithm of an equilibrium contant. Thus, it is desirable in QSAR analysis of biological activity to use the logarithm of some biologically relevant equilibrium constant (e.g., a binding constant) as a measure of potency. The potency of herbicides is often expressed as $\log(1/C_x)$, where C_x is the concentration giving rise to a fixed effect x, e.g., 50% kill or 50% growth inhibition (see Section II.C.1). Assuming Michaelis competitive kinetics at some receptor site, it is reasonable to expect that C_x values will be proportional to inhibitor binding constants[65] as desired. Of course, the actual concentration of a herbicide in the vicinity of its receptor will usually be less than that near the application site due to the several penetration and permeation steps required to reach the vicinity of the receptor. Hence, some terms appearing in an MLR equation for herbicidal activity may be related to penetration and transport.[68]

The significance of an MLR equation can be judged by several statistical measures, which are usually reported along with the equation.[59,69] The square of the multiple linear correlation coefficient, R^2, is the fraction of the potency variance that is accounted for by the MLR equation. Values range from 0.0 to 1.0, with 1.0 indicating perfect correlation. Another measure of the goodness of fit is the standard deviation of the estimate from the measured potency, smaller values indicating better correlations. A statistic that is sensitive to the unwarranted use of too many descriptors is the F value for the regression. It is simply related to the multiple correlation coefficient by the expression

$$\mathrm{F}(n, N-n-1) = \left(\frac{R^2}{1-R^2}\right)^* \left(\frac{N-n-1}{n}\right) \tag{7}$$

where N is the number of compounds included in the regression and n is the number of descriptors used. F values should be compared to tabulated test values available in many standard references. If F exceeds the test value for a high confidence level, one can be confident that the coefficients in the MLR equation all differ significantly from zero. If, however, F is less than the test value, at least one of the terms in the fitting function should be removed since it is decreasing the multiple correlation by its inclusion.[69]

Improvements in these significance measures may be sought by judiciously removing outliers, including or rejecting censored data,[7] remeasuring or recalculating data, or synthesizing and testing new analogues, especially analogues that probe regions of descriptor space that were not well represented in the initial set of analogues.

1. Representative Examples of the Value of MLR in the QSAR Analysis of Herbicides

Some of the earliest applications of MLR to QSAR analysis were for hormonal-type plant growth regulators of the phenoxyacetic class.[9] These early studies demonstrated the importance of log K_{ow} as a correlate of potency and prompted the interpretation that log K_{ow} terms in Equation 6 may be related to the ability of a herbicide to transport through multiple lipophilic barriers.[68] Thus, the presence in Equation 6 of a quadratic term in log K_{ow} may account for the existence of an optimum partition coefficient for transport.

Many studies have demonstrated the value of MLR to concisely summarize large amounts of structure-activity data for herbicides in terms of traditional descriptors such as π, MR, σ, F, and R. Notable among these studies are extensive QSAR analyses of acetolactate synthase inhibition by sulfonylurea herbicides[67] and several studies of Hill reaction inhibition by photosystem II herbicides.

Hill reaction inhibition by phenyl amides depends strongly on log K_{ow}. The optimum log K_{ow} value for inhibition of photosynthetic electron transport in isolated chloroplasts is in the range of 5 to 6, while optimal herbicide activity occurs when log K_{ow} is in the 2 to 3 range.[62] A comparison of these QSARs thus teaches that optimal log K_{ow} for electron transport inhibition is not sufficient for herbicidal activity. Transport from the site of application (e.g., roots) to the site of action (chloroplasts of mature leaves) is also required, and this evidently is most efficient when the K_{ow} is several orders of magnitude lower than that required for optimum activity in the chloroplast. Other authors[63,64] have reached similar conclusions. Thus MLR analysis of structure-activity data has deepened our understanding of requirements for both intrinsic and whole plant activity of photosystem II herbicides and has provided rules of thumb for designing new herbicides with this mode of action. It is also satisfying that the optimum log K_{ow} for whole plant herbicidal activity (2 to 3 range) is near that required for good xylem translocation.[70,71]

The value of MLR analysis for optimizing herbicide related activity has been documented. For example, the Hill reaction activity of linuron analogs has been enhanced,[63] and new active subclasses of *N,N*-dimethyl-*N'*-phenylureas[72] and 1,2,4-triazin-5-ones[73] designed, using directions provided by QSAR analysis. The activity of the *N*-benzylacylamide (PhC[Me][Me]NHC[=O]R) class of herbicides has been related to the size and lipophilicity of the R group.[23] The guidance provided by this MLR analysis resulted in a herbicide (bromobutide) with activity some 100 times greater than the original lead. Ample support for the value of MLR analysis as a lead optimization tool can also be found in medicinal chemistry where many examples can be cited.[74,75]

Often selectivity improvement is just as important as activity optimization. QSAR in the form of MLR analysis can again provide useful guidance. For example, separate MLR equations can be derived for a crop and weed species. These can then be used to optimize selectivity. Such an approach has been exemplified for the selective action of herbicidal *N*-chloroacetyl-*N*-phenylglycine esters on barnyard grass in rice.[76]

Although MLR analysis is usually applied to a well defined class characterized by a common framework (e.g., anilide or triazine herbicides), comparison of separate QSARs for two different classes that bind to the same site can shed light on the nature of the common receptor. For example, strict size limitations common to separate QSARs prompted the observation that the acyl part of herbicidal anilides corresponds to the smaller of the two amino substituents of triazine herbicides.[77] Observations of this kind can be useful in the design of novel hybrid herbicides.

B. PARTIAL LEAST SQUARES (PLS)

In situations where the number of molecular descriptor types to be tested as correlates of biological activity is much larger than the number of molecules for which biological data has been obtained, the partial least square method may be of value.[78] In essence this method combines the large set of descriptor types into two sets: one of reduced size, known as the set of latent variables, that can account for a large fraction of the variance of the biological activity, and another that is of little use in this regard. The least squares procedure yields a linear equation in terms of the reduced set of latent variables. The method is similar to principal components regression (PCR) analysis. However, in PCR the latent variables, once extracted, are related to the response variable (e.g., biological activity) in a separate step, while in PLS the latent variables are extracted subject to a constraint. That constraint maximizes the communality of the extracted variables and the biological activity, thus obviating the need for a subsequent multiple regression step.[79]

The PLS method has been of value in 3D-QSAR approaches where the descriptors consist of electric and steric field values at a large number of grid points (see Section III.E.2). The statistical measures for building confidence in PLS equations are based on the concept of cross-validation. Essentially the data set is randomly divided into training and test sets. The PLS procedure is applied to each training set and the ability to predict the potency of the corresponding test sets is determined.

C. CLUSTER ANALYSIS

Graphical presentation of structure-activity data often provides insights that are not otherwise obvious. For example, a plotting technique known as parameter focusing can assist in the identification of the descriptors most important for controlling bioactivity.[80] In the two dimensional variant, a series of plots are constructed, one for each pair of molecular descriptors (e.g., π and MR). The points are marked according to whether the corresponding molecule is active or inactive. The plot that clusters the active compounds most effectively identifies the two most important parameters. The successful focusing of parameter pairs for a number of herbicide classes has been described.[80] These include bleaching herbicides of the arylpyridazone class, rice selective, barnyard grass active phenoxyacrylates, and pre-emergent triazolidione herbicides.

One advantage of the parameter focusing technique is its ability to handle qualitative or categorical (e.g., active-inactive) herbicidal data, but the method lacks statistical significance measures. A new technique that maintains all of the advantages of parameter focusing, while placing the method on a firmer statistical foundation, is cluster significance analysis (CSA).[81] CSA establishes whether the cluster of active compounds in a given descriptor space (e.g., π vs. MR) is significantly tighter than that generated by an equal number of compounds chosen at ran-dom from the data set of actives and inactives. If the cluster of actives passes the significance test, then the descriptors that define the space are significant determinants of activity.

D. DISCRIMINANT ANALYSIS — ADAPTIVE LEAST SQUARES (ALS)

The adaptive least squares (ALS) method is another method for handling categorical data. The ALS method[82] is an iterative procedure that seeks a single discriminant function that can place a molecule into one of several activity classes. The value of the discriminant function determines the category into which a molecule is predicted to fall (e.g., a value between the cutoffs b_1 and b_2 indicates membership in class 2, etc.). The discriminant function L evaluated for molecule j is formulated as a linear combination of molecular descriptors, X_{ij}

$$L_j = w_0 + w_1 X_{1j} + w_2 X_{2j} + \ldots \tag{8}$$

Initially, all molecules in a given category are assigned the same value for L. The cutoffs b between categories are set at the midpoints between the initial values assigned to the various categories. The weights w_i are then determined to minimize the difference between the values calculated by Equation 8 and those initially assigned. Once the first set of weights are determined, the discriminant function is evaluated for each of the molecules according to Equation 8. On the basis of the L_j values, some molecules may now be assigned to the wrong category. The L_j values for the misassigned molecules are "adapted" to bring them closer to the cutoff, which needs to be crossed in order to re-enter the appropriate category. New weights are then determined and this procedure is iterated until all molecules are assigned to their proper category or a limit of usually 30 iterations is reached.

E. 3D-QSAR

1. Distance Geometry

Despite numerous structure-activity studies within classes of photosystem II inhibitors, little work has been done to correlate activity across class boundaries. For example, only rarely has there been an attempt to find a single MLR equation to correlate biological activity across several herbicide classes, though one relevant example treats photosystem II herbicides from the closely related anilide, urea, and carbamate classes.[64]

The distance geometry method for QSAR analysis[32] promises to accommodate several classes that bind to the same receptor. The product of such an analysis is a receptor image that includes not only the geometrical arrangement of the points to which a ligand might be attracted or repelled, but also, in its full implementation, binding energies associated with the attractive site points. In order to image a binding site, cartesian coordinates and binding energies for each ligand are required. Desired binding modes must also be specified. They consist of pairings of ligand points (usually centered on ligand atoms or on centroids of ligand rings) and receptor site points (usually considered to be associated with functional groups of the receptor).

The distance geometry algorithm provides for rotation about essential single bonds within each ligand. These internal rotations allow interatomic distances within each ligand to vary within bounds. From these bounds and from the assigned binding modes, upper and lower bounds on the distances between receptor site points are calculated. Cartesian coordinates for site points that satisfy these constraints are then determined. The algorithm also proposes alternative binding modes in addition to the desired ones. Finally, pairwise interaction energies between site points and ligand points are varied to minimize the difference of the experimental binding energies and the theoretical binding energies. A theoretical binding energy is calculated as a sum of pairwise interaction energies between ligand points and site points with which they are in contact.

A sample application to the binding site of photosystem II inhibitors is illustrated in Figures 2 through 4. The ligands considered include a truncated plastoquinone (the presumed endogenous ligand), diuron, atrazine, dinoseb, and a methylbenzyl atrazine, MBAT (Figure 2). The desired binding modes (Figure 2) correspond to a recently proposed model.[83] Alternative desired binding modes could have been considered. For the herbicides, experimental binding energies

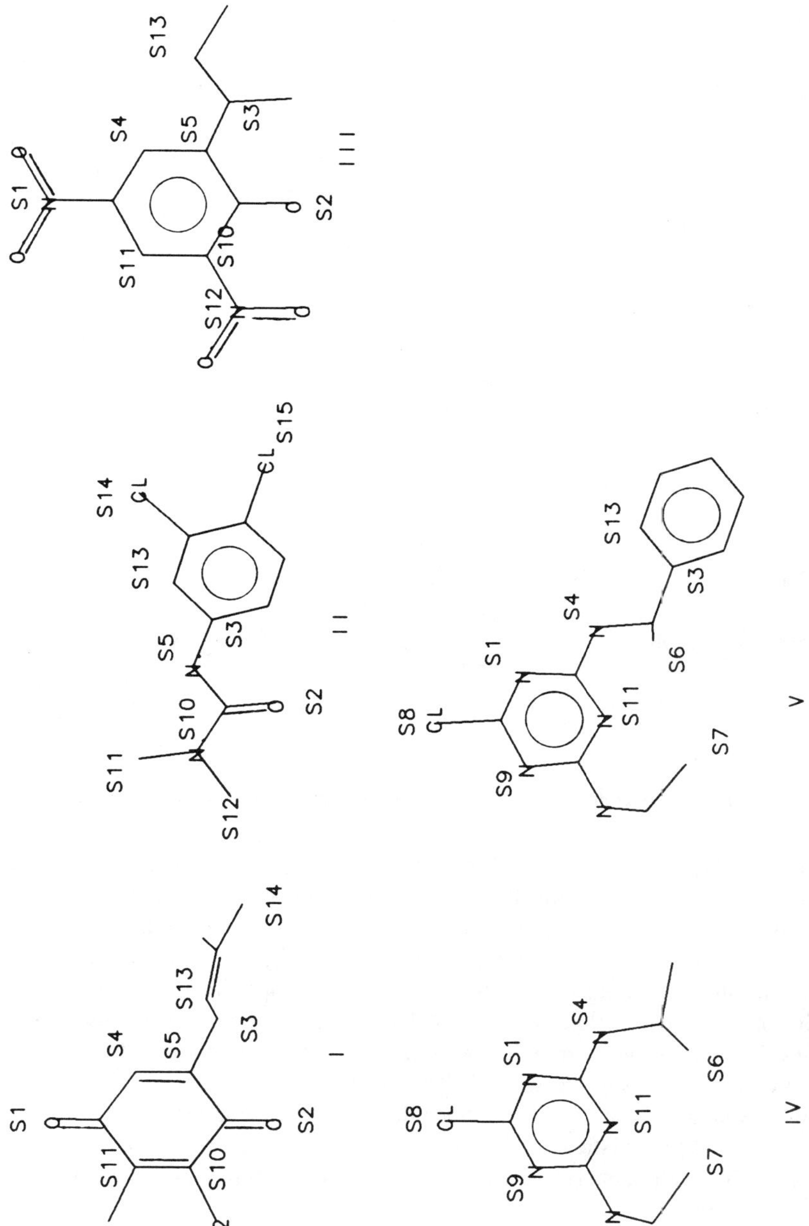

PLASTOQUINONE (I), DIURON (II), DINOSEB (III)

ATRAZINE (IV), MBAT (V)

FIGURE 2. Desired binding modes used to generate binding site of PSII inhibitors. The symbol S1 next to an atomic symbol implies the desire that the atom be in contact with receptor site point S1, etc.

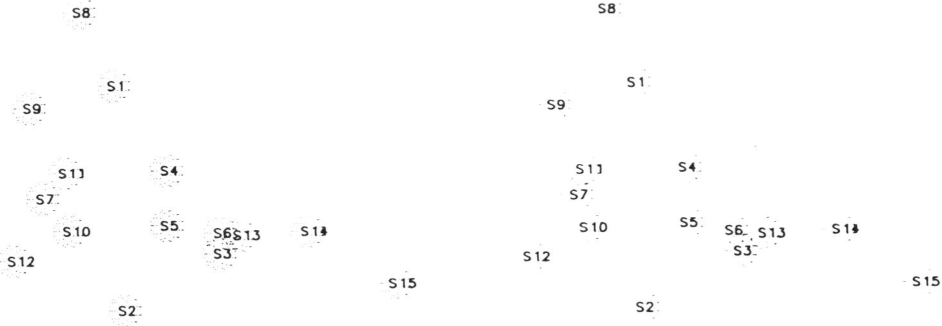

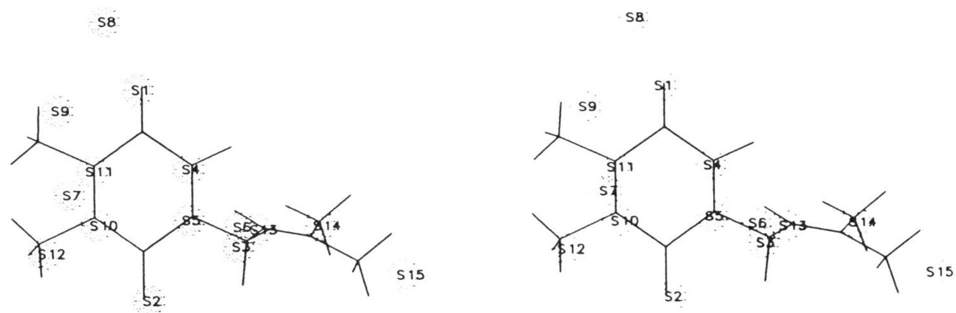

FIGURE 3. Stereoview[95] of 15 point binding site (top) and binding site occupied by plastoquinone (bottom).

were determined from Hill reaction I_{50} values (Table 1). The binding energy for plastoquinone was that determined for ubiquinone at the B-site of bacterial reaction centers.[84]

One of several resultant binding sites consistent with the distance constraints imposed by the desired binding modes is depicted as a stereo pair in Figure 3. Figure 3 (bottom) illustrates how plastoquinone fits into the imaged binding site. Figure 4 (top) illustrates the desired binding mode of diuron. In the cases of atrazine and MBAT, alternative binding modes equally acceptable as the desired modes were found. For atrazine the desired contact site point for the terminal carbon of the *N*-ethyl group was S7. The program has determined that contact with S12 is equally acceptable. Table 1 also includes the calculated binding energies for the photosystem II herbicides. The statistical significance of these binding energies is uncertain, but the overestimate of the binding energy of dinoseb suggests that our model may be missing a key repulsive binding site.

2. Comparative Molecular Field Analysis (CoMFA)

As noted above, steric and electrostatic fields evaluated on a molecule centered grid provide a large number of highly redundant descriptor types (two types for each of a multitude of grid points). The CoMFA method[31] reduces the number of descriptors by combining them according to the PLS procedure (see Section III.B) while simultaneoulsy performing an MLR on the reduced set of "latent" variables. The resultant linear equations can be back transformed

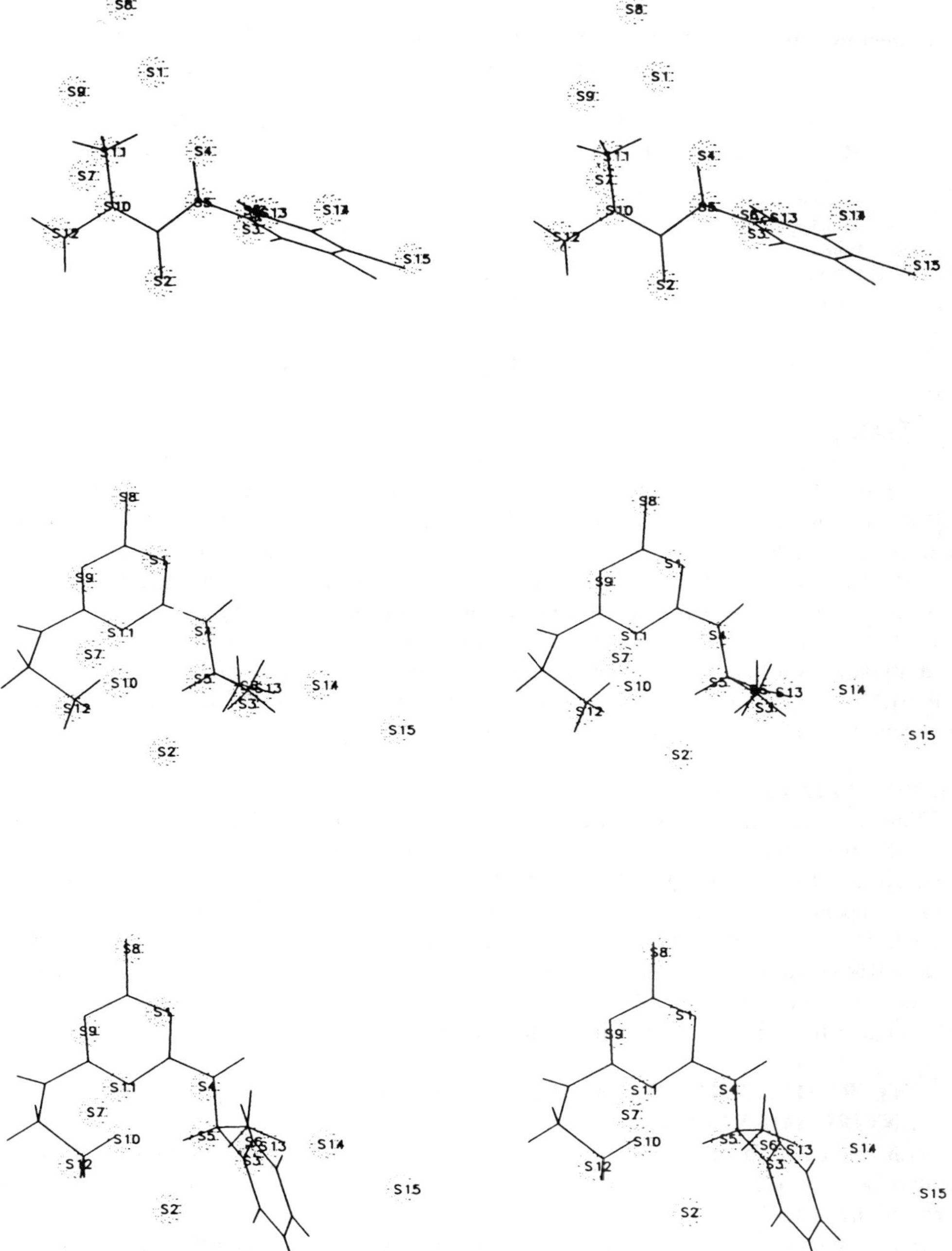

FIGURE 4. Stereoview[95] of binding site occupied by diuron (top), atrazine (middle), and MBAT (bottom).

to field variables on the grid points to visualize the requirements for active compounds on a coefficient contour map. Application of the method requires that ligands be superimposed. This is left to the user, but superposition based upon maximization of overlap of steric and electrostatic fields is recommended if there is no obvious structural feature common to all compounds in the series. Although no applications to herbicides have been reported, this method holds significant promise.

TABLE 1
Experimental and Calculated Binding Energies for Five Molecules Used to Build Image of Binding Site for Photosystem II Herbicides

Molecule	pI_{50}	Reference	Binding energies: Experimental	Binding energies: Calculated
Plastoquinone	—	84	–7.80	–7.81
Atrazine	6.71	94	–9.10	–9.11
Diuron	7.17	94	–9.80	–9.81
MBAT	6.70	94	–9.10	–9.11
Dinoseb	5.10	29	–6.90	–9.80

IV. RECENT AND FUTURE DEVELOPMENTS

A. QSAR SYSTEMS

Academic and commercial developers continue to improve software systems designed to facilitate the discovery of structure-activity trends. Programs especially designed for performing MLR analysis of SAR data are available.[85] ChemStat[86] is a system that collects modeling information, such as charges, and geometric information into a tabular form and provides for MLR analyses. Molecular spreadsheets capable of organizing chemical information by substituents (R groups) provide a convenient starting point for QSAR analysis of large databases.[87] The ADAPT system[88] is a complete system for computer assisted structure-activity studies using pattern recognition techniques. Modules that enable structure-activity rules to be encoded and then used to search large databases continue to appear.[89,90] Recent innovations such as CoMFA are also available[91] commercially.

B. VISUALIZATION

Enhanced visualization tools will facilitate the discernment of structure-activity trends. Visualization enables powerful human pattern recognition skills to complement computer logic. The 3D variants of QSAR (e.g., COMFA and distance geometry) are enlivened, if not totally dependent on, integration with molecular graphics for the results to be fully appreciated. Molecular graphics can also be used to elucidate the significance of various terms in an MLR equation.[50] Techniques such as parameter focusing or cluster significance analysis for viewing multidimensional SAR data in subspaces of reduced dimension (e.g., 1, 2, or 3D) also enable human pattern recognition to play a more active role.

C. INTERFACE WITH DETERMINISTIC MODELS FOR TRANSPORT AND ENVIRONMENTAL FATE

A herbicide's itinerary from point of application to site of action has become well enough understood for deterministic models to have been developed to describe them. Typically the predictions of these models depend upon environmental parameters[54] (e.g., soil pH, cuticle thickness, sap velocity, etc.) and herbicide properties (e.g., log K_{ow} and pK_a). Interfacing these quantitative structure-transport models with the results of QSAR analysis of binding data could enable predictions of whole plant activity from *in vitro* activity, as well as predictions of how activity might vary with species and environmental conditions.

D. ARTIFICIAL INTELLIGENCE

Many modern chemical information systems effectively enable the encoding of SAR information into a set of rules for activity (pharmacophore definition). Such expert systems can be trained to discriminate between active and inactive compounds and can then be used to screen large databases for novel compounds with similar activity. Neural networks can be trained to reproduce information about the activity of a compound from its structure. This information is encoded in the strengths of the connections between computer simulated neurons. Once trained, the neural network can be asked to predict the activity of compounds outside the training set. This method for analyzing structure-activity data may offer the advantage of being able to account for nonlinearities and interactions between structural descriptors in the QSAR more effectively than multiple linear regression.[92] Recently applications of neural nets have been made to structure-activity analysis of several classes of medicinal chemicals.[93] In the cases analyzed, the classification success of the neural network appeared to be superior to that of the ALS method.

V. CONCLUSION

The very discipline of designing an analogue program to yield sound QSARs has benefits, even though a statistically significant relationship may allude the experimenter. Good design minimizes the chance that the optimal compounds in an analogue program will be missed while reducing the number of compounds that need to be synthesized in order to find this optimum. If a valid QSAR can be developed, it provides, at the very least, an excellent summary of structure-activity data; in many instances it will have predictive value. QSAR analysis can thus provide direction to a herbicide optimization program where the goal is to maximize activity within an analog series. Many examples of successful interpolations and, to a lesser extent, extrapolations of QSARs to more active compounds are known. Some of these have been discussed in Section III.A.1 and additional examples can be drawn from other pesticide and pharmaceutical programs.[74] In favorable instances, QSAR analysis has the potential to generate new leads. This possibility becomes less remote when traditional approaches such as MLR are coupled with other computer aided design tools such as molecular graphics and pharmacophore searching. Receptor site images invite the user to design new molecules complementary to the receptor, and pharmacophore searching of 2D and 3D databases can provide a wealth of ideas for new chemistry.

ACKNOWLEDGMENTS

The authors wish to thank Professor Toshio Fujita for helpful comments on this manuscript.

REFERENCES

1. **Box, G. E. P., Hunter, W. G., and Hunter, J. S.,** *Statistics for Experimenters,* John Wiley & Sons, New York, 1978.
2. **Wold, S., Sjostrom, M., Carlson, R., Lundstedt, T., Helberg, S., Skagerberg, B., Wikstorm, C., and Ohman, J.,** Multivariate design, *Analyt. Chem. Acta,* 191, 17, 1986.
3. **Austel, V.,** A manual method for systematic drug design, *Eur. J. Med. Chem., Chim. Ther.,* 17, 9, 1982.
4. **Marsili, M.,** An expert system for chemometrics-based optimization in chemistry, *Tetrahedron Comput. Methodol.,* 1, 71, 1988.
5. **Hansch, C. and Leo, A.,** *Substituent Constants for Correlation Analysis in Chemistry and Biology,* John Wiley & Sons, New York, 1979.
6. **Borth, D. M. and McKay, R. J.,** A difficulty information approach to substituent selection in QSAR studies, *Technometrics,* 27, 25, 1985.
7. **Dekeyser, M. A., Borth, D. M., Moore, C., and Mishra, A.,** Quantitative structure-activity relationships in acaracidal 4H-1,3,4-oxadiazin-5(6H)-ones, *J. Agric. Food Chem.,* 39, 374, 1991.
8. **Topliss, J. G. and Costello, R. J.,** Chance correlations in structure-activity studies using multiple regression analysis, *J. Med. Chem.,* 15, 1066, 1972.

9. **Hansch, C., Muir, R.M., Fujita, T., Maloney, P. P., Geiger, F., and Striech, M.,** The correlation of biological activity of plant growth regulators and chloromycetin derivatives with Hammett constants and partition coefficients, *J. Am. Chem. Soc.,* 85, 2817, 1963.
10. **Levitan, H. and Barker J. L.,** Salicylate: a structure-activity study of its effects on membrane permeability, *Science,* 176, 1423, 1972.
11. **Veith, G. D. and Morris, R. T.,** A rapid method for estimating log P for organic chemicals, *Water Res.,* 13, 43, 1979.
12. MedChem Software, Medicinal Chemistry Project, Pomona College, Claremont, CA.
13. **Ghose, A. K. and Crippen, G. M.,** Atomic physicochemical parameters for three-dimensional-structure-directed quantitative structure-activity relationships. I. Partition coefficients as a measure of hydrophobicity, *J. Comput. Chem.,* 7, 565, 1986.
14. **Bodor, N., Gabanyi, Z., and Wong, C.,** A new method for the estimation of partition coefficients, *J. Am. Chem. Soc.,* 111, 3783, 1989.
15. **Eisenberg, D., and McLachlan, A. D.,** Solvation energy in protein folding and binding, *Nature,* 319, 199, 1984.
16. **Fujita, T. and Iwamura, H.,** Applications of various steric constants to quantitative structure-activity relationships, in *Topics in Current Chemistry,* Boschke, F. L., Ed., Sringer-Verlag, Berlin, 1983, 119.
17. **Connolly, M.,** Computation of molecular volumes, *J. Am. Chem. Soc.,* 107, 1118, 1985.
18. **Dunn, W.J., III,** Molar refractivity as an independent variable in quantitative structure activity studies, *Eur. J. Med. Chem.,* 12, 109, 1977.
19. **Ghose, A. K. and Crippen, G. M.,** Atomic physicochemical parameters for three-dimensional-structure-directed quantitative structure-activity relationships. II. Modeling dispersive and hydrophobic interactions, *J. Chem. Inf. Comput. Sci.,* 27, 21, 1987.
20. **Taft, R.,** in *Steric Effects in Organic Chemistry,* Newman, M. S., Ed., Wiley, New York, 1956, 556.
21. **Kutter, E. and Hansch, C.,** Steric parameters in drug design, monoamine oxidase inhibitors and antihistamines, *J. Med. Chem.,* 12, 647, 1969.
22. **Hancock, C. K., Meyers, E. A., and Yager, B. J.,** Quantitative separation of hyperconjugative effects from steric substituent constants, *J. Am. Chem. Soc.,* 83, 4211, 1961.
23. **Kirino, O., Furuzawa, K., Takayama, C., Matsumoto, H., and Mine, A.,** Quantitative structure-activity relationships of herbicidal *N*-(1-methyl-1-phenethyl)acylamides, *J. Pestic. Sci.,* 8, 301, 1983.
24. **Verloop, A., Hoogenstraaten, W., and Tipker, J.,** Development and application of new steric substituent parameters in drug design, in *Drug Design, Vol. 7, Medicinal Chemistry Series,* Ariens, E. J., Ed., Academic Press, 1976, 165.
25. **Verloop, A.,** The Sterimol approach: further development of the method and new applications, in *Proceedings of 5th International Congress of Pesticide Chemistry,* Vol. 1, Miyamoto, J. and Kearney, P. C., Eds., Pergamon Press, Oxford, 1983, 339.
26. **Kirino, O., Furuzawa, K., Takayama, C., and Mizutani, T.,** Herbicidal activity of *N*-benzylbutanamides. VII. Steric effect of acyl and amine moeity substituents on the herbicidal activity of *N*-benzylbutanamides, *J. Pestic. Sci.,* 9, 345, 1984.
27. **Shaner, D. L., Speltz, L. M., and Szucs, S. S.,** Synthesis and herbicidal activity of pyridazines based on 3-chloro-4-methyl-6-[m-(trifluoromethyl)phenyl]pyridazine, in *Synthesis and Chemistry of Agrochemicals,* ACS, Washington, D.C., 1987, chap. 3, p.84.
28. **Ladner, D. W.,** Structure-activity relationships among the imidazolinone herbicides, *Pestic. Sci.,* 29, 317,1990.
29. **Trebst, A. and Draber, W.,** Structure-activity correlations of recent herbicides in photosynthetic reactions, in *Advances in Pesticide Science, Part 2,* Geissbuhler, H., Ed., Pergamon Press, 1979, 223.
30. **Ohta, H., Jikihara, T., Wakabayashi, K., and Fujita, T.,** Quantitative structure-activity study of herbicidal *N*-aryl-3,4,5,6-tetrahydrophtalimides and related cyclic imides, *Pestic. Biochem. Physiol.,* 14, 153, 1980.
31. **Cramer, R. D., Patterson, D. E., and Bunce, J. D.,** CoMFA:1. Effect of shape on binding of steroids to carrier proteins, *J. Am. Chem. Soc.,* 110, 5959, 1988.
32. **Crippen, G.,** Distance geometry approach to rationalizing binding data, *J. Med. Chem.,* 22, 988, 1979.
33. **Bures, M. G., Black-Schaefer, C.,and Gardner, G.,** The discovery of novel auxin transport inhibitors by molecular modeling and 3-dimensional pattern analysis, *J. Comput. Aided Molec. Design,* 5, 323, 1991.
34. **Kubinyi, H.,** in *Comprehensive Medicinal Chemistry,* Vol. 4, *Quantitative Drug Design,* Hansch, C., Sammes, P. G., Taylor, J. B., and Ramsden, C. A., Eds., Pergamon Press, 1990, 590.
35. **Rouvray, D. R.,** Predicting chemistry from topology, *Sci. Am.,* 255, (3), 40, 1986.
36. **Hall, L. H. and Kier, L. B.,** Molecular connectivity and total response surface, *J. Molec. Struct. (THEOCHEM),* 134, 309, 1986.
37. **Soskic, M. and Sabljic, A.,** Inhibition of the Hill reaction by 3-alkoxy uracil derivatives: Quantitative structure-activity study with topological indices, *Croat. Chem. Acta,* 60, 755, 1987.
38. **Enslein, K.,** Estimation of toxicological endpoints by structure-activity relationships, *Pharm. Rev.,* 36, 131S, 1984.

39. **Hammett, L. P.,** Some relations between reaction rates and equilibrium constants, *Chem. Rev.,* 17, 125, 1935.
40. **Taft, R. W. and Lewis, I. C.,** The general applicability of a fixed scale of inductive effects. II. Inductive effects of dipolar substituents in the reactivities of m- and p-substituted derivatives of benzene, *J. Am. Chem. Soc.,* 80, 2436, 1958.
41. **Swain, C. G. and Lupton, E. C.,** Field and resonance components of substituent effects, *J. Am. Chem. Soc.,* 90, 4328, 1968.
42. **Hansch, C., Leo, A., and Taft, R.W.,** A survey of Hammett substitution constants and resonance and field parameters, *Chem. Rev.,* 91, 165, 1991.
43. **Mulliken, R. S.,** Electronic population analysis on LCAO-MO molecular wavefunctions, *J. Chem. Phys.,* 23, 1833, 1955.
44. **Armstrong, D. R., Perkins, P. G., and Stewart, J. J. P.,** Bond indexes and valence, *J. Chem. Soc., Dalton Trans.,* 838, 1973.
45. **Clark, T.,** *A Handbook of Computational Chemistry,* John Wiley & Sons, New York, 1985.
46. Quantum Chemistry Program Exchange, Department of Chemistry, Indiana University, Bloomington, IN.
47. **Gasteiger, J. and Marsili, M.,** Iterative partial equalization of orbital electronegativity: a rapid access to atomic charges, *Tetrahedron,* 36, 3219, 1980.
48. **Skorczyk, R.,** The calculation of crystal energies as an aid in structural chemistry. I. A semi-empirical potential field model with atomic constants and parameters, *Acta Cryst.,* A32, 447, 1976.
49. **McClellan, A. L.,** *Tables of Experimental Dipole Moments,* Vol. 2, Rahara Enterprises, El Cerrito, CA, 1979.
50. **Hansch, C. and Klein, T.,** Molecular graphics and QSAR in the study of enzyme-ligand interactions. On the definition of bioreceptors, *Acc. Chem. Res.,* 19, 392, 1986.
51. **Gasteiger, J. and Hutchings, M. G.,** Quantitative models of gas phase proton transfer reactions involving alcohols, ethers and their thio analogues. Correlation analysis based on residual electronegativity and effective polarizability, *J. Am. Chem Soc.,* 106, 6489, 1984.
52. **Abdul-Ahad, P. G., Blair, T., and Webb, G. A.,** Quantitative structure-activity study of some enzyme inhibitory quinazolines, *Int. J. Quantum Chem.,* 17, 821, 1980.
53. **Sun, K., Pilgram, K. H., Kleier, D. A., Schroeder, M. E., and Yang, A. Y. S.,** 2′5′;5′2″-terthiophene and heteroarene analogs as potential novel pesticides, in *Synthesis and Chemistry of Agrochemicals (II),* Baker, D. R., Fenyes, J. G., and Moberg, W. K., Eds., American Chemical Society, Washington, D.C., 1991, chap. 29.
54. **Kleier, D. A.,** Phloem mobility of xenobiotics. I. Mathematical model unifying the weak acid and intermediate permeability theories, *Plant Physiol.,* 86, 803, 1986.
55. **Jorgensen, W. L.,** Free energy calculations: a breakthrough for modeling organic chemistry in solution, *Acc. Chem. Res.,* 22, 184, 1989.
56. **Calderbank, A.,** Chemical structure and bioactivity of bipyridinium herbicides, in *Proc. Int. Conf. Pestic. Chem.,* Vol. 5, Tahori, A. S., Ed., Gordon and Breach, 1972, 29.
57. **Soll, H. and Ottmeier, W.,** Inhibitor binding and displacement in plastoquinone depleted chloroplasts, in *Adv. Photosynth. Research, Proc. Int. Congr. Photosynth., 6th,* Vol. 4, Sybesma, C., Ed., Nijhoff/Junk, 1984, 5.
58. **Reynolds, C. A., King, P. M., and Richards, W. G.,** Computed redox potentials and the design of bioreductive agents, *Nature,* 334, 80, 1988.
59. **Martin, Y. C.,** Biological data, in *Quantitative Drug Design,* Martin, Y. C., Ed., Marcel Dekker, New York, 1978, chap. 5.
60. **Martin, Y. C., Bush, E. N., and Kyncyl, J. J.,** Quantitative description of biological activity, in *Comprehensive Medicinal Chemistry,* Vol. 4, Hansch, C., Sammes, P. G., Taylor, J. B., and Ramsden, C. A., Eds., Pergamon Press, Oxford, 1990, 349.
61. **Hansch, C. and Deutsch, E. W.,** Structure-activity relation in amides inhibiting photosynthesis, *Biochim. Biophys. Acta,* 112, 381, 1966.
62. **Kakkis, E., Palmire, V. C., Jr., Strong, C. D., Bertsch, W., Hansch, C., and Schirmer, U.,** Quantitative structure-activity relationships in the inhibition of Photosystem II in chloroplasts by phenylureas, *J. Agric. Food Chem.,* 32, 133, 1984.
63. **Takemoto, I., Yoshida, R., Sumida, S., and Kamoshita, K.,** Quantitative structure-activity relationships of herbicidal *N*′-substituted phenyl-*N*-methoxy-*N*-methylureas, *Pest. Biochem. Physiol.,* 23, 341, 1985.
64. **Mitsutake, K., Iwamura, H., Shimizu, R., and Fujita, T.,** Quantitative structure-activity relationship of Photosystem II inhibitors in chloroplasts and its link to herbicidal action, *J. Agric. Food Chem.,* 34, 725, 1986.
65. **Tischer, W. and Strotmann, H.,** Relationship between inhibitor binding by chloroplasts and inhibition of photosynthetic electron transport, *Biochim. Biophys. Acta,* 460, 113, 1977.
66. **Gardner, G., Pilgram, K. H., Brown, L. J., and Bozarth, G. A.,** *In vitro* activity of sorghum-selective fluorophenyl urea herbicides, *Pest. Biochem. Physiol.,* 24, 285, 1985.

67. **Andrea, T. A., Artz, S. P., Ray, T. B., and Pasteris, R. J.,** Structure-activity relationships of sulfonylurea herbicides, in *Rational Approaches to Structure, Activity and Ecotoxicology of Agro-chemicals,* CRC Press, Boca Raton, FL, 1992, 373.
68. **Penniston, J. T., Beckett, L., Bentley, D. L., and Hansch, C.,** Passive permeation of organic compounds through biological tissue: a non-steady state theory, *Mol. Pharmacol.,* 5, 333, 1969.
69. **Bevington, P. R.,** *Data Reduction and Error Analysis for the Physical Sciences,* McGraw-Hill, New York, 1969.
70. **Briggs, G., Bromilow, R. H., and Evans, A. A.,** Relationships between lipophilicity and root uptake and translocation of non-ionised chemicals in barley, *Pestic. Sci.,* 13, 495, 1982.
71. **Hsu, F. C., Marxmiller, R. L., and Yang, A.,** Study of root uptake and xylem translocation of cinmethylin and related compounds in detopped soybean roots using a pressure chamber technique, *Plant Physiol.,* 93, 1573, 1990.
72. **Camilleri, P., Bowyer, J. R., Gilderson, T., Odell, B., and Weaver, R. C.,** Structure activity relations in Hill inhibition of substituted phenylureas, *J. Agric. Food Chem.,* 35, 479, 1987.
73. **Draber, W.,** Can quantitative structure activity analysis and molecular graphics assist in designing new inhibitors of photosystem II?, *Z. Naturforsch.,* 42c, 713, 1987.
74. **Fujita, T.,** The extrathermodynamic approach in drug design, in *Comprehensive Medicinal Chemistry,* Vol 4, *Quantitative Drug Design,* Hansch, C., Sammes, P. G., Taylor, J. B., and Ramsden, C. A., Eds., Pergamon Press, 1990, chap. 21.
75. **Martin, Y. C.,** A practitioner's perspective of the role of quantitative structure-activity analysis in medicinal chemistry, *J. Med. Chem.,* 24, 229, 1981.
76. **Fujinami, A., Satomi, T., Mine, A., and Fujita, T.,** Structure-activity study of herbicidal *N*-chloroacetyl-*N*-phenylglycine esters, *Pestic. Biochem. Physiol.,* 6, 287, 1976.
77. **Shimizu, R., Iwamura, H., and Fujita, T.,** Quantitative structure-activity relationships of photosystem II inhibitory anilides and triazines. Topological aspects of their binding to the active site, *J. Agric. Food Chem.,* 36, 1276, 1988.
78. **Wold, S.,** Multivariate design, *Analyt. Chim. Acta,* 191, 17, 1986.
79. Sybyl 5.4 Theory Manual, Tripos Associates, St. Louis, MO, chap. 6.3.4, 1991.
80. **Magee, P. S.,** Parameter focusing — a new QSAR technique, in *IUPAC Pesticide Chemistry: Human Welfare and the Environment,* Miyamoto, J. and Kearney, P. C., Eds., Pergamon Press, Oxford, 1963, 251.
81. **McFarland, J. W. and Gans, D. J.,** On the significance of clusters in the graphical display of structure-activity data, *J. Med. Chem.,* 29, 505, 1986.
82. **Moriguchi, I., Katsuichiro, K., and Matsushita, Y.,** Adaptive least squares method applied to structure-activity corelation of hypotensive *N*-alkyl-*N″*-cyano-*N′*-pyridylguanidines, *J. Med. Chem.,* 23, 20, 1980.
83. **Gardner, G.,** A stereochemical model for the active site of photosystem II herbicides, *Photochem. Photobiol.,* 49, 331, 1989.
84. **Stein, R. R., Castellvi, A. L., Bogacz, J. P., and Wraight, C. A.,** Herbicide-quinone competition in the acceptor complex of photosynthetic reaction centers from Rhodopseudomonas sphaeroides: a bacterial model for photosystem II herbicide activity in plants, *J. Cell Biochem.,* 24, 243, 1984.
85. QSAR: Medicinal Chemistry Project, Pomona College, Claremont, CA.
86. ChemStat: Chemical Design, Ltd., Oxford, England.
87. MOLECULAR SPREADSHEET: Molecular Design Ltd., San Leandro, CA.
88. ADAPT: Molecular Design Ltd., San Leandro, CA.
89. MACCS-3D, POWERSEARCH: Molecular Design Ltd., San Leandro, CA.
90. ALADDIN, MERLIN, GENIE: Daylight Chemical Information Systems, New Orleans, LA.
91. CoMFA: Tripos Associates, St. Louis, MO.
92. **Andrea, T. A. and Kaleyah, H.,** Application of neural network technology to quantitative structure activity problems, *J. Med. Chem.,* 34, 2824, 1991.
93. **Aoyama, T., Suzuki, Y., and Ichikawa, I.,** Neural networks applied to structure-activity relationships, *J. Med. Chem.,* 33, 905, 1990.
94. **Gardner, G.,** unpublished data.
95. CHEMX: Chemical Design, Ltd., Oxford, England.

Chapter 6

FORMULATIONS AND ADJUVANTS

Per Kudsk and Jens C. Streibig

TABLE OF CONTENTS

0-8493-6603-8/93/$0.00+$.50

I. INTRODUCTION

If not formulated, many herbicides, insoluble in water, are biologically inactive and impossible to apply with conventional application devices. Even water-soluble compounds may be inactive when applied as unformulated material. Formulating herbicides is therefore crucial for enhancing the effect of commercial herbicides.

Among the formulation constituents, the adjuvants are probably the most important. An adjuvant has been defined as "an ingredient in a pesticide prescription which aids or modifies the action of the principal ingredient."[1] The selection of adjuvants to optimize the activity and selectivity of herbicide formulations normally takes place within the chemical companies, and the exact composition of a formulation is proprietary information. Various formulations, particularly off-patent herbicides, however, are often available in the market, and assessment of their herbicidal efficacy is therefore of general interest.

Adjuvants already present in a formulation are termed formulation adjuvants, whereas the commercial adjuvants, added to the spray solution along with the herbicides by the end user, are called spray adjuvants.[1] The literature on adjuvants mainly deals with the effect of these spray adjuvants. In this chapter we just use the term adjuvants.

Although adjuvants are sometimes used for specific purposes, e.g., to reduce drift, foaming, or evaporation or as buffering agents, the majority of adjuvants are used to promote herbicide activity. This enhancing effect on herbicide activity has been known since the introduction of the organic herbicides almost 50 years ago.[2] Early research focused on adjuvants that optimized activity and selectivity of herbicides such as diuron[3,4] and atrazine,[5] herbicides otherwise possessing little or no foliar activity, and relatively expensive foliar-active herbicides such as glyphosate.[6,7] Some recent work is more directed to elucidating the mechanisms by which adjuvants enhance biological activity and demonstrating relationships between the physicochemical properties of adjuvants and pesticides.[8-11] Despite this basic research we still know very little about the mechanism of adjuvant action. The selection of an adjuvant is therefore still largely a matter of trial and error and not a matter of rational approach.[12]

The above-mentioned studies have mainly focused on the effect of adjuvants on uptake. Although it is generally accepted that adjuvants primarily promote uptake and have no effect on translocation per se,[1] it should be stressed that in practice we are concerned with the overall effect and not just uptake.

The purpose of this chapter is to discuss the use of bioassays to evaluate the overall effect of formulations and adjuvants. We will use the concept of parallel dose-response curves, presented in Chapter 3, to rank formulations and adjuvants in terms of relative potency. The mode of action of adjuvants has been dealt with in several recent reviews,[13-15] and this topic is not discussed in this chapter.

We will focus on the effect of formulations and adjuvants as promoters of herbicide activity and as modifiers of herbicide performance under contrasting climatic conditions.

II. ASSESSING THE POTENCY OF FORMULATIONS AND ADJUVANTS

A. THE CONCEPT OF PARALLEL DOSE-RESPONSE CURVES

In most cases the dose-response curves for a herbicide and an adjuvant or blank formulation can be described as in Figure 1A. While the effect of a herbicide can be described by a logistic dose-response curve (see Chapter 3), the adjuvant or blank formulation is virtually without any effect on plant growth at dose rates used under normal conditions. Adjuvants can, however, have effects on the cellular and membrane level.[16-18] We can therefore assume that when applying a formulated herbicide or a herbicide-adjuvant mixture it is only the herbicide molecule that acts in the plants. All formulants and other constituents are biologically inactive

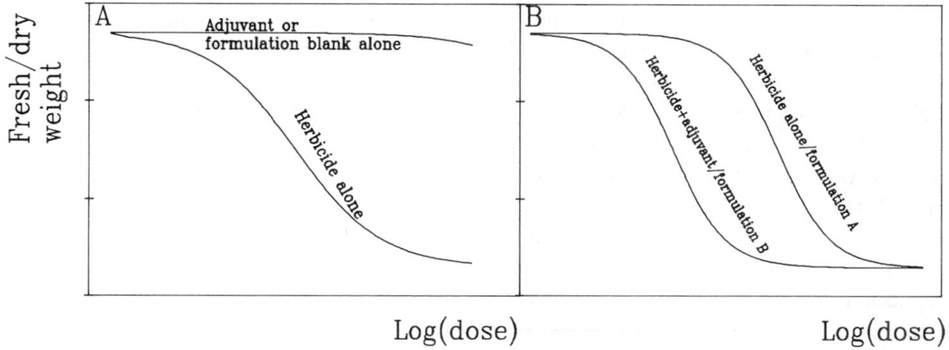

FIGURE 1. Schematic dose-response curves of (A) a herbicide and an adjuvant/formulation blank and (B) a herbicide applied alone and in mixture with an adjuvant or two formulations of the same herbicide.

or at least have no visual effect on plant growth. Although biologically inert when applied alone, formulation constituents and adjuvants can influence herbicide activity by promoting spray retention, herbicide penetration into the plant, or translocation in the plant, and thus increase the amount of herbicide reaching its site(s) of action. Consequently, the dose-response curves for different herbicide formulations or a herbicide applied alone and mixed with an adjuvant must be parallel (Figure 1B). The active ingredient is the same, i.e., the mode of action is similar; the only difference is the amount of herbicide reaching the site(s) of action. In other words, the applied dose is the same but the efficacy of the applied dose (effective dose) depends upon the formulation or adjuvant.

The dose-response curve for a standard preparation, for example, a commercial formulation without additional adjuvant, can be described by[19]

$$U = \frac{D - C}{1 + \exp\{-2[a + b\log(z)]\}} + C, \quad b < 0 \tag{1}$$

U is the plant response (fresh or dry weight), z is the herbicide dose, D and C denote the upper and lower limit at zero and high doses, and a and b are parameters describing the horizontal location and the slope of the curve around the point of inflection, which is antilog ($-a/b$) and corresponds with ED_{50} (see Chapter 3 for further details). A important property of Equation 1 is that the four parameters can be biologically interpreted.

The dose-response curve for a test formulation or a commercial formulation mixed with an adjuvant can then be described by

$$U = \frac{D - C}{1 + \exp\{-2[a + b\log(Rz)]\}} + C, \quad b < 0 \tag{2}$$

The only difference between Equations 1 and 2 is the parameter R. R is called the relative potency and expresses the ratio of the doses of the standard and test preparations giving the same effect. R is constant and independent of dose rate and response level (see Chapter 3).

The effect of formulations or adjuvants is normally compared at a few preset herbicide dose rates.[20-22] By assuming the relationship in Figure 1B, it is obvious that the differences of effects must be dose dependent. The differences in effect at low and high doses are relatively small; maximum differences are found in the midregion. The choice of dose rate is therefore decisive for the difference of effects and, consequently, for the conclusions drawn from such an experiment. Often the enhancement of herbicide activity by adjuvants is so pronounced that the best herbicide-adjuvant mixtures will give almost full effect, although a dose rate is carefully selected to give 50% growth reduction for the herbicide applied alone. If this happens it is impossible to distinguish between the adjuvants. This problem depends largely on the steepness of the dose-response curve; the steeper the curve the more difficult it is to select the proper dose rates.

As the observed difference of effects depends on the dose rate, conclusions about the intrinsic effect of formulations and adjuvants are specific and not general. A more general way to assess the efficacy of formulations and adjuvants is to apply the concept of parallel dose-response curves and then summarize the experiment by relative potencies, which are independent of dose rates. The relative potency of a herbicide-adjuvant combination quantitatively expresses how much a herbicide dose can be reduced without loss of efficacy when adding the adjuvant. If the price relation between the herbicide and the adjuvant is known, it is possible to determine the most cost-effective combination of the two components. This aspect will be discussed later.

In all experiments shown in the following, the responses were transformed to stabilize the variance. A transform-both-sides technique was used (see Chapter 3) and a logarithmic transformation appeared to be appropriate in all experiments. A test for lack of fit, by comparing analysis of variance and nonparallel line regression and subsequently parallel line and nonparallel line regression for the same data, shows whether the hypothesis of parallel dose-response curves holds.

B. FORMULATIONS

Various aspects have to be taken into consideration when formulating a herbicide. Practical problems such as storage stability, packaging, ease of mixing in water, and problems with operator exposure when handling the product have to be considered. In this context, however, we will only focus on the biological activity.

The costs of formulating a herbicide are usually insignificant relative to the costs of developing a herbicide (see Chapter 2). The number of papers reporting benefits from adding adjuvants to commercial herbicides may therefore seem surprising. One reason for this could be that most companies only market few formulations of each herbicide. Weed flora, climatic conditions, and application techniques differ widely around the world. Consequently, it is unlikely that one or a few formulations should be optimal for every purpose. As pointed out by Turner,[12] pesticides have a short patent life and it is therefore crucial from an economic point of view to get a new herbicide on sale as soon as possible, despite shortcomings of the formulation.

The increasing costs for developing new herbicides and the toxicological and ecotoxicological demands have resulted in a reduction in the number of new products. More attention is now being paid to products about to run off-patent. Companies often compete on improved formulations rather than new active ingredients. Recent examples are phenmedipham and glyphosate in some European countries.

The assumption of parallel dose-response curves provides a biological and economic test for performance of different formulations. Examples from recent work on formulations of fluazifop-*p*-butyl (Figure 2A) and glyphosate (Figure 2B) are summarized in Table 1. The old EC formulation of fluazifop-*p*-butyl was compared with a new oil-based EW formulation containing less organic solvents. The performance of the new and environmentally more acceptable EW formulation on barley (*Hordeum vulgare* L.) was significantly better than the old EC formulation. In a second assay, an experimental glyphosate formulation containing 120 g a.i. l^{-1} specifically developed for control of annual weed species, performed significantly better on *Stellaria media* L. than the standard formulation Roundup (360 g a.i. l^{-1}). It is obvious from Figures 2A and 2B, that the assumption of parallel dose-response curves for different formulations of the same active ingredient was reasonable. This concept has also been used to evaluate various phenmedipham formulations.[23,24]

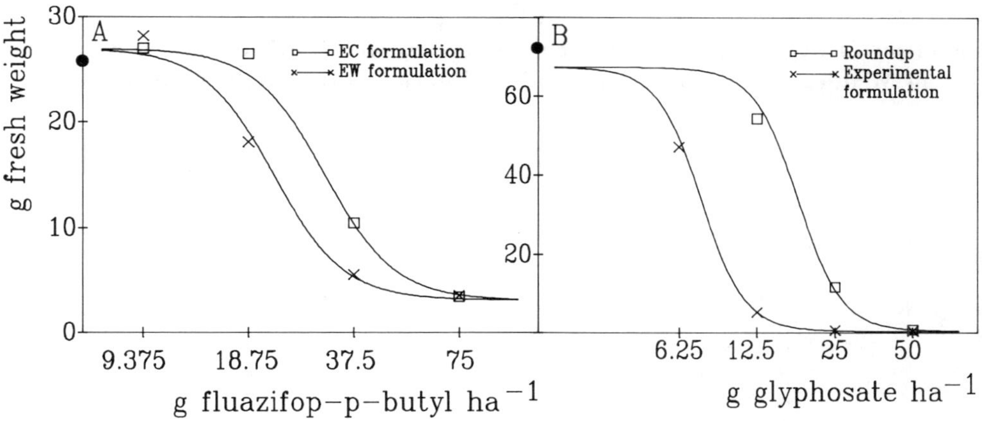

FIGURE 2.. Dose-response curves and observed mean fresh weights when applying (A) fluazifop-*p*-butyl as an EC or EW formulation to barley and (B) glyphosate formulated as Roundup and an experimental formulation to *Stellaria media* L. (untreated control: ●).

TABLE 1
Summary of Regression of Fluazifop-*p*-butyl Formulations on Barley and Glyphosate Formulations on *Stellaria media*

Fluazifop-*p*-butyl

D g pot^{-1}	C g pot^{-1}	a	b
26.95 (24.30 to 29.60)	3.12 (2.74 to 3.50)	7.38 (5.02 to 9.74)	–4.95 (–6.42 to –3.48)

Formulation	Relative potency
EC formulation	1.00
EW formulation	1.40 (1.20–1.60)

Test for lack of fit: $F_{1,29,0.95} = 1.54$

Glyphosate

D g pot^{-1}	C g pot^{-1}	a	b
67.37 (40.57 to 94.17)	0.37 (0.20 to 0.54)	7.35 (4.67 to 10.03)	–5.85 (–7.60 to 4.10)

Formulation	Relative potency
Roundup	1.00
Experimental formulation	2.37 (2.05 to –2.70)

Test for lack of fit: $F_{1,17,0.95} = 0.25$

Note: Figures in parentheses are approximate 95% confidence intervals.

Besides using the relative potency to rank formulations and select candidates for further research and development, the results can also be used in an economic context. The relative potency of 2.37 in the glyphosate experiment indicates that with the new formulation, the amount of active ingredient can be reduced 42% (1/2.37) without loss of efficacy. Based on results from similar experiments with other important annual weed species, an average relative potency can be calculated. Combining the information on biological activity with production

TABLE 2
Summary of Regression of Tribenuron on *Sinapis alba* L. when Applied Alone and in Mixture with 0.1% of Various Nonyl-phenol-polyethoxylate Surfactants

D g pot^{-1}	C g pot^{-1}	a	b
65.54 (59.33 to 71.75)	1.42 (0.75 to 2.09)	–1.90 (–2.07 to –1.73)	–1.39 (–1.56 to –1.22)

Surfactant	Relative potency
No surfactant	1.00
0.1% NP4[a]	1.49 (1.24–1.74)
0.1% NP6	1.24 (1.03–1.45)
0.1% NP9.5	1.72 (1.42–2.01)
0.1% NP15	1.48 (1.23–1.73)
0.1% NP30	0.99 (0.82–1.16)
Test for lack of fit: $F_{5,38,0.95} = 1.35$	

Note: Figures in parentheses are approximate 95% confidence intervals.

[a] The figure indicates the average number of ethylene oxide units.

costs of formulations and the active ingredient, it is simple to calculate the profitability of marketing the two formulations and then pick out candidate formulations for further work. Comparing the relative biological activity with formulation costs is particularly important for herbicides that are difficult and expensive to formulate, for example phenmedipham.

C. ADJUVANTS

The development of herbicide formulations is normally done by the chemical companies before marketing. The array of adjuvants available in the market provides the user with an opportunity to improve biological activity of commercial formulations. As for formulations, the relative potency of various herbicide-adjuvant combinations can answer the user's question: "How much can I reduce the herbicide dose when adding an adjuvant without losing effect?" The selection of adjuvants is still largely empirical. Therefore, a reliable and efficient bioassay technique to assess adjuvants for herbicides is still an important requirement. The parallel line assay concept meets this demand. We will look at three important aspects in adjuvant evaluation. First, to optimize herbicide activity a proper adjuvant has to be selected. Second, the adjuvant dose has to be optimized. The third aspect we will address is the use of adjuvants as modifiers of herbicide activity. Adjuvants can help overcome environmental impediments and thereby improve reliability.

1. Assessment of Adjuvants

In experiments with fluazifop-*p*-butyl, sethoxydim, and alloxydim-sodium the parallel line assay was used to rank adjuvants belonging to different adjuvant classes.[25] An experiment on *Sinapis alba* L. comparing the potency of the methyl ester of the sulfonylurea herbicide tribenuron (formerly DPX-L5300) with a series of nonyl-phenol-polyethoxylate surfactants is summarized in Table 2.

The relative potencies in Table 2 show that all surfactants except NP30 significantly enhanced herbicide activity compared to the standard with no surfactant. The maximum activity was obtained with NP9.5; the activity of the herbicide in mixture with NP4 and NP15 was not, however, significantly different from the activity observed with NP9.5. The relationship between the activity of tribenuron and the hydrophilic-lipophilic balance (HLB) of the nonyl-phenol-polyethoxylate surfactants is shown in Figure 3. This illustrates that HLB

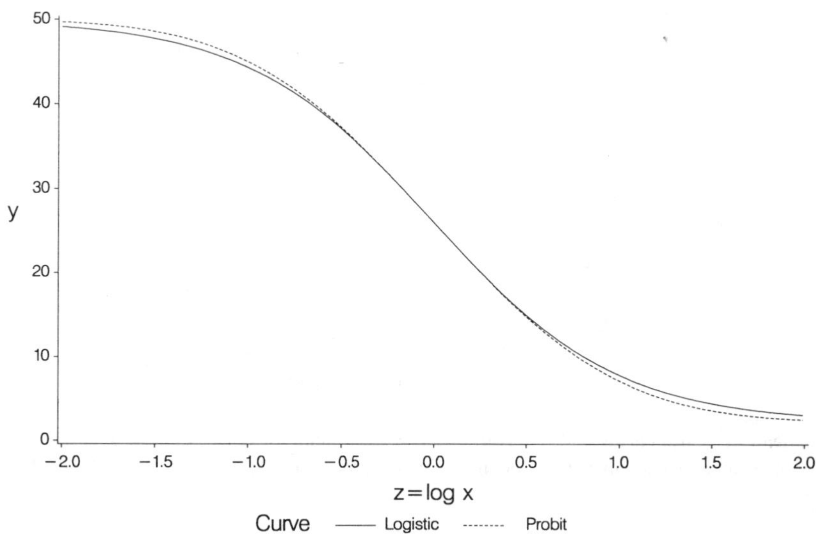

FIGURE 3. Relationship between HLB and relative potencies of tribenuron on *Sinapis alba* L. applied in mixture with 0.1% of a series of nonyl-phenol-polyethoxylate surfactants. The relative potency of tribenuron applied without surfactant is 1.00.

is an important factor governing herbicide activity. Green and Brown,[26] working with three other sulfonylurea herbicides, also found that HLB was an important surfactant property affecting the phytotoxicity of sulfonylurea herbicides. The existence of such a relationship had previously been postulated,[21,27,28] but always based on results with a single herbicide dose. As pointed out in Section II.A, conclusions based on relative potencies are independent of dose rate and response level and, hence, more general than conclusions based on only one dose rate. The relationship established in studies with only one dose rate apply only to that particular dose rate. Other dose rates may not change the ranking of the surfactants, i.e., change the optimum HLB value, but perhaps affect the magnitude of the differences between surfactants. In contrast, Figure 3 is based on the relative horizontal displacement of the parallel dose-response curves, and consequently, the intrinsic effects of the members of this series of surfactants are independent of effect level and dose rate.

The results in Table 2 can be interpreted in economic terms to calculate which combination of herbicide and surfactant is most cost-effective, whereas results from experiments based on one dose cannot be used to calculate cost-effectiveness.

In practice, herbicides are often tank mixed with other pesticides or fertilizers. For fungicides and insecticides, whose formulations often contain surfactants and other formulation constituents, we could expect that some products may have an influence on the activity of the herbicide. Similarly, inorganic salts have been shown to enhance phytotoxicity of certain herbicides.[7,29-31] Consequently, the effect of a herbicide applied in mixture with a fertilizer could be modified.

Fungicides and insecticides normally only have minor effects on the growth of weed plants. They can, like adjuvants and blank formulations, be considered biologically inert when applied alone. Hence we can assume parallel dose-response curves when a herbicide is applied alone and in mixture with a fungicide or insecticide, respectively. Dose-response curves for wild oat (*Avena fatua* L.) with difenzoquat applied alone and in tank mixture with the fungicide, propiconazole, and the insecticide, fenvalerate, are shown in Figure 4. While the insecticide did not affect the activity of difenzoquat, the fungicide significantly enhanced difenzoquat activity (Table 3).

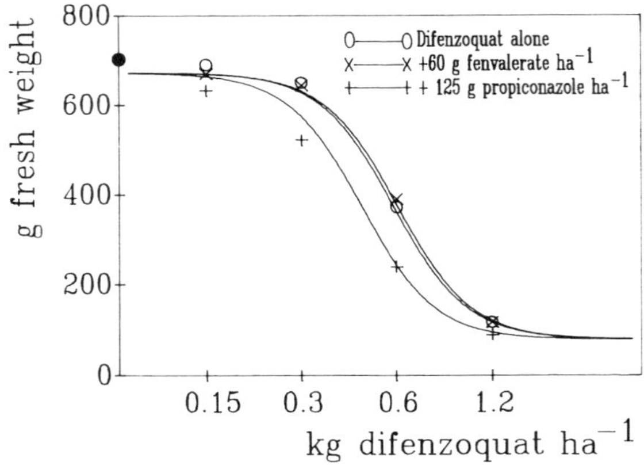

FIGURE 4. Dose-response curves and observed mean fresh weights for difenzoquat to wild oat either alone or in mixture with the insecticide, fenvalerate, and the fungicide, propiconazole (untreated control: ●). (Modified from Kudsk et al.[32])

TABLE 3
Summary of Regression of Difenzoquat on Wild Oat when Applied Alone and in Tank Mixture with the Insecticide Fenvalerate and the Fungicide Propiconazole

D g pot^{-1}	C g pot^{-1}	a	b
672.8 (646.3 to 699.3)	79.1 (62.8 to 95.4)	–0.95 (–1.15 to –0.75)	–4.30 (–5.14 to –3.46)

Tank mixture	Relative potency
Difenzoquat alone	1.00
+60 g fenvalerate ha^{-1}	1.03 (0.94–1.11)
+125 g propiconazole ha^{-1}	1.30 (1.19–1.41)
Test for lack of fit: $F_{2,20,0.95} = 2.18$	

Note: Figures in parentheses are approximate 95% confidence intervals.

Modified from Kudsk et al.[32]

This difference may be related to formulation. Fenvalerate is a water-based soluble concentrate formulation, while propiconazole, at that time, was formulated as an emulsifiable concentrate containing the organic solvent, xylene. Propiconazole belongs to a group of ergosterol inhibitors that are also known to inhibit monooxygenases in plants.[33] It is not known whether monooxygenases are involved in the degradation of difenzoquat in wild oat. Gressel,[34] however, suggested that the effect of propiconazole on difenzoquat activity could be caused by an inhibition of a monooxygenase mediated metabolism of difenzoquat. Whether the synergizing effect of propiconazole is due to the formulation itself or the inhibition of the metabolism, the result is the same — more active difenzoquat reaches the site of action. Thus the parallel line assay is applicable in both cases.

The parallel line assay was also used to show an antagonistic effect of maneb on the activity of phenoxyalkanoic acid herbicides and chlorsulfuron.[35]

Tank mixes of herbicides with other pesticides, particularly fungicides, has become more common during the last 10 years. Despite this, the potential of other pesticides to act as adjuvants has not been thoroughly investigated. In some countries application of herbicides in combination with other pesticides is as common as application with adjuvants.

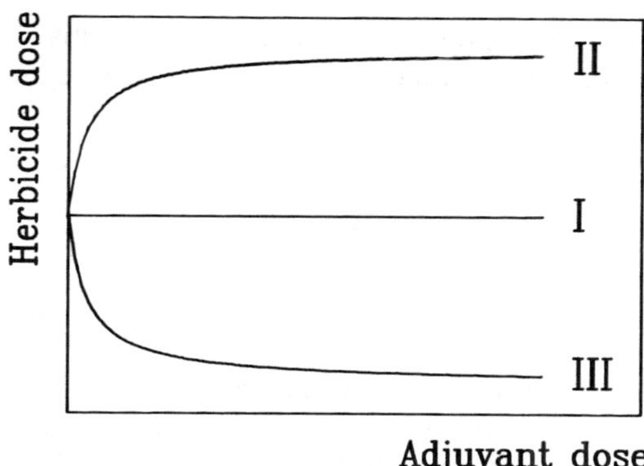

FIGURE 5. Schematic illustration of the three situations that can occur when increasing doses of an adjuvant is added to a herbicide. I denotes additivity, II antagonism, and III synergism. For further details see the text.

2. Adjuvant Dose and Herbicide Activity

In a number of previous studies a progressive enhancement of herbicide activity was found with increasing adjuvant dose.[3,4,20,21,27,36,37] Jansen et al.[20] examined the effect of 63 surfactants on the activity of 4 herbicides on 2 plant species. They found examples of a progressive reduction of herbicide activity with increasing surfactant dose. None of these studies, however, attempted to describe the relationship between herbicide activity and adjuvant dose.

Basically, the influence of an adjuvant on the activity of a herbicide calls for the same reference models as do mixtures of herbicides (see Chapter 7). If an adjuvant is biologically inactive, we have a limiting case of a reference model for assessing joint action of herbicides. The isoboles in Figure 5 illustrate the three situations found when a herbicide is mixed with a biologically inert adjuvant.

If the dose-response curves are identical, whatever adjuvant dose is added, the isobole will look like I in Figure 5. We have an additive action of the compounds,[38] i.e., the dose of the adjuvant does not influence the dose of the herbicide required to produce a given effect. If an adjuvant reduces the activity of the herbicide, i.e., the dose-response curve is displaced to the right, then the isobole will proceed upward like II and antagonism occurs.[38] If the adjuvant increases the activity of the herbicide, i.e., the dose-response curve is displaced to the left, as in Figure 1B, then an increase in adjuvant dose reduces the herbicide dose required to produce a given effect. In this case the isobole will proceed downward like III and we have synergism.[38]

Hewlett[39] has suggested that the effect of a synergist (isobole III, Figure 5), can be described by a rectangular hyperbola. If z_{herb} is the dose of the herbicide and z_{adj} is the dose of the adjuvant then the equation

$$z_{\text{herb}} = \frac{m_1 + m_2}{m_3 + z_{\text{adj}}} \tag{3}$$

can describe the relationship (m_1, m_2, and m_3 are parameters with positive values). Hewlett used ED_{50} or ED_{90} values to derive Equation 3. As mentioned earlier, the relative potency expresses the ratio between the herbicide doses giving the same effect. Consequently, the ED_{50} or ED_{90} values can be replaced by the relative potencies. As the ED_{50} value of a herbicide-adjuvant mixture is estimated as the ED_{50} value of the herbicide alone divided by the relative potency, the relationship in Equation 3 can be expressed as

$$\frac{1}{R} = \frac{m_1 + m_2}{m_3 + z_{\text{adj}}} \tag{4}$$

The joint action ratio (JAR), expressing the ratio between relative potency of the herbicide applied alone and in mixture with an indefinitely large dose of adjuvant, can be calculated as

$$\mathrm{JAR} = 1 + \frac{m_2}{m_1 m_3} \tag{5}$$

Alloxydim-sodium was mixed with six doses of a mineral oil adjuvant. The relationship between the relative potency of the herbicide and the adjuvant dose was linear on a double-logarithmic scale.[40] With seven herbicide-adjuvant combinations we found that the promotion of herbicide activity by increasing adjuvant doses could be described by a linear relationship on either an arithmetic, semilogarithmic, or double-logarithmic scale.[41] These results, from experiments with relatively few adjuvant doses, indicate that the relative potencies were located at different parts of a sigmoid curve. We therefore tentatively suggested that the relationship between the relative potency of a herbicide (R) and the dose of an adjuvant (z_{adj}) could be described by a sigmoid curve very much similar to that of a dose-response curve for a herbicide (Equation 1):

$$R = \frac{D - C}{1 + \exp\left\{-2\left[a + b\log\left(z_{\mathrm{adj}}\right)\right]\right\}} + C, \quad b > 0 \tag{6}$$

The parameters D, C, a, and b denote the same as in Equation 1, the only difference being that $b > 0$. The sigmoid curve implies that at a certain adjuvant dose, describing the upper limit D, no further increase in herbicide activity can be obtained by increasing the adjuvant dose. This is a reasonable assumption from a biological point of view. As adjuvants themselves can be phytotoxic, they may at very high doses actually inhibit herbicide uptake and movement and hence reduce the overall effect.[13] This effect is to be expected, particularly with systemic herbicides, but it has also been found with the contact herbicide, paraquat.[42] Being a monotonic increasing function, the sigmoid relationship in Equation 6 cannot describe this reduction in herbicide efficacy at high adjuvant doses. This problem will be discussed later in this section.

As pointed out in Section II.A, the symmetric logistic curve used in this chapter has one important property in which it surpasses most other dose-response relationships; the parameters can be biologically interpreted. As we have no previous experience in describing the relationship between the relative potency and adjuvant dose with Equation 6, we did not know whether a symmetric curve was appropiate or not. The data were therefore also fitted to an asymmetric curve suggested by Finney:[19]

$$R = \frac{D - C}{1 + \exp\left\{-2\left[a + b\log\left(z_{\mathrm{adj}}\right)\right]\right\}^{\beta}} + C, \quad b > 0 \tag{7}$$

β being an extra parameter conveying asymmetry. One disadvantage of Equation 7 is that the a and b parameters no longer have any obvious biological meaning. Regressions with Equations 6 and 7 were checked against each other by a test for lack of fit.

The effect of increasing doses of an adjuvant on the phytotoxicity of glyphosate, formulated as Roundup, and fluazifop-*p*-butyl, formulated as Fusilade 5EC, was assessed on barley.

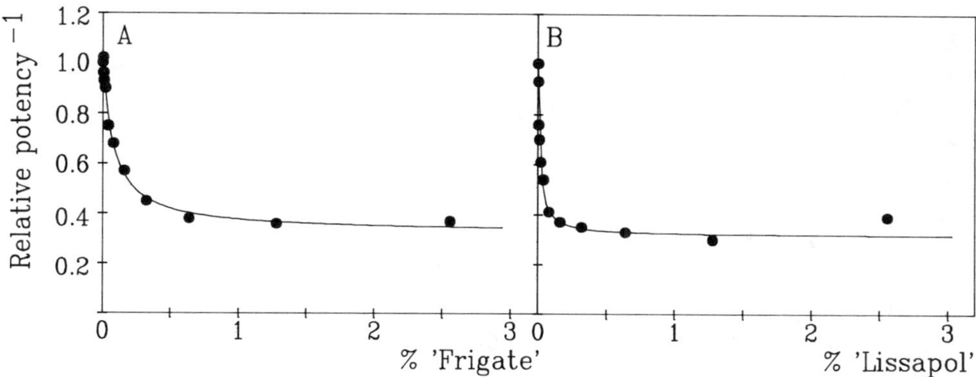

FIGURE 6. Inverse relative potencies of (A) glyphosate on barley in mixture with increasing doses of Frigate (tallow amine polyethoxylate with an average of 15 ethylene oxide units) and (B) fluazifop-*p*-butyl on barley in mixture with increasing doses of Lissapol (nonyl-phenol-polyethoxylate with an average number of 8 ethylene oxide units) calculated from Equation 4. The estimated parameters with approximate 95% confidence intervals in parenthesis were $m_1 = 0.3272$ (0.0368), $m_2 = 0.0532$ (0.0157), and $m_3 = 0.0765$ (0.0222) for glyphosate plus Frigate and $m_1 = 0.3171$ (0.0399), $m_2 = 0.0093$ (0.0034), and $m_3 = 0.0135$ (0.0051) for fluazifop-*p*-butyl plus Lissapol.

The herbicides were applied alone or in mixture with 11 doses of the commercial adjuvants Frigate and Lissapol.

The estimated curves using Hewlett's hyperbola (Equation 4) is shown in Figures 6A and 6B. The joint action ratio was 3.12 for glyphosate and Frigate and 3.17 for fluazifop-*p*-butyl and Lissapol, respectively. This indicates that at very large adjuvant doses it was possible to reduce the herbicide dose to approximately one third without losing efficacy.

Equation 6 fitted to the same data shown in Figures 7A and 7B. The tests of lack of fit revealed that the asymmetric curve (Equation 7) did not describe the relationship significantly better than the symmetric curve (Equation 6). It is obvious that the estimated curves fitted very closely to the relative potencies, indicating that the logistic curve adequately described the relationship between herbicide activity and adjuvant dose.

The *D* parameter denotes the upper limit of the curve and expresses the maximum activity of the herbicide at very large adjuvant doses. Thus the *D* parameter is the same as Hewlett's joint action ratio. According to the confidence intervals the *D* parameters were not significantly different from the estimated joint action ratios, i.e., both methods estimated the same maximum enhancement of the herbicide activity by the adjuvants. Furthermore, the estimated *D* parameters were in no instances significantly different from the maximum relative potencies in Table 4.

The maximum change in herbicide phytotoxicity to small increments of adjuvant dose is around the point of inflection midway between the upper and lower asymptote. The corresponding dose is a kind of ED_{50}.

The *b* parameter is proportional to the slope in the vicinity of the point of inflection. It describes the rate of increase in herbicide activity when increasing the dose of the adjuvant. The *b* parameter is correlated with the *D* and *C* parameters, and a comparison of *b* parameters between assays is therefore only possible if *D* and *C* are not significantly different between assays. In the present experiments *D* and *C* were not significantly different and the *b* parameters can therefore be compared. Although *b* was somewhat higher in the fluazifop-*p*-butyl experiment, this difference was not significant. Thus the maximum response to small increments in adjuvant dose was similar for the two herbicides.

Figures 7A and 7B reveal that small doses of Lissapol improved the activity of fluazifop-*p*-butyl, whereas corresponding doses of Frigate had no effect on the efficacy of glyphosate. This is also reflected in the ED_{50} values, being 0.06% Lissapol and 0.2% Frigate. The

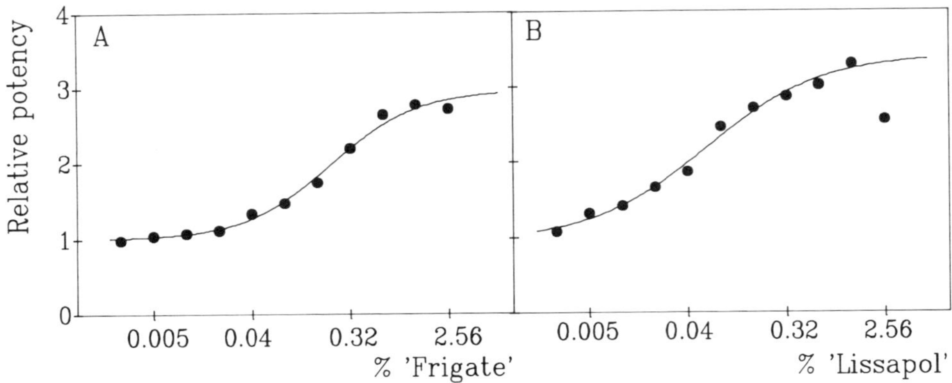

FIGURE 7. Relative potencies of (A) glyphosate on barley in mixture with increasing doses of Frigate and (B) fluazifop-*p*-butyl on barley in mixture with increasing doses of Lissapol calculated from Equation 6. The estimated parameters with approximate 95% confidence intervals in parenthesis were D = 2.96 (0.36), C = 1.00 (0.09), a = 0.86 (0.51), and b = 1.28 (0.45) for glyphosate plus Frigate and D = 3.39 (0.46), C = 0.97 (0.13), a = 1.18 (0.61), and b = 0.96 (0.33) for fluazifop-*p*-butyl plus Lissapol. For adjuvant nomenclature see legend to Figure 6.

TABLE 4
Summary of Regression of Chlorsulfuron on *Sinapis alba* L. when Applied Alone and in Mixture with the Nonionic Surfactant Citowett[a] at High and Low Humidity

D g pot^{-1}	C g pot^{-1}	a	b
101.29 (91.64 to 110.94)	4.07 (2.65 to 5.49)	−1.31 (−1.46 to −1.16)	−1.59 (−1.83 to −1.35)

Humidity	Adjuvant	Relative potency[b]	
35% rh	No adjuvant	1.00	
85% rh	No adjuvant	5.42 (4.34–6.59)	
35% rh	0.1% Citowett	4.90 (3.92–5.87)	1.00 (0.80–1.20)
85% rh	0.1% Citowett	7.95 (6.40–9.50)	1.62 (1.31–1.94)

Test for lack of fit: $F_{3,38,0.95} = 1.69$

[a] Isooctyl-phenol-polyethoxylate with an average of 6 ethylene oxide units.

[b] In the right column the relative potencies of chlorsulfuron plus 0.1% Citowett are expressed relative to the effect of this tank mixture at 35% rh.

Note: The figures in parenthesis are approximate 95% confidence intervals.

recommended concentration is 0.1% Lissapol to fluazifop-*p*-butyl and 0.5% Frigate to glyphosate. While the 0.5% Frigate results in an effect close to the maximum effect, 0.1% Lissapol only yields a little more than 50% of maximum effect. Thus, increasing the recommended adjuvant dose to fluazifop-*p*-butyl, when controlling volunteer cereals, allows for a reduction in the herbicide dose without loss of efficacy. Fluazifop-*p*-butyl is also used to control perennial grass weeds. The results above cannot be extended to perennial grasses, as plant species as well as different requirements for translocation of the herbicide might have an influence on the optimum adjuvant dose.

In both experiments reduced activity was observed when increasing the adjuvant dose from 1.28 to 2.56% (Figures 7A and 7B). As mentioned earlier, a reduced herbicide phytotoxicity at very large adjuvant doses can occur, particularly with systemic herbicides like fluazifop-*p*-butyl and glyphosate. Large adjuvant doses can cause localized necrosis and thereby restrict herbicide translocation in the plant. Equation 6 is a monotonic function that cannot describe

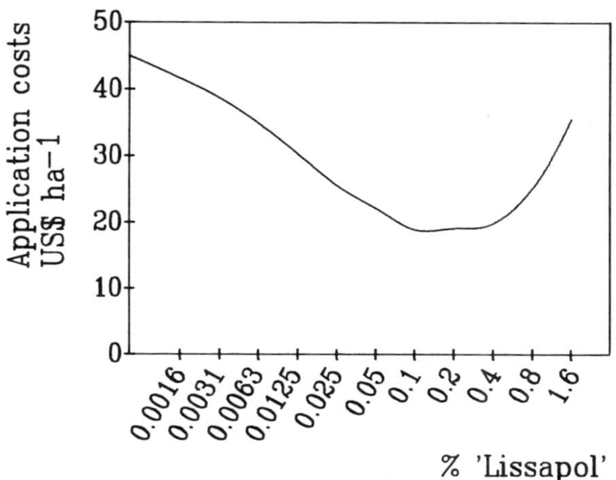

FIGURE 8. Application costs (herbicide plus adjuvant) when applying fluazifop-*p*-butyl to barley in mixture with increasing doses of Lissapol at a volume rate of 150 l ha^{-1}. The relative potencies of fluazifop-*p*-butyl were calculated using Equation 6 and the regression parameters in legend to Figure 7. The recommended herbicide dose without adjuvant was fixed at 1 l ha^{-1} commercial product (125 g fluazifop-*p*-butyl l^{-1}). The prices were 45 and 13.5 $/l (U.S.) herbicide and adjuvant, respectively.

these effects. In the fluazifop-*p*-butyl experiment the decrease in the relative potency when increasing the Lissapol dose from 1.28 to 2.56% was significant. Consequently, the result at 2.56% was omitted when estimating the rectangular hyperbola in Figure 6B and the dose-response curve in Figure 7B. In the glyphosate experiment the difference in the relative potency between the two highest adjuvant doses was not significant, and the result at 2.56% was included in the regressions.

The application of the logistic curve to describe the effect of increasing adjuvant dose on herbicide efficacy needs further validation before it can be generally accepted. More herbicides, systemic as well as nonsystemic, adjuvants, and plant species have to be included. If, however, the logistic curve turns out to be reasonable, it provides a better understanding of the relationship between herbicide activity and adjuvant dose. Furthermore, the biological interpretation of parameters makes the logistic curve superior to Hewlett's hyperbola.

The curves in Figures 7A and 7B can be used to determine the most cost-effective combination of herbicide and adjuvant. The Lissapol dose should be kept between 0.1 and 0.4% to obtain the most cost-effective mixture of fluazifop-*p*-butyl and adjuvant (Figure 8). Although the fluazifop-*p*-butyl dose can be reduced even further when increasing the Lissapol dose beyond the recommended 0.1%, the total costs will not be reduced.

3. Adjuvants and Environmental Factors

The increasing public concern for the agricultural use of pesticides has stimulated the interest in reducing herbicide use. A widespread use of reduced herbicide doses in a number of European countries is a result of this public concern and economical considerations by the farmer. In Denmark extensive field research has shown that it is often possible to substantially reduce the recommended herbicide dose, particularly in cereals.[43] One reason is that the recommended dose is often based on the rate required to control the most difficult to control weed species. If a farmer knows the species at his field and the required dose to control each species, then the dose can be lowered. Another reason that the dose can be reduced without significant loss of efficacy is that the recommended dose rate often corresponds to an effect on the lower, flat part of the dose-response curve (see Figure 1A). One reason for recommending rather high doses is that substantial reduction in the effective dose (see Section II.A), for example due to unfavorable climatic conditions, will result in only

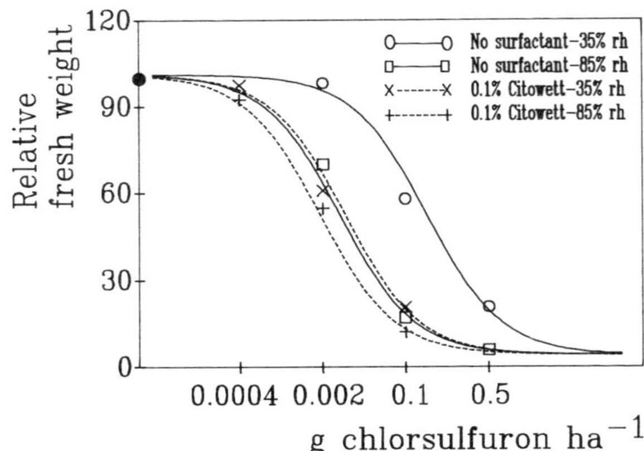

FIGURE 9. Dose-response curves and mean relative fresh weights of *Sinapis alba* L. for chlorsulfuron alone and in mixture with 0.1% Citowett at constant temperature but contrasting humidities (untreated control: ●).

a minor reduction in efficacy. Changes in the effective dose have no significant effect on herbicide activity on the lower, flat part of the dose-response curve. Consequently, recommended dose rates will provide the required effect under most conditions, because the recommendations tend to reflect the worst rather than the normal case. This approach is no longer acceptable to the public in a number of countries. Instead the dose should be adjusted to give exactly the required effect and no more under the prevailing conditions.[43]

The goal of current research is to find a dose that results in an effect near the point where the dose-response curve flattens out. We can, however, only control the applied dose, or what we could call the apparent dose. The effect, however, depends on how much herbicide reaches the site of action, i.e., the effective dose. The effective dose is dependent on a variety of factors, for example, environmental conditions and application techniques. Although the apparent dose is held constant, the effective dose will vary. In graphical terms, this means that environmental factors can shift the horizontal position of the dose-response curve along the dose axis.

Variations in the effective dose can have a pronounced impact on the efficacy when applying reduced herbicide doses, intended to give an effect in the vicinity of the point where the dose-response curve flattens out. In this range even small changes in dose rate significantly influence herbicide efficacy. In practice, effects of reduced doses are therefore more variable than those of recommended doses. The increased use of reduced doses has put the adjuvants into a new perspective as modifiers of herbicide efficacy. Adjuvants can overcome the constraints imposed by adverse environmental conditions. As will be shown in the following, the parallel line assumption may also be applicable in such studies.

In our work we have focused mainly on the ability of adjuvants to overcome the influence of humidity and rain on herbicide activity. The dose-response curves for chlorsulfuron from an experiment in controlled environment chambers, examining the influence of humidity on foliar activity, is shown in Figure 9. The temperature was held constant at 22.2°C througout the experiment. The herbicide was administered alone and in tank mixture with 1% (v/v) of the nonionic surfactant, Citowett. Besides increasing the phytotoxicity of chlorsulfuron, the surfactant reduced the impact of humidity. An increase in relative potency from 1.00 to 5.42 was observed with chlorsulfuron alone when the relative humidity increased from 35 to 85%, whereas the relative potency was only increased to 1.62 when chlorsulfuron was administered with the surfactant (Table 4). Although the influence of humidity was statistically significant, 0.1% Citowett considerably reduced the effect of humidity. Similar results were found with other sulfonylurea herbicides[44,45] and bentazon.[46] In all our studies with sulfonylurea herbicides, the effect of temperature on herbicide activity was only slightly reduced by Citowett.

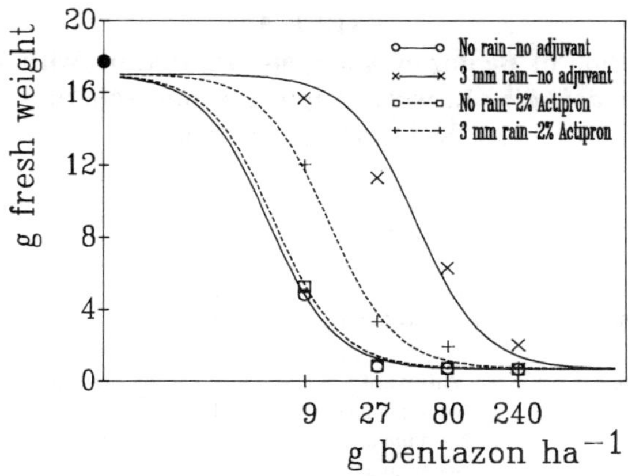

FIGURE 10. Dose-response curves and observed mean fresh weights of *Sinapis alba* L. for bentazon alone and in mixture with 2% Actipron under rain-free conditions or when 3 mm rain was applied 1/2 h after herbicide application (untreated control: ●).

High humidity usually facilitates foliar absorption of herbicides.[47] This is partly attributed to an increased hydration of the cuticle and an extended droplet drying period.[48] The observed effect of Citowett, which reduced the influence of humidity, could be caused by the hygroscopic properties of surfactants.[49] At low humidity, surfactant molecules on the leaf surface could bind water, thereby extending the droplet drying time.[49] Surfactant molecules, penetrating the cuticle could attract water, causing swelling of the cuticle, and thus easing uptake.[50]

Another factor, which may have a substantial effect on the activity of foliar-applied herbicides, is rain occurring shortly after application. Adjuvants can improve rainfastness of herbicides (Figure 10). With no rain, Actipron, a mineral oil adjuvant, did not improve the phytotoxicity of bentazon on *Sinapis alba* L. However, the effect of bentazon, applied with the mineral oil, was significantly better than bentazon applied alone when rain occurred 1/2 h after herbicide application. Rain 3 h after application had no significant influence on bentazon with the adjuvant, whereas the phytotoxicity was significantly reduced if bentazon was applied alone (Table 5). A similar effect of a nonionic surfactant on the rainfastness of various sulfonylurea herbicides has been reported.[51]

The results presented in this section clearly demonstrate that adjuvants can overcome environmental impediments and thereby reduce variability of performance. This is an aspect particularly important when considering the use of reduced herbicide doses. The main advantage of assuming parallel dose-response curves is that we can distinguish between influence of the adjuvant and the environmental factor on herbicide activity, as illustrated in Tables 4 and 5. Comparing the influence of humidity on a herbicide applied alone and in mixture with an adjuvant at a preset herbicide dose rate can lead to erroneous conclusions, because the increase in herbicide activity by an adjuvant often masks the influence of environmental factors. The comparisons of effect with and without adjuvant are done on different sections of the dose-response curve. This can be avoided by describing the whole dose-response curve and comparing the horizontal distances between the curves instead of the vertical distances.[52]

Adjuvants might also have a potential to overcome the effect of soil moisture stress on the activity of foliar-applied herbicides. This is an aspect we have not looked at so far.

It should be mentioned that although we have focused only on the use of adjuvants to reduce variability in performance, similar differences can be observed between formulations. For example, pronounced differences in rainfastness has been observed between an ester and an amine formulation of 2,4-D.[53] The parallel line assay assumption can also be applied in this context. For instance, we have found significant differences in rainfastness between a potassium salt and an ester formulation of bromoxynil (unpublished data).

TABLE 5
Summary of Regression of Bentazon Alone and in Mixture with 1% of the Mineral Oil Actipron[a] on *Sinapis alba* L. when Appling 3 mm Rain at Various Times after Herbicide Application

D g pot^{-1}	C g pot^{-1}	a	b
17.03	0.66	–4.41	–2.22
(14.50 to 19.56)	(0.54 to 0.78)	(–4.95 to 3.87)	(–2.53 to –1.91)

Rain regime	Adjuvant	Relative potency[b]	
No rain	No adjuvant	1.00	
Rain after 1/2 h	No adjuvant	0.10 (0.07–0.12)	
Rain after 1 h	No adjuvant	0.22 (0.16–0.29)	
Rain after 3 h	No adjuvant	0.61 (0.42–0.80)	
No rain	1% Actipron	0.91 (0.63–1.19)	1.00 (0.69–1.31)
Rain after 1/2 h	1% Actipron	0.39 (0.27–0.52)	0.43 (0.30–0.57)
Rain after 1 h	1% Actipron	0.50 (0.35–0.64)	0.55 (0.38–0.70)
Rain after 3 h	1% Actipron	0.76 (0.52–0.99)	0.84 (0.57–1.09)

Test for lack of fit: $F_{8,90,0.95} = 0.87$

[a] 98% mineral oil and 2% emulsifier.

[b] In the right column the relative potencies of bentazon plus 1% Actipron are expressed relative to the effect of this tank mixture without rain.

Note: Figures in parenthesis are approximate 95% confidence intervals.

III. CONCLUSIONS

The selection of formulations and adjuvants is still largely empirical. The use of proper assay systems is therefore imperative to provide general and not specific conclusions. We find that dose-response curves covering the whole dose range from no effect to complete kill should be used. For ease of interpretation the concept of parallel dose-response curves should be tested. As shown in this chapter, this assumption has proven reasonable for herbicide formulations and tank mixtures with adjuvants.

The increased public interest in agricultural pesticide use has already lead to a demand for a reduction of use. Reduced doses is one avenue to take to reduce the total amount of herbicides used. In this context adjuvants are important because of their potential to enhance or modify activity, minimizing the variability in herbicide performance.

In the future, formulations and adjuvants may become even more important. A proper formulation or adjuvant may promote the foliar activity of residual herbicides. An improved foliar activity may reduce the required dose rate, and thereby perhaps delay the evolution of resistant weed populations, because the reduced residual activity decreases selection pressure. An example could be the relatively persistent sulfonylurea herbicides, chlorsulfuron and metsulfuron, which are foliar-active at extremely low doses.

Another important development in some countries could be a reduced number of active ingredients available in the market. This is partly because the number of new compounds introduced are becoming scarce due to increasing costs for development, registration, etc., and partly because some of the "older" compounds are banned for toxicological and ecotoxicological reasons. With a reduced number of herbicides, improvements of activity to control more weed species and to extend selectivity in crops become relevant. Formulations and adjuvants are important in achieving this goal.

ACKNOWLEDGMENTS

Appreciation is provided to ICI Denmark for providing the series of nonyl-phenol-polyethoxylate surfactants.

REFERENCES

1. **Foy, C. L.,** Adjuvants: terminology, classification, and mode of action, in *Adjuvants and Agrochemicals,* Vol. 1, Chow, P. N. P., Grant, C. A., Hinshalwood, A. M., and Simundsson, E., Eds., CRC Press, Boca Raton, FL, 1989, chap. 1.
2. **Foy, C. L.,** Adjuvants for agrochemicals: introduction, historical overview, and future outlook, in *Adjuvants and Agrochemicals,* Vol. 2, Chow, P. N. P., Grant, C. A., Hinshalwood, A. M., and Simundsson, E., Eds., CRC Press, Boca Raton, FL, 1989, chap. 21.
3. **McWhorter, C. G.,** Effects of surfactants on the herbicidal activity of foliar sprays of diuron, *Weeds,* 11, 265, 1963.
4. **Bayer, D. E. and Drever, H. R.,** The effects of surfactants on efficiency of foliar-applied diuron, *Weeds,* 13, 222, 1965.
5. **Dexter, A. G., Burnside, O. C., and Lavy, T.L.,** Factors influencing the phytotoxicity of foliar applications of atrazine, *Weeds,* 14, 222, 1966.
6. **Wyrill, J. B. and Burnside, O. C.,** Glyphosate toxicity to common milkweed and hemp dogbane as influenced by surfactants, *Weed Sci.,* 25, 275, 1977.
7. **Turner, D. J. and Loader, M. P. C.,** Effects of ammonium sulphate and other additives on the phytotoxicity of glyphosate to *Agropyron repens* (L.), *Weed Res.,* 20, 139, 1980.
8. **Silcox, D. and Holloway, P. J.,** Foliar absorption of some nonionic surfactants from aqueous solutions in the absence and presence of pesticidal active ingredients, in *Adjuvants and Agrochemicals,* Chow, P. N. P., Grant, C. A., Hinshalwood, A. M., and Simundsson, E., Eds., CRC Press, Boca Raton, FL, 1989, chap. 12.
9. **Holloway, P. J., Stock, D.,Whitehouse, P., and Grayson, B. T.,** Rational approaches to selection of surfactants for optimizing uptake of foliage-applied agrochemicals, in *Brighton Crop Protection Conference—Weeds,* BCPC, Croydon, U. K., 1989, 225.
10. **Schönherr, J. and Bauer, H.,** Analysis of effects of surfactants on permeability of plant cuticles, in *Adjuvants and Agrichemicals,* Foy, C. L., Ed., CRC Press, Boca Raton, FL, 1992.
11. **Riederer, M. and Schönherr, J.,** Effects of surfactants on water permeability of isolated plant cuticles and on the composition of their cuticular waxes, *Pestic. Sci.,* 29, 85, 1990.
12. **Turner, D. J.,** Additives for use with herbicides, a review, *J. Plant Prot. Tropics,* 1, 77, 1984.
13. **Hull, H. M., Davis, D. G., and Stolzenberg, D. G.,** Action of adjuvants on plant surfaces, in *Adjuvants for Herbicides,* Hodgson, R. H., Ed., Weed Science Society of America, Champaign, IL, 1982, chap. 3.
14. **Norris, R. F.,** Action and fate of adjuvants in plants, in *Adjuvants for Herbicides,* Hodgson, R. H., Ed., Weed Science Society of America, Champaign, IL, 1982, chap. 4.
15. **McWhorter, C. G.,** The physiological effects of adjuvants on plants, in *Weed Physiology, Vol. II: Herbicide Physiology,* Duke, S. O., Ed., CRC Press, Boca Raton, FL, 1985, chap. 6.
16. **Miller, G. M. and St. John, J. B.,** Membrane-surfactant interactions in lipid micelles labelled with 1-anilo-8-naphthalenesulfonate, *Plant Physiol.,* 54, 527, 1974.
17. **St. John, J. B., Bartels, P. G., and Hamilton, K. C.,** Surfactant effects on isolated plant cells, *Weed Sci.,* 22, 233, 1974.
18. **Watson, M. C., Bartels, P. G., and Hamilton, K. C.,** Action of selected herbicides and Tween 20 on oat (*Avena sativa*) membranes, *Weed Sci.,* 28, 122, 1980.
19. **Finney, D. J.,** Bioassay and the practice of statistical inference, *Int. Stat. Rev.,* 47, 1, 1979.
20. **Jansen, L. L., Gentner, W. A., and Shaw, W. C.,** Effects of surfactants on the herbicidal activity of several herbicides in aqueous spray systems, *Weeds, 9*, 381, 1961.
21. **Jansen, L. L.,** Relation of structure of ethylene oxide ether-type nonionic surfactants to herbicidal activity of water-soluble herbicides, *J. Agri.. Food Chem.,* 12, 223, 1964.
22. **Jansen, L. L.,** Effects of structural variations in ionic surfactants on phytotoxicity and physical-chemical properties of aqueous sprays of several herbicides, *Weeds,* 10, 117, 1965.
23. **Streibig, J. C., Thonke, K. E., and Kudsk, P.,** The effect of formulations on herbicidal activity of phenmedipham, in *Adjuvants for Agrochemicals,* Chow, P. N. P., Grant, C. A., Hinshalwood, A. M., and Simundsson, E., Eds., CRC Press, Boca Raton, FL, 1989, chap. 9.

24. **Streibig, J. C.,** The effects of adjuvants and formulations on herbicide phytotoxicity, *7. Int. Cong. of Pest. Chem. Abstr.*, IUPAC, Hamburg, 1990, 5A-02.
25. **Kudsk, P., Thonke, K. E., and Streibig, J. C.,** Method for assessing the influence of additives on the activity of foliar-applied herbicides, *Weed Res.*, 27, 425, 1987.
26. **Green, J. M. and Brown, P. A.,** Influence of surfactant properties on nicosulfuron, DPX-E9636, and thifensulfuron performance in corn, *7. Int. Cong. Pest. Chem. Abstr.*, IUPAC, Hamburg, 1990, 5A-04.
27. **Smith, L. W., Foy, C. L., and Bayer, D. E.,** Structure-activity relationships of alkyl-phenol ethylene oxide ether non-ionic surfactants and three water-soluble herbicides, *Weed Res.*, 6, 233, 1966.
28. **Becher, P.,** The emulsifier, in *Pesticide Formulations*, Van Walkenburg, J. W., Ed., Marcel Dekker, New York, 1973, chap. 2.
29. **Crafts, A. S. and Reiber, H. S.,** Studies on the activation of herbicides, *Hilgardia,* 16, 487, 1945.
30. **Wills, G. D.,** Effect of inorganic salts on the toxicity of dalapon and MSMA to purple nutsedge, *Weed Sci. Soc. Am.. Abstr.*, 84, 1971.
31. **Turner, D. J. and Loader, M. P. C.,** Effects of ammonium sulphate and related salts on the phytotoxicity of dichlorprop and other herbicides used for broadleaved weed control in cereals, *Weed Res.*, 24, 67, 1984.
32. **Kudsk, P., Thonke, K. E., and Streibig, J. C.,** The phytotoxicity of difenzoquat to wild oat as influenced by other pesticides, in *Adjuvants for Agrochemicals,* Chow, P. N. P., Grant, C. A., Hinhalswood, A. M., and Simundsson, E., Eds., CRC Press, Boca Raton, FL, 1989, chap. 18.
33. **Koller, W.,** Isomers of sterol synthesis inhibitors. Fungicidal effects and plant growth regulator activity, *Pestic. Sci.,* 18, 129, 1987.
34. **Gressel, J,** Synergizing herbicides, *Rev. Weed Sci.,* 5, 49, 1990.
35. **Kudsk, P.,** The influence of maneb on the effect of the phenoxy herbicides and chlorsulfuron, in *3. Danish Plant Protection Conference/Weeds,* 1987, 267.
36. **McWhorter, C. G.,** Effect of surfactant concentration on Johnsongrass control with dalapon, *Weeds,* 11, 83, 1963.
37. **Foy, C. L. and Smith, L. W.,** Surface tension lowering, wettability of paraffin and corn leaf surfaces, and herbicidal enhancement of dalapon by seven surfactants, *Weeds*, 13, 15, 1965.
38. **Hewlett, P. S. and Plackett, R. L.,** *The Interpretation of Quantal Responses in Biology,* University Park Press, Baltimore, 1979.
39. **Hewlett, P. S.,** Measurement of the potencies of drug mixtures, *Biometrics,* 25, 477, 1979.
40. **Streibig, J. C. and Thonke, K. E.,** The effect of a surfactant on alloxydim-sodium and sethoxydim potency, in *Application and Biology,* Monogr. No. 28, BCPC, Croydon, U. K., 1985, 147.
41. **Streibig, J. C. and Kudsk, P.,** The influence of adjuvants on phytotoxicity of alloxydim-sodium, fluazifop-butyl and DPX-L5300, in *Adjuvants and Agrochemicals,* Foy, C.L., Ed., CRC Press, Boca Raton, FL, 1992.
42. **Bland, P. D. and Brian, R. C.,** Surfactants and the uptake and movement of paraquat in plants, *Pestic. Sci.,* 6, 419, 1975.
43. **Kudsk, P.,** Experiences with reduced doses in Denmark and the development of the concept of factor-adjusted doses, in *Brighton Crop Protection Conference—Weeds,* 1989, 545.
44. **Thonke, K. E.,** Temperaturens of luftfugtighedens indflydelse på 3 forskellige bladherbiciders effekt, in *NJF-Seminar No. 58,* 1984, 6.1.
45. **Kudsk, P., Olesen, T., and Thonke, K. E.,** The influence of temperature, humidity and simulated rain on the performance of thiameturon-methyl, *Weed Res.,* 30, 261, 1990.
46. **Nalewaja, J. D., Pudelko, J., and Adamczewski, K. A.,** Influence of climate and additives on bentazon, *Weed Sci.,* 23, 504, 1975.
47. **Hull, H. M.,** Leaf structure as related to absorption of pesticides and other compounds, *Residue Rev.*, 31, 1, 1970.
48. **Price, C. E.,** The effect of environment on foliage uptake and translocation of herbicides, *Aspects Appl. Biol.,* 4, 157, 1983.
49. **Price, C. E.,** Penetration and translocation of herbicides and fungicides in plants, in *Herbicides and Fungicides Factors Affecting their Activity,* McFarlane, N. R., Ed., Special Publication No. 29, The Chemical Society, London, 1976, 42.
50. **Foy, C. L. and Smith, L. W.,** The role of surfactants in modifying the activity of herbicidal sprays, in *Pesticidal Formulations Research,* Van Walkenburg, J. W., Ed., ACS Adv. Chem. Ser. 86, American Chemical Society, Washington D.C., 1969, 55.
51. **Kudsk, P., Mathiassen, S. K., and Kristensen, J. L.,** The rainfastness of five sulfonylurea herbicides on *Sinapis alba* L., *Meded. Fac. Landbouwwet. Rijksuniv. Gent,* 54/2a, 327, 1989.
52. **Streibig, J. C.,** Herbicide bioassays, *Weed Res.,* 28, 479, 1988.
53. **Behrens, R. and Elakkad, M. A.,** Influence of rainfall on the phytotoxicity of foliarly-applied 2,4-D, *Weed Sci.,* 29, 349, 1981.

Chapter 7

HERBICIDE MIXTURES

Jerry M. Green and Jens C. Streibig

TABLE OF CONTENTS

0-8493-6603-8/93/$0.00+$.50

I. INTRODUCTION

Historically, researchers have improved herbicide performance with mixtures. Because weed flora normally consists of many species with varying herbicide sensitivity, mixtures have mainly used herbicides with complementary spectrums to achieve complete control. Recently, more emphasis has been placed on mixing for additive or synergistic activity, reducing rates, and controlling resistant weeds. Researchers traditionally determined these mixtures by empirical study, but today a more efficient and quantitative method is needed. This chapter presents such a method that combines bioassay and mixture modeling.

The number of potential herbicide mixtures is so large that to test empirically all possibilities is impossible. New registrations, herbicide-resistant crops, safeners, adjuvants, and mixtures with other pesticides continue to increase and make mixtures more complex. Today the most efficient strategy to develop new mixtures uses bioassays to

1. Quantify single herbicide activity on crop(s) and weeds,
2. Select likely mixture partners, and
3. Assay mixtures to determine joint action.

Predictive models can use this information, with appropriate economic and environmental criteria, to select the specific mixtures most likely to succeed.

Herbicide mixtures are used commonly in every crop.[1] Herbicide use has evolved rapidly from high rates of single herbicides to low rates of three-, four-, and five-way mixtures that are mixed with one or two adjuvants that vary according to weed spectrum and environmental conditions. A fungicide, insecticide, or fertilizer also may be included. Mixtures are the most common way to improve efficacy and reduce costs.[2,3]

Many crops in the U.S. and Europe are treated with more than 150 different mixtures each season.[4,5] In 1986, Europeans applied 156 herbicides in 500 different mixtures.[4] Such complexity forces the market to simplify product use by developing premixtures. Now some new herbicides are being introduced as mixture components only.[6,7]

Mixture considerations include tank mixtures, premixtures, and chemicals applied sequentially. Chemicals with long soil residual activity can affect herbicides applied later.[8] Product labels commonly restrict use if certain chemicals previously were applied.*

Herbicides may interact physically or chemically in the spray solution[9] or biologically in the plant. The three basic types of joint action are as follows:

1. Chemicals alone are inactive but the mixture is active;
2. One chemical is active but the inactive(s) affects it;
3. Both chemicals are active.

Joint action is not easy to assess. We must choose a reference model according to the type of joint action, and base this model on the activity of the herbicides applied alone. This chapter primarily addresses models with two herbicides, while Chapter 6 addresses mixtures where only one chemical is a herbicide.

Herbicides can act at the same or different site(s) in the plant. Their intrinsic activity is expressed in the slope and shape of their response curves. The mechanisms of their joint action can be grouped broadly into the following categories:[10,11]

1. **Biochemical** — one herbicide changes the amount of the other that reaches its site of action through uptake, translocation, or metabolism;

* Accent® and Classic® Product Labels, Agricultural Products, E.I. du Pont de Nemours and Company, Wilmington, DE, 1989.

2. **Competitive** — binding at the active site is changed;
3. **Physiological** — the biological effect of either counteracts or synergizes the overall effect;
4. **Chemical** — herbicides react chemically.

Ideally, a mixture model should evolve from biological concepts and define an expected response. To be consistent with how herbicides act alone, the model should revert to the response equation when only one herbicide is present. Joint action should be independent on the response equation type.

Morse[12] reviewed the literature and concluded that many investigators confused "additivity of effects" with "additivity of doses". She clarified some basic principles for determining mixture responses, but these principles have not been extensively adopted.[11] Many investigators overly interpret small tests with only one or two rates and confuse effects due to dose with joint action.

For example, 2 g ha^{-1} DPX-E9636 {*N*-([4,6-dimethoxypyrimidin-2-yl]amino-carbonyl)-3-(ethylsulfonyl)-2-pyridinesulfonamide}, common name rimsulfuron, controls *Setaria faberi* Herrm., but it requires 100 g ha^{-1} to injure most corn hybrids. At 2 g ha^{-1}, the safener, DPX-D5293 (*N*-[amino-carbonyl]-2-chlorobenzene-sulfonamide), reduces *Setaria* control but does not change the corn response (Figure 1A). At 100 g ha^{-1} DPX-E9636, the reverse occurs (Figure 1B). Viewed separately, such results often are interpreted incorrectly as species-specific joint action.

In fact, the safener reduces DPX-E9636 activity on both species similarly by shifting the response curve. The effect on herbicide rate is constant, but the biological effect depends on the shape and slope of the response curve. Consequently, several rates must be tested to separate joint action from dose response curve effects.

II. TERMINOLOGY

The herbicide literature reveals a Babylonian confusion of mixture models and terminology.[4,12] Terminology is confused partly because the meaning of interaction in statistics is not the same as in biology. We wish to distinguish clearly between these two uses.

A. INTERACTION

Statistically, mixture interactions are determined with factorial analysis of variance, testing whether a herbicide effect remains unchanged in mixture. Table 1 defines additivity of effects. The difference between any two columns and rows of corresponding rates is always the same. Interaction occurs when this difference is not constant. Additivity of effects is independent of the dose. If there is no interaction, the response surface is

$$y = F(x_1) + G(x_2) \tag{1}$$

where y is the response and x_1 and x_2 are the herbicide rates. The response surface for interaction is the sum of two functions

$$y = F(x_1, x_2) \tag{2}$$

where $F(x_1,x_2) \neq F(x_1) + G(x_2)$.

Such a definition is biologically unacceptable for the assessment of joint action. For illustration, consider the case of a single herbicide as a hypothetical mixture with itself. In this case, herbicide 1 is the same as herbicide 2, making them interchangeable. Thus,

A

B

FIGURE 1. The joint action of the herbicide DPX-E9636 and the safener DPX-D5293 on corn, *Zea mays* L.'Cargill 937', and *Setaria faberi*: (A) 2 g ha^{-1} DPX-E9636 and 1000 g ha^{-1} DPX-D5293, (B) 100 g ha^{-1} DPX-E9636 and 1000 g ha^{-1} DPX-D5293.

TABLE 1
A Hypothetical Factorial Experiment without Interaction

Herbicide 1 ($g\ ha^{-1}$)	Herbicide 2 ($g\ ha^{-1}$)		
	0	**100**	**200**
		% Control	
0	0	10	50
100	25	35	75
200	50	60	100

100 g ha^{-1} of herbicide 1 with 100 g ha^{-1} of herbicide 2 yields the same as 200 g ha^{-1} of either herbicides. This seldom occurs with additivity of effects because of the asymptotic response curves (Chapter 3).

B. JOINT ACTION

A better way to define herbicide mixture response is to use the dose-response curves. Additivity in this context is the additivity of doses. Additivity assumes that herbicides in mixture can be substituted for each other at equivalent biological rates without changing the mixture response. This exchange corresponds to the relative potency (Chapter 3). Thus, joint action requires knowledge of the response curves, but interaction does not.[12,13]

Additivity of doses is harder to comprehend because we are trained to think in terms of effects. The following analogy to money exchange may help clarify.

For illustration, assume the exchange ratio between the Danish Kroner (DKr) and the U.S. dollar ($) is 7 DKr to $1. If we have 700 DKr and $100 and want to spend in the U.S., we have the equivalent of $200. If we want to spend in Denmark, we have the equivalent of 1400 DKr. The value is the same in both countries, and the exchange ratio corresponds to relative potency. Exchanging money back and forth without losing value is analogous to additivity. However, antagonism results with exchange commissions.

To avoid confusion, we recommend using "joint action" instead of "interaction" to describe herbicide-herbicide mixture response.

III. JOINT ACTION MODELS

The different methods to analyze herbicide joint action can be grouped into three model types: additive dose models, multiplicative survival models, and empirical models. We will review these models for two herbicide mixtures. Models with three or more herbicides are mathematically more complex but do not introduce any differences in principle.[14,15]

Analogous mixtures exist in pharmacology.[13,16,17] To a large extent, we have adopted this terminology and that of Morse[12] and Green and Bailey.[15]

Herbicide mixture models should be based on the single response equations and revert to them when one herbicide is present. Most models use probit or logit transformations.[13,16] These transformations were developed for quantal responses, but often are used for quantitative responses such as fresh or dry weights.[4,12] Joint action is independent of the response equation type. Any dose-response relationship can be used as long as it adequately describes the response. Chapter 3 describes how linear response curves are obtained from intrinsically nonlinear curves.

A. ADDITIVE DOSE MODEL (ADM)

Figure 2 graphically defines the three types of herbicide joint action: additivity, antagonism, and synergism.

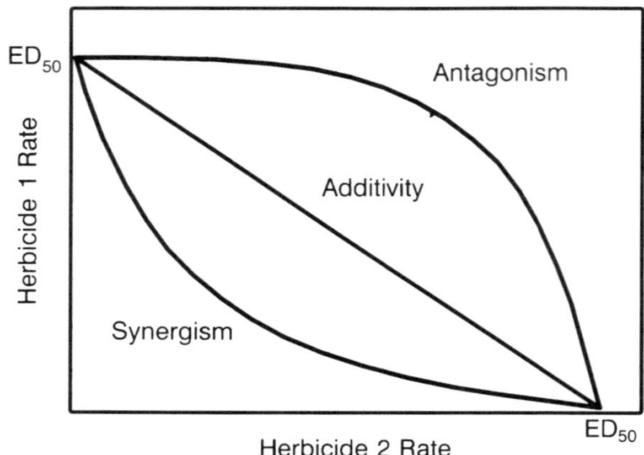

FIGURE 2. Isobole plot defining the three types of herbicide-herbicide joint action: additivity, antagonism, and synergism. ED_{50} is the 50% control rate.

Additivity occurs when the same effect is obtained when one herbicide is substituted for another at equivalent biological rates.[16-18] If the herbicide response curves are parallel, additivity gives linear isoboles. Equivalence is based on relative potency.[19] Antagonism occurs when the mixture has less than expected activity and the isobole bends outward. Synergism is the reverse.

Much of the confusion in weed science,[12,15] toxicology, and pharmacology[13] is because these definitions are not standardized. Lack of standardization results in different interpretations.

The isoboles intercept the axes at rates that give the same response (Figure 2). At any response, the mathematics are

$$x_{1M}\big|x_1 + x_{2M}\big|x_2 = 1 \tag{3}$$

where x_1 and x_2 are the herbicide rates if applied separately, and x_{1M} and x_{2M} are the corresponding mixture rates.

Another way to describe additivity is to use the relative potency, $R = x_1/x_2$. Substituting x_1 with Rx_2 in Equation 3 we get

$$x_1 + Rx_2 = Rx_{2M} \tag{4}$$

The case where mixture x_m consists of fixed proportions p of x_1 and $(1–p)$ of x_2, Equation 4 leads to

$$x_m\left[p + R(1 - p)\right] = Rx_2 = x_1 \tag{5}$$

The rate of a mixture x_m thus can be expressed by a biological equivalent rate of either herbicide. Furthermore, the relationship between the relative potency of the mixture R_m and that of the herbicide applied alone is

$$R_m = p + R(1 - p) \tag{6}$$

Equations 4 through 6 can be used if the response curves of the herbicides alone and in fixed ratios are parallel (Figure 3A and 3B).[20] It is easy to determine if response curves are parallel when they are transformed to logits (Figure 3B) (see Chapter 3).

For parallel response curves ADM can be tested by simultaneously fitting response curves for the herbicides applied separately:

$$y_1 = F(x_1) \tag{7}$$

$$y_2 = F(Rx_2) \tag{8}$$

and the response curves for the mixtures

$$y_m = F\{x_m[p + R(1-p)]\} \tag{9}$$

Equation 9 is reduced to Equation 7 if $p = 1$ and to Equation 8 if $p = 0$.

Table 2 illustrates ADM for a mixture of two sulfonylureas, the methyl esters of thifensulfuron, and tribenuron. The dry weight of *Sinapis alba* L. with thifensulfuron and tribenuron alone ($i = 1$ and 2) and in 67% mixture of thifensulfuron ($i = 3$) is described with ADM assuming parallel response curves, with upper limit D and lower limit C:

$$y_i = \frac{C + (D - C)}{1 + \exp\left(b \log(\mathrm{ED}_{50}) - \log[px_i + R(1-p)x_i]\right)} \tag{10}$$

where $p = 1$ or $p = 0$ for the herbicides alone and $p = 0.67$ for the mixture. ED_{50} is for thifensulfuron and R is the relative potency.

In Table 2 the relative potency for the mixture was 1.22 when the model was assumed to describe three parallel curves. The hypothetical relative potency according to Equation 6 was 1.11. Using Equation 10, the mixture response does not give a significant test for lack of fit; hence, the mixture follows ADM (Figure 4).

If response curves are not parallel, R will vary as the distance between the lines changes. The mixture response curves will not follow the same functional relationship as the herbicides alone (Figure 3C).

For many years investigators have taken the straight line isobole as ADM. However, this occurs only when the response curves are parallel. The joint action of chlorsulfuron and linuron mixtures does not satisfy ADM, and the herbicides do not have parallel response curves.[21] Thus, mixtures may have response curves different from their components.

One mixture model that does not require parallel response curves is the equivalents model (EQM).[15] EQM is mathematically complex but corresponds to ADM. For illustration, log-logit response equations, which linearlize the response, for each herbicide are

$$Y_1 = \log\left(\frac{y_1}{100 - y_1}\right) = a_1 + b_1 \log(x_1) \tag{11a}$$

$$Y_2 = \log\left(\frac{y_2}{100 - y_2}\right) = a_2 + b_2 \log(x_2) \tag{11b}$$

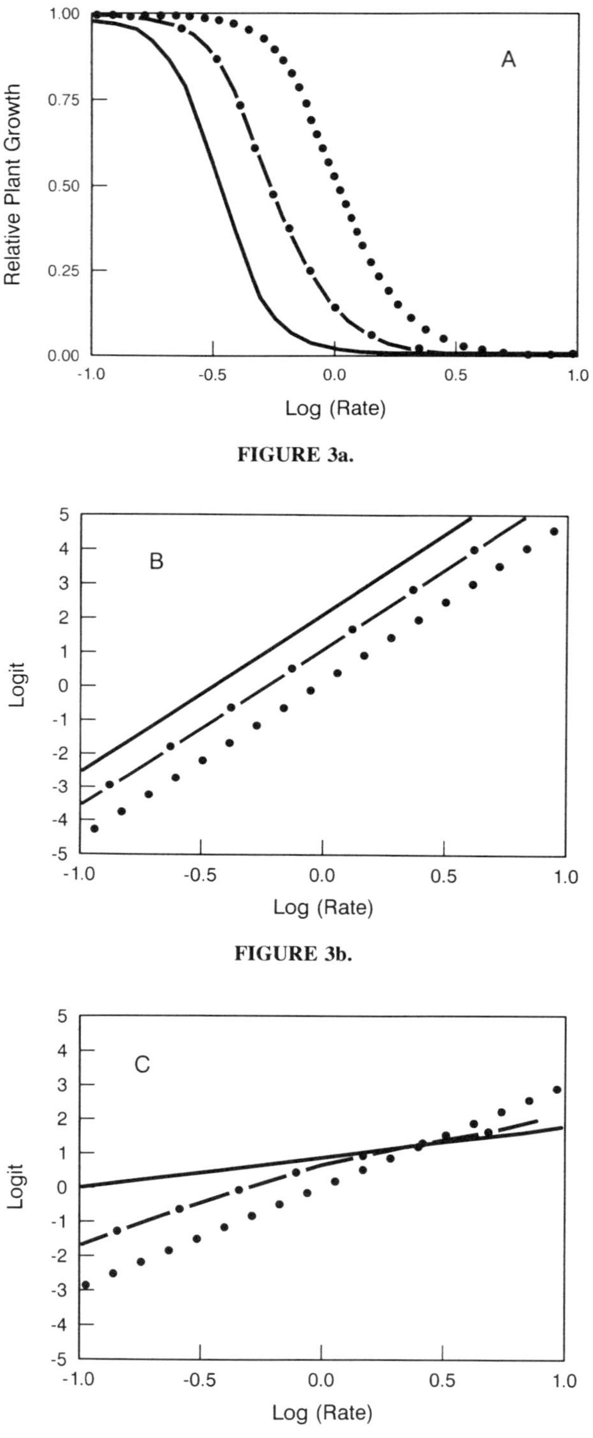

FIGURE 3a.

FIGURE 3b.

FIGURE 3c.

FIGURE 3. Comparison of response curves using ADM: (A) response curves for herbicide 1 (—) and 2 (••••) applied alone and in a 60:40 mixture (- • - • -). (B) logit-transformed responses from parallel response curves of A. (C) logit-transformed responses with nonparallel response curves.

TABLE 2
Summary of Regressions for Thifensulfuron, the Reference Herbicide, and Tribenuron Applied Alone and in a 2:1 Mixture on *Sinapis alba*

Thifensulfuron					
D g pot^{-1}	C g pot^{-1}	b	ED_{50} g ha^{-1}	*R*	R_{mix}
		Parallel lines[a]			
18.15	4.04	−4.220	0.207	1.320	1.222
(1.28)	(0.49)	(0.776)	(0.033)	(0.215)	(0.199)
		Parallel lines and ADM[a]			
18.26	4.05	−4.154	0.195	1.296	
(1.28)	(0.47)	(0.734)	(0.029)	(0.212)	

Note: *R* is the relative potency of thifensulfuron and tribenuron; R_{mix} is the relative potency of thifensulfuron and the 2:1 mixture. The test for lack of fit between the two models was $F_{1,46} = 0.56$.

[a] Standard deviations in parentheses.

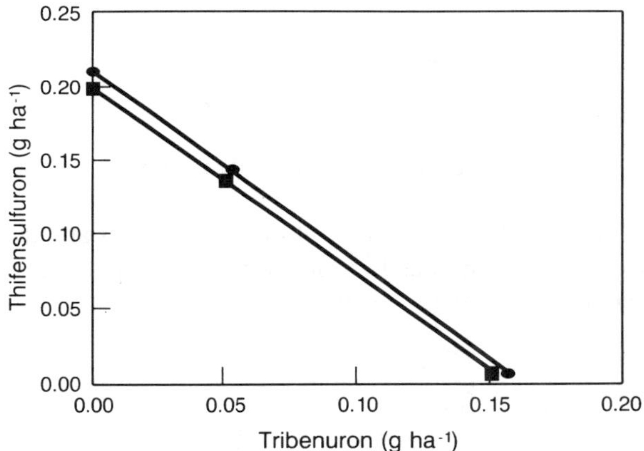

FIGURE 4. ED_{50} isoboles based upon regressions in Table 2. Parallel curves (●); ADM (■).

where y_1 and y_2 are the percent responses, Y_1 and Y_2 are the logit transformations, and x_1 and x_2 are the herbicide rates. The parameters a_1 and a_2 are the intercepts, and b_1 and b_2 are the slopes of the response equations.

For an additive mixture, the model is

$$Y = p\left[a_1 + b_1 \log\left(x_1 + x_{1(2)}\right)\right] + q\left[a_2 + b_2 \log\left(x_2 + x_{2(1)}\right)\right] \tag{12}$$

where Y is the mixture response expressed as logit, $x_{1(2)}$ and $x_{2(1)}$ are the equivalent rates of each herbicide in terms of the other (Figure 5).

The log-logit response equation easily calculates the biologically equivalent rates of each herbicide at any one response level, here Y_1:

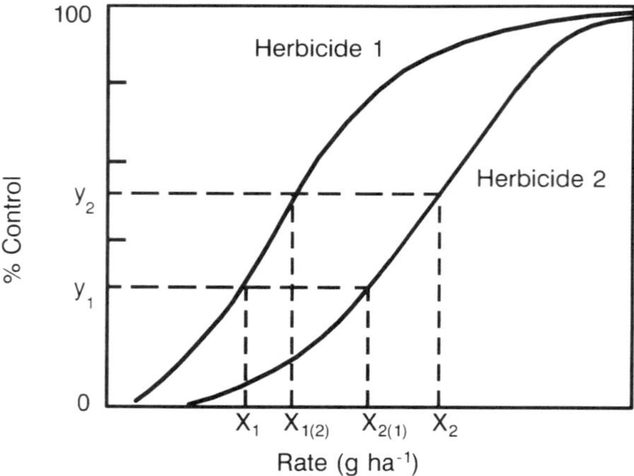

FIGURE 5. EQM terms defined graphically with response curves. y_1 and y_2 denote the % control derived from the logit of control in Equations 11a and 11b. The rate axis is a logarithmic scale.

$$x_{1(2)} = \exp\left(\frac{Y_1 - a_1}{b_1}\right) \tag{13a}$$

$$x_{2(1)} = \exp\left(\frac{Y_1 - a_2}{b_2}\right) \tag{13b}$$

For parallel response curves, the biologically equivalent rates are a constant ratio that corresponds to the relative potency.

Each herbicide is converted to the equivalent rate of the other and thus incorporated into each other's response equation. The p and q terms weight the response equations in proportion to their relative biological activity ($p + q = 1$):

$$p = \frac{x_1 + x_{2(1)}}{x_1 + x_2 + x_{1(2)} + x_{2(1)}} \tag{14a}$$

$$q = 1 - p \tag{14b}$$

Weighting allows the model to use nonparallel response equations. When only one herbicide is present, the weighing terms are 0 and 1 and the model Equation 11 is reduced to a single equation.

This model quantifies deviations from additivity with the joint action parameter (θ) before the equivalent rate terms:

$$Y = p\left[a_1 + b_1 \log\left(x_1 + \theta x_{1(2)}\right)\right] + q\left[a_2 + b_2 \log\left(x_2 + \theta x_{2(1)}\right)\right] \tag{15}$$

The θ parameter quantifies herbicide joint action. For the reference ADM, θ equals 1. The mixture is synergistic when θ is >1 and antagonistic when θ is <1. When the response curves are parallel, this definition corresponds exactly to Figure 2.

TABLE 3
6 × 7 Factorial Test to Evaluate the Joint Action of Chlorimuron and Mefluidide on *Setaria viridis*

Mefluidide (g ha^{-1})		Chlorimuron (g ha^{-1}) 0	2	4	8	16	32	64
		% Control						
0	Observed	0	7	11	14	22	28	38
	EQM	0	5	9	14	22	33	45
10	Observed	1	27	26	33	33	37	50
	EQM	2	19	21	25	30	39	49
20	Observed	4	33	37	43	44	48	50
	EQM	5	31	34	37	40	46	54
40	Observed	18	34	48	49	51	59	61
	EQM	12	43	49	53	56	59	63
80	Observed	26	48	54	61	66	70	80
	EQM	24	53	61	67	71	74	76
160	Observed	44	62	70	82	78	84	90
	EQM	43	63	70	75	80	84	86

Note: The data are the means of 10 replicates weighed and converted to % control (Green, unpublished). Parameters (with 95% confidence intervals) were $a_1 = -3.43$ (–3.07 to –3.80), $b_1 = 0.78$ (0.67 to 0.89), $a_2 = -6.63$ (–6.07 to –7.19), $b_2 = 1.25$ (1.10 to 1.39), and $\theta = 6.9$ (4.7 to 9.9).

Two θ parameters can be defined to differentiate the effect of the first herbicide on the second and vice versa. An F-test can determine if two joint action parameters are statistically justified.

Green and Bailey[15] compared EQM to 6 other models on a mixture of alachlor and atrazine that acted additively. To show EQM on nonadditive mixture, we analyzed the joint action of the ethyl ester of chlorimuron and mefluidide on *Setaria viridis*.

Mefluidide and chlorimuron are highly synergistic. Table 3 summarizes the parameters and their 95% confidence intervals. The θ value of 6.9 results in sharply inward bending contours (Figure 6). Mixtures of mefluidide and sulfonylureas are often synergistic.[22]

Some herbicide mixtures are more antagonistic than EQM can model as described in Equation 15. These antagonistic mixtures can be modeled similarly to herbicide safener mixtures by including a rate factor (RF) to reduce the herbicide rate.[15] The two proposed rate factors are the asymptotic rate factor:

$$ARF = \frac{1 + x_2}{1 + \delta x_2} \tag{16}$$

and the proportionate rate factor:

$$PRF = \frac{1}{1 + \delta x_2} \tag{17}$$

where x_2 is the rate of the antagonist and the parameter δ quantifies its activity.

When the second chemical is a nonherbicide, the rate factor is placed immediately before the herbicide rate. These factors imply small differences in mechanisms that whole plant tests usually cannot distinguish. The model with PRF is

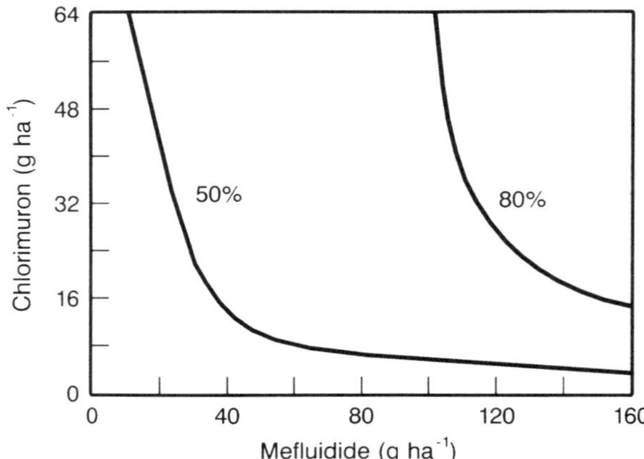

FIGURE 6. Synergistic mixture response of mefluidide and chlorimuron on *Setaria viridis*. Percent denotes control level.

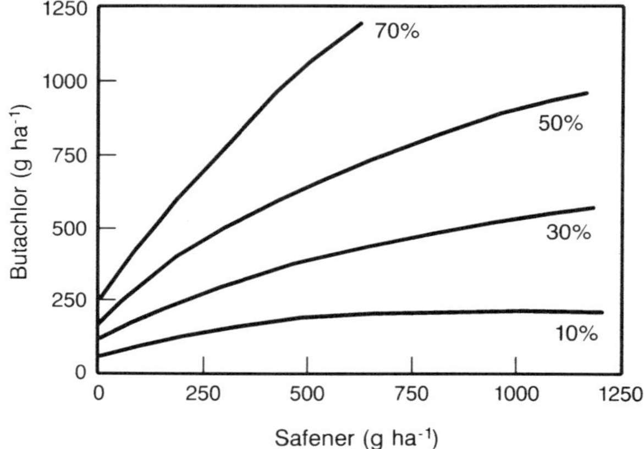

FIGURE 7. Mixture response of a herbicide safener, phenyl-2-chloro-4-trifluoromethyl-5-thiazole-carboxylate, and the herbicide, butachlor, on rice. Percent denotes control level.

$$Y = p\left[a_1 + b_1 \log\left(PRFx_1\right)\right] + q\left[a_2 + b_2 \log\left(PRFx_2\right)\right] \tag{18}$$

Rate factors are used within the response equation and are thus sensitive to its slope and shape and asymptotic to 0 and 100%. For interpretation, rate factors quantify the shift of the herbicide response curve.

To illustrate such modeling, Figure 7 uses a published example with the safener (phenyl-2-chloro-4-trifluoromethyl-5-thiazole-carboxylate) and butachlor on rice.[23] The safener does not injure rice, but it dramatically reduces butachlor injury. This strong antagonism can be readily modeled by combining the PRF factor and EQM responses.

A rate factor also can be used with EQM. When θ reaches 0, EQM Equation 15 is simplified to

$$Y = p\left[a_1 + b_1 \log\left(x_1\right)\right] + q\left[a_2 + b_2 \log\left(x_2\right)\right] \tag{19}$$

Incorporating a modified rate factor term, θ becomes negative and EQM evolves to

$$Y = p\left[a_1 + b_1 \log\left(x_1/\left(1-\theta x_{1(2)}\right)\right)\right] + q\left[a_2 + b_2 \log\left(x_2/\left(1-\theta x_{2(1)}\right)\right)\right] \tag{20}$$

With a rate factor, EQM can model extremely antagonistic mixtures.

B. MULTIPLICATIVE SURVIVAL MODEL (MSM)

If two herbicides act in entirely different ways, a contrasting type of joint action models may be useful. These models are called random kill models,[24] multiplicative survival models (MSM),[12] or simple independent action.[13,16] Their basis also can be herbicide response curves, but they usually require that the responses be expressed as a proportion of a hypothetical maximum value Q. In weed science MSM has often been confused with ADM.[25]

Suppose that a mixture contains two herbicides, and if administered separately, yield responses P_1 and P_2. If each herbicide acts independently, a proportion P_2 of the treated plants will not respond to the first herbicide but would respond to the second herbicide

$$P = P_1 + P_2\left(1 - P_1\right) \tag{21}$$

If it is symmetrical in P_1 and P_2, then

$$P = 1 - \left(1 - P_1\right)\left(1 - P_2\right) \tag{22}$$

This natural form for quantal responses is more problematic for quantitative responses, because the hypothetical maximum value Q has to be estimated from the data (Chapter 3). Use of percentage of untreated control does not have the same meaning. In quantitative bioassays, the percentage is based on the mean response, while in quantal assays it refers to proportions.

If the dose-response curves for two herbicides are

$$q_1 = F\left(x_1\right) \quad q_2 = G\left(x_2\right) \tag{23}$$

then for any mixture satisfying MSM it is

$$q_m = F\left(x_1\right)G\left(x_2\right) \tag{24}$$

The rather elementary concept of MSM poses more complicated statistical problems than that of simple similar action.[16] For example, Figure 8A shows herbicides with parallel response curves applied separately and in fixed ratio satisfying MSM. The response curve for a mixture is not parallel and thus has a different functional relationship (Figure 8B). MSM isoboles depend on the response level, while ADM isoboles do not (Figure 9).[21]

Many statistical methodologies surround MSM.[25,26] These methods often determine different joint action for different preset rate combinations of the same mixture components. MSM does not define one joint action.

Mixture studies often use MSM with a factorial analysis of variance.[27,28] These methods and that of Drury[24] can be linked to MSM by taking the logarithm of the effect

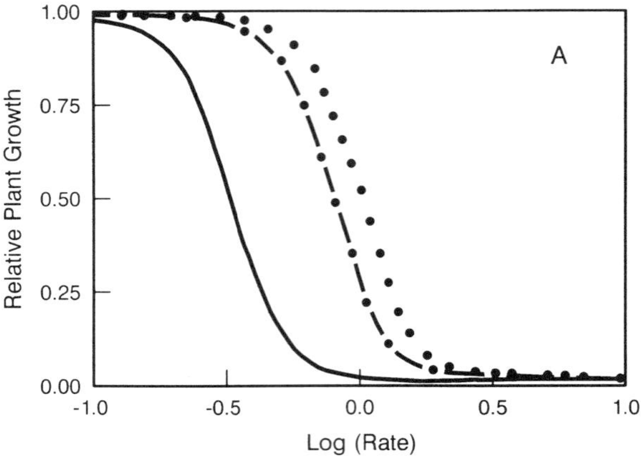

FIGURE 8a.

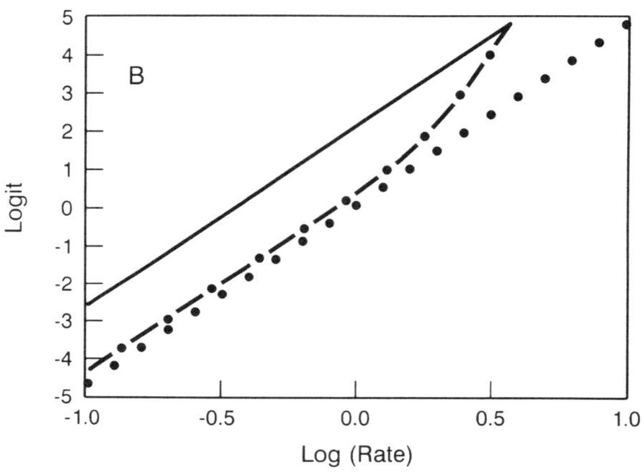

FIGURE 8b.

FIGURE 8. Comparison of response curves using MSM: (A) response curves for herbicide 1 (—) and 2 (••••) applied alone and in a 60:40 mixture (- • - • -), (B) logit-transformed responses.

$$\log(q_m) = \log[F(x_1)] + \log[G(x_2)] \tag{25}$$

If interaction disappears when using Equation 25, then the mixture satisfies MSM.

Table 1 illustrates this type of mixture. A 100 g ha^{-1} rate of herbicide 1 controls 25%, while 100 g ha^{-1} of herbicide 2 controls 10%. According to Equation 22, the expected mixture response is 32.5% for MSM. The actual mixture response for additive effect is 35%

Hewlett and Plackett[13] developed a general theory that combined simple similar (ADM) and independent action (MSM). Their theory does not require parallel response curves as a basis for similar action. The theory is based on quantal responses and has not yet been used in herbicide studies.

Their biological concepts are based on chemicals acting selectively through a site of action (Table 4). When two herbicides act at the same site, their joint action is similar. If they act at different sites, they are dissimilar. Interaction occurs irrespective of similar or dissimilar joint action. In this context, interaction is concerned with a physiological system and not effects per se.

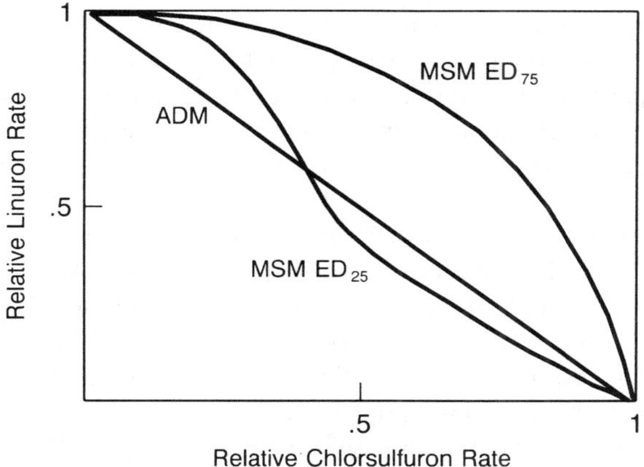

FIGURE 9. Comparison of ADM and MSM isoboles at different response levels on *Sinapis alba*.[21] The axes have been scaled so that the doses of the separate herbicides are unity.

TABLE 4
The Four Types of Joint Action According to Hewlett and Plackett[13]

	Similar	**Dissimilar**
Noninteractive	Simple similar (e.g., ADM)	Independent (e.g., MSM)
Interactive	Complex similar	Dependent

C. EMPIRICAL REGRESSION

Polynomial equations are often used to analyze mixture results with linear regression techniques.[25,29] Polynomials are difficult to interpret biologically and do not make good predictor models. However, they are well founded in statistics and are often the first approach for scientists. Polynomials are based on rate terms, and their interactions do not provide information on joint action. High degree polynomials often give unrealistic interpolation within and extrapolations outside the data range.[30]

Full polynomials usually have too many parameters, but these can be reduced by various stepwise procedures. These procedures usually fit the data well, but they often produce a cumbersome array of parameters and a plot that is overly sensitive and impossible to interpret.

Piepel and Cornell[31] proposed a linear regression model called a mixture amount model that is based on proportions and total herbicide amount instead of rate. For two herbicides where the proportions are z_1 and z_2 and A is the total amount, then

$$y = a_1z_1 + a_2z_2 + b_1z_1A + b_2z_2A + a_{12}z_1z_2 + b_{12}z_1z_2A \tag{26}$$

If y is a logit transformation and A is replaced by log(A), the model reduces to a log-logit model when only one herbicide is present. The terms, a_{12} and b_{12}, quantify joint action.[32]

Computer technology called neural networks or back propagation networks[33] also can analyze mixture results. These methods generate plots but no mathematical model.

IV. EXPERIMENTAL DESIGN FOR JOINT ACTION

Proper experimental design minimizes the work and resources needed to answer an experimental question. Mixture assays are often more complex than single herbicide assays, making proper design more important. Although extensive literature exists for mixture design,[31,34] little exists for biological experiments to determine joint action.[35]

The key design questions are the selection of herbicides, their rates alone and in mixtures, replication number, treatment number, and response to be measured. Design must be related to the method of analysis. However, the joint action model is not usually known when the assay is designed.

Factorial designs provide enough data for analysis but are usually inefficient. For example, 5 × 5 factorial has four or less nonzero treatments to fit the dose-response parameters. This is less than the five rates normally considered a practical minimum.[36-38] In contrast, this 5 × 5 factorial has 16 mixture treatments to determine one joint action parameter in EQM.

Another inefficiency with factorials is selecting rates for single herbicides so the effects range between 0 and 100% but do not predominately give 0 or 100% in their corresponding mixtures. The midregion carries the most information, and 0 and 100% effects do not help quantity response curves or joint action (Chapter 3). Incomplete factorials or design with one or two mixture ratios are usually more efficient.

The optimum experimental design for MSM is more complex. Individual constituents are treated as ADM, but a different principle is needed because MSM is bivariate, while ADM is univariate.[35] Many scientists use factorial designs so results can be viewed in tables and analyzed with Colby's equation.[25,28]

Colby's equation is useful as a first approximation and attractive because of its simplicity. However, it unrealistically assumes that herbicides act independently and sequentially. This "interaction of effects" is dependent on the rates chosen, because the dose-response curve has asymptotic upper and lower limits (Chapter 3). If applied to a response curve as a hypothetical mixture of a herbicide with itself, Colby's equation almost always determines synergism and antagonism depending on the response curve.

ADM is different in that rates, not effects, are considered. Herbicide rates with similar effects can be substituted for one another to maintain efficacy depending on cost, environmental safety, or other factors. Computer models can evaluate many herbicide combinations with it and devise complex mixtures optimally suited to defined performance criteria.

Unless *a priori* requirements govern, the mixture ratios should be evenly spaced between those for the herbicides alone.[16] "Evenly spaced" is best defined by using the ED_{50} isobole method. Finney[16] advocates even spacing on the basis of $\log(ED_{50})$ while Hewlett advocates spacing on the basis of ED_{50}.[35]

If one herbicide is R times as potent as the other, its ED_{50} will be shifted a distance $\log(R)$ to the left. With ADM, a mixture in the proportion p and $(1-p)$ yields a response curve at a distance $O\log(r)$ from the first herbicide, where O is some fraction between 0 and 1. According to Equation 6

$$\log\left[p + R(1-p)\right] = O\log(R) \tag{27}$$

where

$$p = \frac{R - R^{O}}{R - 1} \tag{28}$$

If v mixtures are used then for an evenly spaced distribution along the isobole (Figure 10)

$$O_i = \frac{i}{v + 1} \tag{29}$$

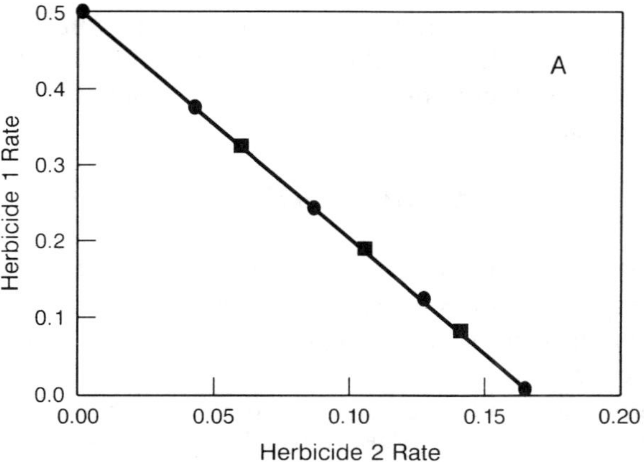

FIGURE 10A.

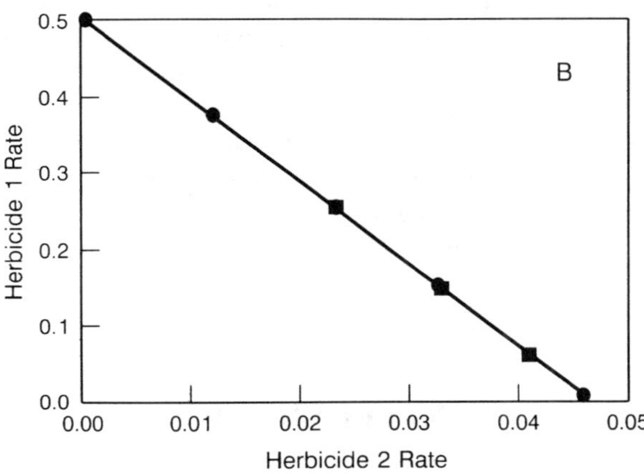

FIGURE 10B.

FIGURE 10. Examples of designing mixture experiments so resulting isoboles are properly described. Ratio selection of herbicide in mixtures using ED_{50}. (A) Relative potency of 3 with spacing based on ED_{50} (●) or log(ED_{50}) (■). (B) Relative potency of 11 with spacing based on ED_{50} (●) or log(ED_{50}) (■).

If an approximate R is known and O is chosen half way between the herbicides separately active, then p can be calculated. Figure 10 illustrates that the difference in spacing with log(R) or R disappears with greater relative potencies.

Rates need to be selected so both the single herbicide and mixture responses cover most of the response range, particularly the ED_{50}. The more rates tested, the less number of replications needed. Statistical precision of parameters need several degrees of freedom, but beyond that replications and treatments are only needed to appraise the experimental variation.

Choose one meaningful performance criteria and definitely measure it! Overzealous researchers mistakenly evaluate several responses, i.e., root and shoot length, dry and fresh weights for the roots and shoots, and multiple visual observations on several weeds and the crop. Quality is often lost, and all reponses are likely to be highly correlated. Discussion gets complicated and focuses on small differences between responses of dubious significance.

The experimental question with mixtures is not always concerned with reference models for joint action. In practice, the many users do not care about a joint action model but concern themselves only with the mixture performance. For these tests, a mixture can be considered a treatment, and one does not need to include multiple rates of all herbicides and mixtures. Such tests only evaluate a specific mixture under one condition and do not lend well to extrapolation to other rates.

V. FUTURE DIRECTIONS

Herbicide mixtures have grown in importance and complexity, and we must now accurately model their performance. Bioassays to assess joint action are much easier under controlled environment conditions than in the field. Computer models can readily assess bioassay results and quantify joint action.

This joint action information along with field dose-response information allows realistic projections of mixture performance. These projections with economic and environmental criteria can focus field development efforts on mixtures most likely to succeed.

Expert computer systems will play a greater role in the analysis, prediction, and prescription of herbicide mixtures. We need to develop the quantitative information to support these systems. These systems will use dose-response and joint action information to reduce herbicide rates and increase performance.[39,40] Such mixtures will maximize herbicide efficiency and help meet the goals in some countries to reduce herbicide rates.[41-43]

Because we cannot empirically test all possible mixtures, we need bioassays to give us an efficient, fast way to quantify herbicide performance alone and their joint action. This information allows us to project realistic mixture performance with ADM-type models and to select potentially useful mixtures. Together, bioassays and models give us the most efficient means to optimize mixture performance.

REFERENCES

1. **Green, J. M.,** Herbicide antagonism at the whole plant level, *Weed Technol.,* 3, 217, 1989.
2. **Barrett, M. and Witt, W. W.,** Maximizing pesticide use efficiency, in *Energy in World Agriculture Handbook,* Helsel, Z., Ed., Elsevier Science Publishers, Amsterdam, 1987, 10, 235.
3. **Green, J. M. and Amuti, K. S.,** Maximizing the performance of antagonistic mixtures, in *International Symposium of Herbicide Safeners,* Hatzios, K. K. and Komives, T., Eds. Budapest, Hungary, 1991, in press.
4. **Bödeker, W., Altenburger, R., Faust, M., and Grimme, L. H.,** Methods for the assessment of mixtures of plant protection substances (pesticides): mathematical analysis of combination effects in phytopharmacology and ecotoxicology. *Nachrichtenbl. Dtsch. Pflanzenschutzdienstes (Braunschweig),* 42, 70, 1990.
5. **Delvo, H. W.,** Herbicide use in wheat and soybean production, *Proc. North Cent. Weed Control Conf.,* 42, 28, 1987.
6. **Gerwick, B. C. and Kleschick, W. A.,** DE-498: a new broadspectrum herbicide for soybeans and other crops, *Proc. North Cent. Weed Sci. Soc.,* 45, 113, 1990.
7. **Green, J. M., Leek, G. L., Strachan, S. D., Palm, H. L., and Rowe, S. W.,** DPX-79406 — a new postemergence herbicide product for corn, *Abstr. Weed Sci. Soc. Am.,* 30, 4, 1990.
8. **Johnson, M. D., Stahlberg, L. A., Porpiglia, P. J., Pruss, S. R., and Gillespie, G. R.,** Interaction between postemergence applications of CGA-136872 and soil-applied insecticides in corn, *Proc. North Cent. Weed Sci. Soc.,* 44, 40, 1989.
9. **O'Donovan, J. T., O'Sullivan, P. A., and Caldwell, C. D.,** Basis for antagonism of paraquat phytotoxicity to barley by MCPA dimethylamine, *Weed Res.,* 23, 165, 1983.
10. **Ezra, G., Rusness, D. G., Lamoureux, G. L., and Stephenson, G. R.,** The effect of CDAA (*N,N*,-diallyl-2-chloroacetamide) pretreatments on subsequent CDAA injury to corn (*Zea mays* L.). *Pestic. Biochem. Physiol.,* 23, 108, 1985.
11. **Hatzios, K. K. and Penner, D.,** Interactions of herbicides with other agrochemicals in higher plants, *Rev. Weed Sci.,* 1, 1, 1985.

12. **Morse, P. M.,** Some comments on the assessment of joint action in herbicide mixtures, *Weed Sci.,* 26, 58, 1978.
13. **Hewlett, P. S. and Plackett, R. L.,** *The Interpretation of Quantal Response in Biology,* University Park Press, Baltimore, 1979.
14. **Berenbaum, M. C.,** The expected effect of a combination of agents: the general solution, *J. Theor. Biol.,* 114, 431, 1985.
15. **Green, J. M. and Bailey, S. P.,** Herbicide interactions with herbicides and other agricultural chemicals, in *Methods of Applying Herbicides,* McWhorter, C. G. and Gebhardt, M. R., Eds., Weed Science Society of America, Champaign, IL, 1987, 37.
16. **Finney, D. J.,** *Probit Analysis.* 3rd. ed. Cambridge University Press, Cambridge, 1971.
17. **Hewlett, P. S.,** Measurement of the potencies of drug mixtures, *Biometrics,* 25, 477, 1969.
18. **Akobundu, I. O., Sweet, R. D., and Duke, W. B.,** A method of evaluating herbicide combinations and determining herbicide synergism, *Weed Sci.,* 23, 20, 1975.
19. **Tammes, P. M. L.,** Isoboles, a graphic representation of synergism in pesticides, *Neth. J. Plant Pathol.,* 70, 73, 1964.
20. **Streibig, J. C.,** Joint action of root-absorbed mixtures of auxin herbicides in *Sinapis alba* L. and barley (*Hordeum vulgare* L.), 27, 337, 1987.
21. **Streibig, J. C.,** Joint action of root-absorbed mixtures of DPX-4189 and linuron in *Sinapis alba* L. and barley, *Weed Res.,* 23, 3, 1983.
22. **Tautvydaus, K. J. and Hargroder, T. J.,** Synergistic growth retardation of grasses with mefluidide/PGR combinations, *Proc. Plant Growth Reg. Soc. Am.,* 10, 51, 1985.
23. **Howe, R. K. and Lee, L. F.,** 2-4-Disubstituted-5-thiazole-carboxylic acids and derivatives, U.S. Patent 4199506, 1980.
24. **Drury, R. E.,** Physiological interaction, its mathematical expression, *Weed Sci.,* 28, 573, 1980.
25. **Nash, R. G.,** Phytotoxic interaction studies — techniques for evaluation and presentation of results, *Weed Sci.,* 29, 147, 1981.
26. **Flint, J. L., Cornelius, P. L., and Barrett, M.,** Analyzing herbicide interactions: a statistical treatment of Colby's Method, *Weed Technol.,* 2, 304, 1988.
27. **Gowing, D. P.,** Comments on tests of herbicide mixtures, *Weeds,* 8, 379, 1960.
28. **Colby, S. R.,** Calculating synergistic and antagonistic responses of herbicide concentrations. *Weeds,* 15, 20, 1967.
29. **Campbell, T. A., Genter, W. A., and Danielson, L. L.,** Evaluation of herbicide interactions using linear regression modeling. *Weed Sci.,* 29, 378, 1981.
30. **Streibig, J. C. and Thonke, K. E.,** Fitting equations to herbicide bioassays: using response surfaces in factorial designs, *Ber. Fachg. Herbologie,* 24, 173, 1983.
31. **Piepel, G. F. and Cornell, J.A.,** Models for mixture experiments when the response depends on the total amount, *Technometrics,* 27, 219, 1985.
32. **Rummens, F. H. A.,** An improved definition of synergistic and antagonistic effects, *Weed Sci.,* 23, 4, 1975.
33. **Rumelhart, D. E. and McClelland, J. L., Eds.,** *Parallel Distributed Processing.* Vols. 1 and 2, MIT Press, Cambridge, MS, 1986.
34. **Cornell, J. A.,** *Experiments with Mixtures,* John Wiley & Sons, New York, 1981.
35. **Abdelbasit, K. M. and Plackett, R. L.,** Experimental design for joint action, *Biometrics,* 38, 171, 1982.
36. **Becker, N. G.,** Models and designs for experiments with mixtures, *Aust. J. Stat.,* 3, 195, 1978.
37. **Finney, D. J.,** *Probit Analysis. A Statistical Treatment of the Sigmoid Response Curve,* 2nd ed., Cambridge University Press, 1952, 318.
38. **Nash, R. G. and Jansen, L. L.,** Determining phytotoxic pesticide interactions, *J. Environ. Qual.,* 2, 493, 1973.
39. **Combellack, J. H.,** Efficient utilisation of herbicides, *Proc. 9th Australian Weeds Conf.,* 552, 1990.
40. **Green, J. M.,** Maximizing herbicide efficiency with mixtures and expert systems, *Weed Technol.,* 5, 894, 1991.
41. **Haas, H.,** Danish experience in initiation and implementing a policy to reduce herbicide use, *Plant Prot. Q.,* 4, 38, 1989.
42. **Baandrup, M. and Ballegaard, T.,** Three years of field experience with an advisory computer system applying factor-adjusted doses, in *Brighton Crop Protection Conference — Weeds,* 1989, 555.
43. **Kudsk, P.,** Experiences with reduced herbicide doses in Denmark and the development of the concept of factor-adjusted doses, in *Brighton Crop Protection Conference —Weeds,* 1989, 545.

Chapter 8

NO-OBSERVABLE-EFFECT LEVEL (NOEL)

Wilfried Pestemer and Petra Günther

TABLE OF CONTENTS

0-8493-6603-8/93/$0.00+$.50

I. INTRODUCTION

The terms "no-effect level" (NEL) and "no-observable-effect level" (NOEL) originate from toxicology. The values of NEL and NOEL mark the border line between effect and no effect of a substance. In herbicide research these terms are used in connection with dose-response curves describing subtoxic doses. The shape of a dose-response curve often can be described by a sigmoid curve that is symmetrically on log dose (see Chapter 3). This curve can be described by a log-logistic equation.[1-5] A typical dose-response curve is shown in Figure 1. Often an initial stimulation of response at very low doses can be observed.[4,6-10] At higher doses the curve decreases until it is almost linear. Generally, comparisons of efficiency of different herbicides are most distinct in this linear range.[5,9,11-13] The lower asymptote of the curve can be zero, depending on the testing procedure and the response variable, but it is usually somewhat higher.[3]

Generally it is assumed that xenobiotics like herbicides will not influence plant growth at very low concentrations. Increasing the dose beyond a specific value (NEL), results in a reduction of plant growth.[14] The problem, in this context, is to determine the lowest dose that produces negative effects. Detection limits are decreasing the more refined the methods of measurement are becoming. For example, changes in the rate of photosynthesis or enzyme activity in plants are detectable at considerably lower doses than is an effect on the fresh weight.[8] Thus, very low doses, although not affecting growth, might affect the metabolism of the plant, but these effects are not detected by the method used.[15-18] The real NEL can therefore not be determined.[19,20] Consequently, it is more convenient to speak of "no-observable-effect level" or "no-observed-effect level", as they are closely connected with the testing procedure.[18,21] Usually NOEL is defined as the highest dose having no effect on the growth of the test organism.[22,23]

A statistically significant difference at a certain probability level may not in all cases be of practical relevance, for example, if the obtained results, such as stimulation of growth, are not particularly important to the problem in question. This is expressed in the so called "no-observed-adverse-effect level" (NOAEL), where only relevant negative effects on the organism are taken into consideration.[22]

In herbicide research an increasing demand for prior consideration of side effects has made the assessment of NOEL important. New herbicides with high selectivity and biological activity, e.g., sulfonylureas, have put NOEL into new perspective,[10] because a safe crop rotation often depends on a 99% dissipation of the herbicide from the upper soil layer.[24]

II. STATISTICAL METHODS FOR DETERMINATION OF THE NO-OBSERVABLE-EFFECT LEVEL

In literature two basically different methods for determining a NOEL are described. A multiple comparison test between treatment means can be used to detect doses that have a significant effect. The highest dose that does not cause a significant reduction is called the NOEL.[23] Alternatively, NOEL can be obtained from a dose-response curve fitted to data by regression analysis. In this case a dose that causes a specific reduction of response (commonly 5 or 10% in herbicide research) is determined from the regression line and is denoted NOEL or toxicity threshold.[25-27] Both methods have advantages and disadvantages and call for specific experimental designs.

A. MULTIPLE COMPARISON TESTS

In multiple comparison procedure, the means of responses at various doses are compared. The highest dose that does not produce a significant effect is equivalent to NOEL.[23]

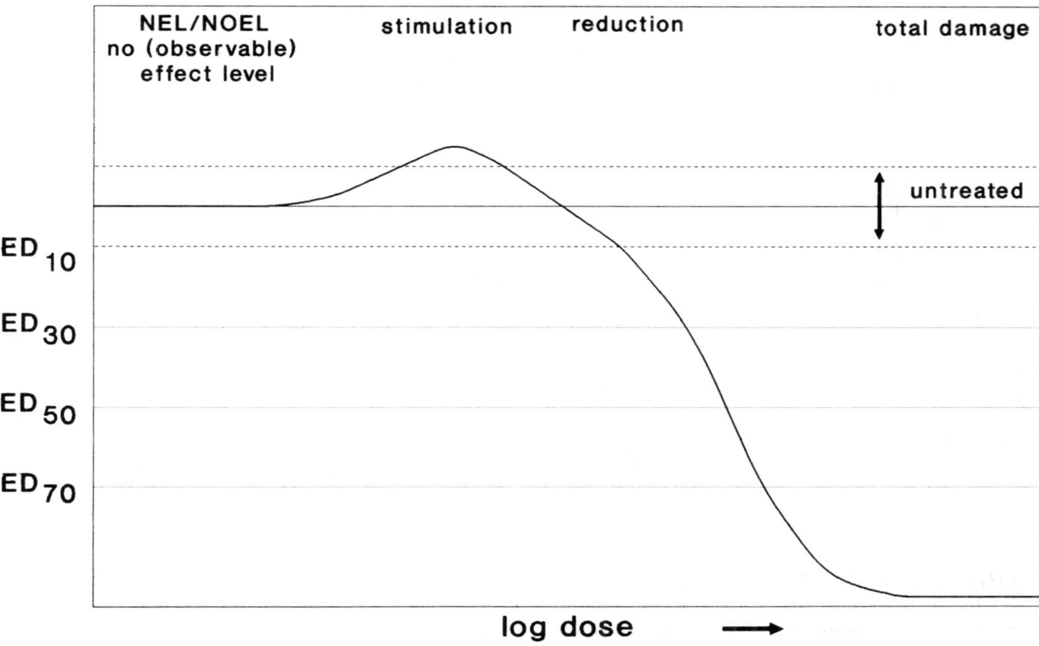

FIGURE 1. Schematic relationship between plant response and herbicide dose with range of variation of the untreated plants.

1. Influence of the Experimental Design on the NOEL

The experimental design and number of replications have a marked influence on the size of differences that can be deemed significantly different. The variability of data originates, besides the random variation, from many sources, e.g., genetic variability of the test organisms. This affects responses particularly in the low dose range.

Minor differences cannot be statistically detected in an experiment with great inherent variation. Test systems with low variability in plant material are preferred (e.g., weeds are not always suitable as test plants because of their considerable genetic variability). If a response variable is to be used because it is easy to measure, e.g., fresh weight, despite its relatively great variability, it can be compensated for by increasing the number of replications.[28] However, the increased demand for labor and facilities also must be taken into account.

Methods for calculating an appropriate number of replications are given elsewhere.[29,30] The number of replications necessary to calculate a significance is influenced by the number of treatments, the specific size of the true difference, the chosen level of confidence (Type I error, i.e., the probability of deeming a difference of means significant when there is no true difference), and the variability of the test system.[31]

With a small number of replications in an assay, a fairly severe inhibition of the test plants is needed to produce a statistically significant difference. Results from an experiment with chlorotoluron and blackgrass (*Alopecurus myosuroides*) based on 8 doses in 20 replications are shown in Figure 2.[32] From the data, 5, 10, or 15 replicates per treatment were selected, so four data sets with similar standard deviations and treatment means were obtained. These data were analyzed by a multiple comparison procedure (Dunnet test) to determine the NOEL. The minimum significant differences of this test for the four samples are shown in Figure 2 together with a logistic curve fitted to the same data using all 20 replicates; moreover, the ED_{10} is shown (see Section II.B.2). It is obvious, that NOEL obtained by the Dunnet test

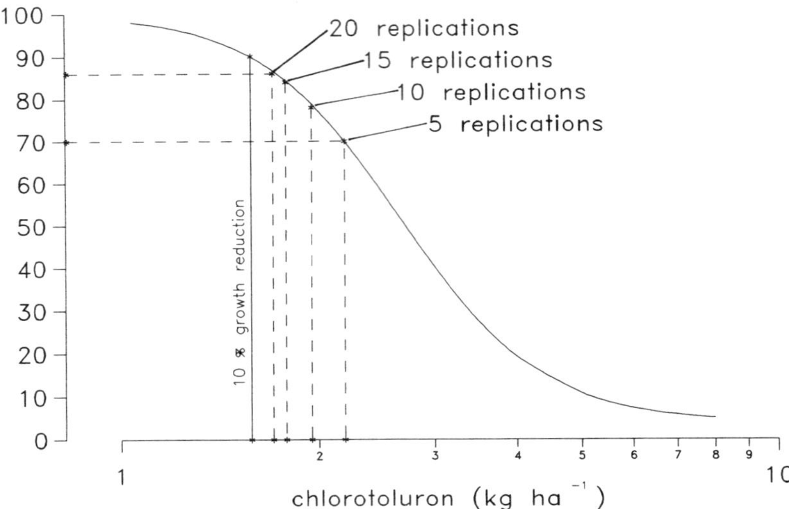

FIGURE 2. NOELs calculated by multiple comparison tests with different number of replications. (Data from Duefer, B., Ursachen der Minderwirkung von Harnstoffderivaten gegenüber *Alopecurus* auf hochgradig verseuchten Standorten in Norddeutschland, Ph.D. thesis, University of Göttingen, 1991.)

differed widely depending on the number of replications. Using 5 replicates, the minimum significant difference corresponded to ca. 30% growth reduction. Even with all 20 replications, the 10% level was not reached.

Thus a statistically significant effect may not be of any practical relevance, because even nonsignificant differences can lead to economic losses.[31,33,34] Gaylor[20] stated that a large nonsignificant difference may be biologically important. But if it is not reproducible, the experiment should be repeated using a more appropriate experimental design.

In contrast, given a sufficient number of replications even very small differences can be significant.[35,36] Elias[18] stated that a number of 50 to 60 animals is needed to establish significance for <7% tumor incidence within one treatment. The question must be asked whether a significant difference obtained in an experiment has any practical implications in relation to the objective of the assay.[18,28,34,36] A certain degree of experience is required when interpreting the significance of effects from these kind of experiments.[21]

2. Disadvantages of Multiple Comparison Procedures

Although it is widely used in toxicological and biological research, a number of authors have criticized NOEL obtained from multiple comparison procedures due to problems with the statistical assumptions and Type II error, i.e., the probability of overlooking a true difference between the treatment means.[16,31,37-39]

Dawkins[38] stated that in the multiple comparison procedure some of the assumptions do not apply to dose-response assays. The fundamental assumption of multiple comparison tests is that the same response is expected for all treatments. This assumption is not true for herbicide assays because different responses are expected with different doses.[4,31] In addition, the slope of the dose-response curve can have considerable influence on the determination of the NOEL and should be taken more into consideration than the sheer mean values.[19,40]

Often too little attention is paid to an interesting point, the so-called risk or Type II error.[19,22,31] This type of error is especially important when determining the NOEL, because doses that definitely will not cause any damage are of interest. The Type II error is difficult to estimate but increases when the level of confidence (Type I error) decreases. On the other hand, the level of confidence (usually 5%) merely supplies information about the percent of cases where significance is claimed but no real difference exists. Consequently, the actual probability should be calculated for marginally nonsignificant values.[31] If the probability is close to the chosen confidence level, a more appropriate experimental design may produce a significant difference.

B. DOSE-RESPONSE RELATIONSHIPS

Because of the inherent disadvantages of multiple comparison procedures, various authors have suggested that the NOEL be determined as the threshold dose that causes a specific inhibition.

Dose-response curves cover a range of often equally spaced doses on a logarithmic scale.[41] A wide range of doses has been suggested for herbicide bioassays,[4] the emphasis being on the range of small effects, if the NOEL has to be determined.

Details of the mathematics and statistics of dose-response relationships are given in Chapter 3. We will only outline some of the aspects with particular reference to NOEL. A mathematical description is possible either by linear regression after transforming the data, nonlinear regression, or graphical methods.

1. Linear Regression

One method for determining NOEL is direct reading from a regression line. This can lead to erratic values depending on the transformation used to linearize the curves. The curves differ only slightly from a straight line in the steep middle section, but they can differ considerably at the upper and lower limits of the curve.[42,43] Consequently, if responses are either very small or very great, deviations from the regression line may occur.[3,44] Transformations such as probit and logit, however, are widely used in herbicide research. For example, Pestemer and Auspurg[45] estimated NOEL from a probit line by calculating a least significant damage by multiple t-test and reading the dose causing this inhibition from the probit line.

Another method, the so-called "broken stick model", was used by Beckett and Davis.[46] They calculated two regression lines, one for doses causing no effects, parallel to the x-axis, and one for the higher doses giving a slope differing from zero. The intersection of the lines is then used as an "upper critical limit".

An interesting suggestion for the determination of NOEL was made by Lowy et al.[47] It is based on the assumption that, by chance, no response may occur even at high doses. The probability of no response can be calculated using the F-value of the difference to the untreated control.[48] The probability of finding an equal or higher F-value, by chance, is equivalent to the probability of finding a significant difference that in fact does not exist, or of finding no effect to occur. The values for each dose can be transformed to probit values using the normal distribution (see Chapter 3). Plotted against the log-scaled doses (Figure 3), it results in a linear relationship to which a regression line can be fitted. From the regression line a dose can be related to every probability of finding no difference, e.g., the dose where in 90% of all cases no effect will be found. In this way it is possible to determine a definite probability for NOEL.

As seen in Figure 3, extrapolation beyond the tested dose range is often necessary. This procedure is critical because extrapolations involve great uncertainties.[40,50] According to Finney,[44] an extrapolation of a linear regression line is not to be recommended. Brown[42] stated that extrapolation should only be used if the data indicate a clear linear dose-response relationship.

Another disadvantage of linear transformations is the lack of confidence limits, which otherwise would give valuable information on the precision of the estimated value. The

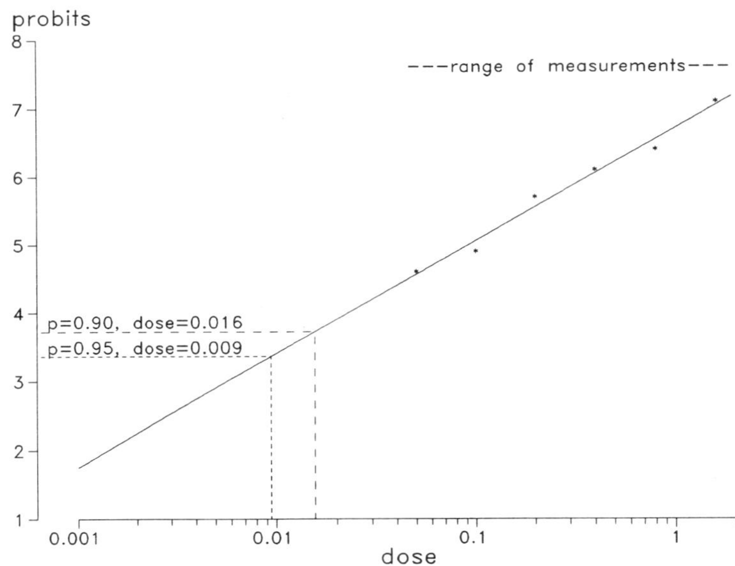

FIGURE 3. Illustration of the method of Lowy et al.[47] using constructed data. The probits correspond to the probability that no damage occurs at the corresponding doses.

transformations were initially developed for quantal data, and the calculation of confidence limits is only permitted with such data.[14] A comparison between NOEL extrapolated from dose-response relationships and NOEL obtained from multiple comparison procedures and safety factors, however, revealed that lower and therefore safer values were obtained using the extrapolation procedure.[39]

2. Nonlinear Regression

Because of the disadvantages of using linear regression methods on an essentially nonlinear curve, a nonlinear regression is more suitable.[4,5,10,51] Numerous approaches to find appropriate mathematical equations and regression procedures have been discussed elsewhere[3-5] (see also Chapter 3 for further details). The heterogeneity of variances could affect the predicted curve and thus the NOEL estimated from data. In dose-response experiments, the standard deviations of responses may differ greatly due to the wide range of responses from virtually no effect at small doses to complete kill at high doses. Therefore a transformation or weighting of responses should be carried out, e.g., logarithm or square root transformations.[4] A comparison of logistic curves fitted to original and logarithmic transformed data derived from the blackgrass experiment (Figure 2) is shown in Figure 4. The curves differ, particularly in the range of small responses resulting in an ED_{10} value (assumed NOEL) of 1.57 kg ha^{-1} for the original data and 1.15 kg ha^{-1} for the transformed data. Calculations of data from various assays have shown that ED_{10} values derived from curves fitted to original, logarithmic, or square root transformed data generally differed to the same relative extent as in Figure 4. An example is shown in Table 1.[52]

If the variances of responses at the lower doses are markedly larger than in the range of higher doses, as usually found in dose-response assays, the responses in the NOEL range will influence the shape of the curve more than the responses at higher doses. Thus the curves will fit better to the data at the upper limit when they are fitted to original data (Figure 5). This may be an advantage if the NOEL is to be estimated and will normally not affect the estimated ED_{50} value (Table 1). This procedure, however, is not correct from a statistical point of view if the whole dose-response curve is to be described.

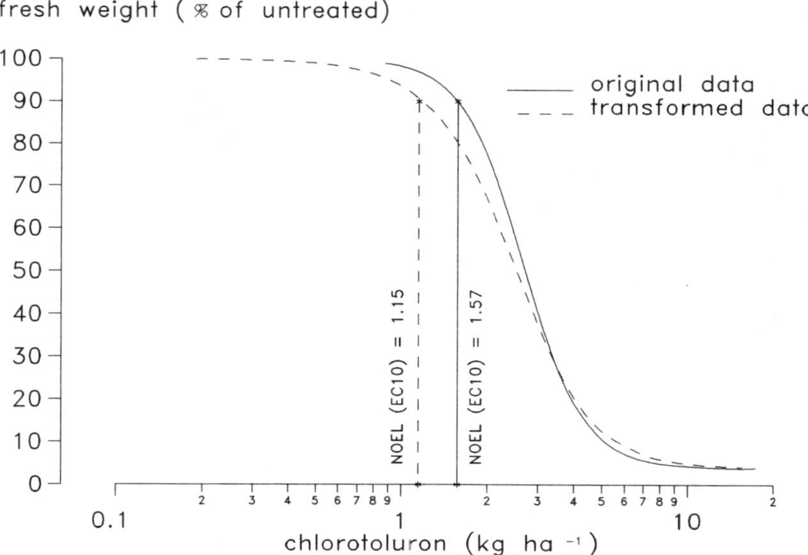

FIGURE 4. Comparison of two dose-response curves and the calculated NOELs (ED_{10}) obtained from original and logarithmic transformed data (data from Duefer[32]).

TABLE 1
Effective Doses (µg kg⁻¹) Derived from Dose-Response Curves for Metsulfuron-Methyl Based on Original and Transformed Fresh Weight Data[52]

Effective doses	Original data	Square root transformed	Logarithm transformed
ED_{10}	0.19	0.15	0.13
ED_{30}	0.44	0.40	0.40
ED_{50}	0.77	0.77	0.76

If even the lowest dose causes damage to the plants, a NOEL can be obtained from a graph connecting all data points drawn on a logarithmic scale.[14] This procedure may produce erratic NOEL values, because the controls of zero dose cannot be properly defined on a logarithmic scale.

3. Determination of a NOEL from a Dose-Response Curve

For determination of NOEL the shape and slope of the dose-response curve is important. From the estimated dose-response curves, doses causing inhibition can be calculated.[16,37,49,53] In this case, a multiple comparison test is not necessary.[35]

The NOELs obtained by regressions are not influenced to the same extent by the size of the experiment as are those from multiple comparison procedures. The precision and reproducibility of the obtained curves, however, increases with the number of replications. This is indicated by the comparison of four logistic curves fitted to the blackgrass data using 5, 10, 15, or 20 replicates. The parameters did not differ significantly (Table 2), but the confidence limits of the curves were dependent on the number of replications (Figure 6). Generally, the greater the number of replications, the smaller the confidence intervals. If the NOEL is obtained from a dose-response curve at a specific level of growth inhibition, the confidence limits should be taken into consideration.

A logistic dose-response curve with ED_{10} (0.18 µg kg^{-1}) and its corresponding confidence limits is shown in Figure 7. It is obvious that the range of confidence limits covers a part

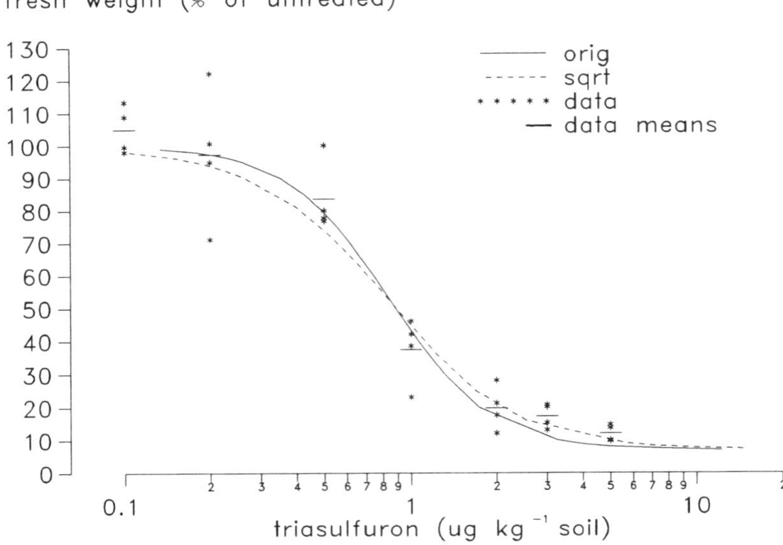

FIGURE 5. Dose-response curves obtained from original and square root transformed data shown along with the observed data.

TABLE 2
Parameters Describing Logistic Dose-Response Curves Obtained from Data with Similar Means and Variation, but Different Numbers of Replications

Number of replications	ED_{50}	Slope	Lower asymptote	Upper asymptote
5	4.01	4.10	41.4	1039.2
10	4.01	4.09	40.6	1038.3
15	4.16	4.20	41.6	1039.5
20	4.05	4.12	41.3	1039.8

of the curve that represents up to about 15% growth inhibition and a dose range from 0.12 to 0.23 μg kg^{-1}. This indicates that, in the worst case, more than 10% inhibition might occur. If the slope of the curve is very steep or if the inherent variation of data is great, the confidence intervals may cover even more than 15% growth inhibition. Potential risks can be minimized by choosing a lower ED value, e.g., ED_5,[27] or by considering the upper confidence limit (0.12 μg kg^{-1} in this example) to be the NOEL.[39]

C. SPECIAL PROBLEMS IN DETERMINING THE NOEL

One problem inherent in the mathematical analysis of dose-response relationships is the stimulation that often occurs with low herbicide doses (Figure 1). This stimulation is not described by the logistic model. Another important aspect is the choice of a suitable response variable and the duration of exposure, as both can affect the NOEL. As shown in the previous section, the number of replications has not a pronounced influence on the NOEL; problems may arise more frequently if there is a considerable variation in the data or outliers.[54]

1. Stimulation

Many examples of stimulation of plant growth at low doses of herbicides can be found in the literature.[6] The phenomenon is known to occur in dose-response relationships between

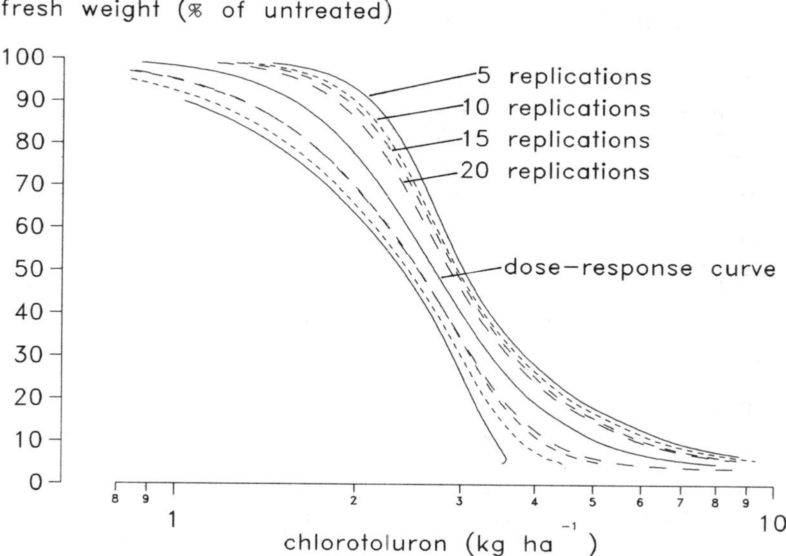

FIGURE 6. Confidence limits for logistic dose-response curves derived by using different numbers of replications (data from Duefer[32]).

organisms and chemicals of all kinds.[7,55] A general explanation for this stimulation may be found in the responses of the control mechanisms, which regulate the adaptation of biological processes at the subcellular level.[56] Following a change in environmental conditions (e.g., climate, nutrients, herbicides, and other xenobiotics) the organism responds with increased activity to compensate for possible further changes. This can result in, e.g., a higher growth rate. This over-compensation is retarded until a specific "preferred growth rate" is restored. This is an optimal adaptation to environmental conditions. Thus, the response measured after a specific duration of exposure is the result of an increase and decrease of growth rate during the experiment. Stimulation is caused by a higher frequency of increased growth rates, while a preponderance of decreased growth rates results in inhibition. The point of the dose-response curve where inhibition is first measured is considered to be the "saturation point" of the control mechanism. The NOEL is the highest dose that causes no reaction (fluctuation) in the growth rate.

Stimulation is important for determination of a NOEL from regression analysis in that it affects the shape of the curve close to the upper asymptote.[3,4] The effect of stimulation must therefore be incorporated into the model when justified by data. In certain cases, e.g., when determining NOEL in relation to crop rotation, a stimulation of growth is not critical. In this case only the reduction of growth or yield is taken into account.[10,57]

By transformations of responses, e.g., probit or logit used by various authors,[58-60] stimulation of growth renders the model incorrect. This is not critical if only a comparison in the linear range of the curve near the ED_{50} is of interest, or if only large effects are important. If the impact of a herbicide on the environment (e.g., for the assessment of side effects) is of interest, then inhibition and stimulation must be taken into consideration, because any disturbance of the environmental balance can be potentially harmful.

In the case of nonlinear regressions, the upper asymptote of a logistic curve can be determined using the data from the untreated controls and/or the stimulating doses.[5] This method may, however, give problems when used to derive a NOEL by using specific inhibition from the curve, because the upper limit is no longer the untreated control but a value based on subtoxic doses of the herbicide. Streibig[9] suggests that doses masking the

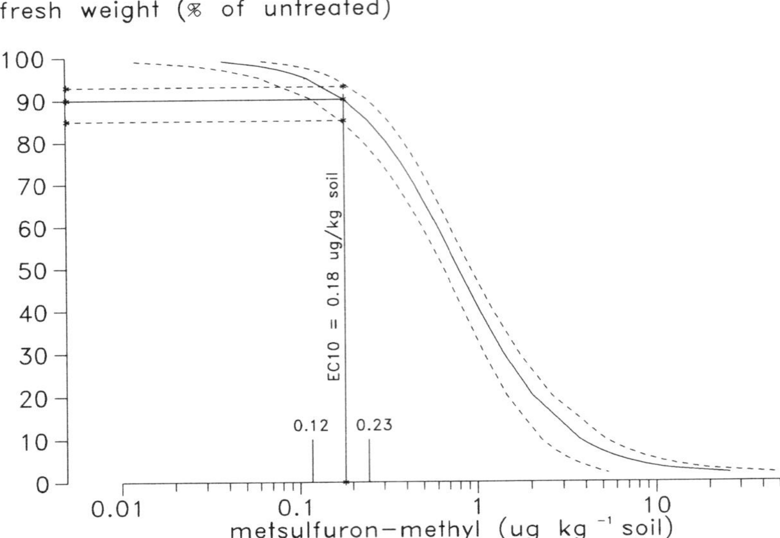

FIGURE 7. Confidence limits of a logistic curve. The NOEL, defined as ED_{10}, is shown together with the confidence limits on the ordinate and the potential range of response on the abscissa.

stimulating range should be used when fitting logistic dose-response curves. Otherwise, to obtain realistic doses for small responses, the upper level could be used as the value of the untreated control. Welp et al.[14] suggested, however, to choose the maximum response as upper limit of the curve in an assay using soil microorganisms, because the ED_{10} values will be lower than those obtained using the untreated control as reference.

The model approach described by Brain and Cousens[4] may offer an even better description of the data by including additional parameters to describe stimulation. This model seems to be suitable for determining a NOEL in the range of growth stimulation given a sufficient number of replications.

2. Influence of Experimental Design and Response Variable

Basically, the choice of response variables to determine the NOEL depends on the purpose of the investigation. For example, possible loss of crop yield is of primary consideration when investigating problems related to crop rotation. Minor damage in the early stages of growth is often compensated by the plants during the growth period. On the other hand, a merely "cosmetic" damage can result in great loss of marketable value of horticultural crops.[33]

An example of the determination of a NOEL for rotational crops with logistic dose-response curves is shown in Figure 8.[61] A dose measured in soil that reduced the fresh weight of the treated plants by 10% relative to the untreated control was defined as NOEL. In this experiment the fresh weight of the edible part of the plant was chosen as response variable. In order to keep possible risks to a minimum, responses offering high sensitivity, simple measurement, and low variability should be used.

The influence of response variable on the NOEL in a specific experiment is shown with sunflowers (*Helianthus annuus*) treated with various doses of triasulfuron in water culture.[62] After 14 days the fresh weight of shoots and true leaves were measured. The root response was expressed as dry weight. As shown in Figure 9, the shape of the logistic curves depends on the response variable. Root and whole shoot growth were not as strongly affected as the growth of the true leaves. The curve fitted to the fresh weight of the true leaves had a much lower limit than the other two curves. The NOEL, defined as a 10% growth reduction, was

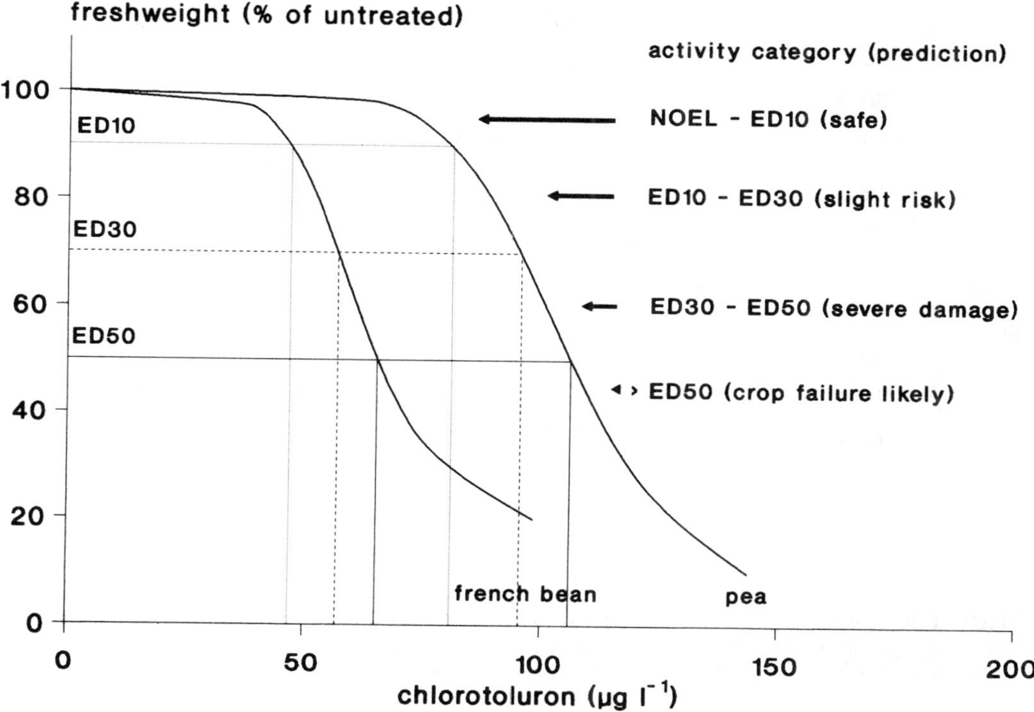

FIGURE 8. Application of logistic dose-response curves for prediction of potential crop damage. (Data from Günther, P., Persistenz: der bestimmende Faktor für das Verhalten und die Wirkung von Herbiziden im Boden, in *Proc. EWRS Symp. on Herbicidal Activity and Selectivity,* Wageningen, 1988, 281.)

affected by the choice of response variable. The NOEL of true leaves, shoots, and roots were reached at 0.077, 0.084, and 0.11 $\mu g\ l^{-1}$, respectively. Measurements of biochemical and physiological processes would perhaps have been even more sensitive and are often chosen for ecotoxicological hazard assessment, e.g., plasmalemma permeability,[63,64] ethylene release,[65] oxygen production,[66] or fluorescence of algae.[67] NOEL values measured in algae cultures, e.g., by oxygen evolution, often depend on the cell density at the beginning of the test. The assay will show higher sensitivity using lower cell densities.[68]

The calculation of a NOEL for an ecosystem is difficult, because model systems, e.g., closed systems containing crop plants, weeds, insects, and different soil organisms,[69] limit the number of doses. Even multispecies tests, which are occasionally used for such purposes,[70] are very laborious and do not sufficiently cover the situation of ecosystems. Therefore single species tests with important and/or sensitive organisms belonging to the ecosystem are commonly used, and the established values are extrapolated to the whole systems. Some authors do not permit this extrapolation from one level of biological organization to a higher one.[40,50] Sloof et al.,[71] however, found that the reaction of acute single species tests were as sensitive as model ecosystems. The authors concluded that more information is obtained by testing some important species in short-time assays than by one experiment using a model ecosystem.

3. Duration of Exposure

Exposition of an organism to a chemical over a long period may produce effects at much lower concentrations compared to short term assays.[72] Thus, in toxicology, chronic experiments over the whole lifetime of a test organism are carried out to determine the NOEL. For example,

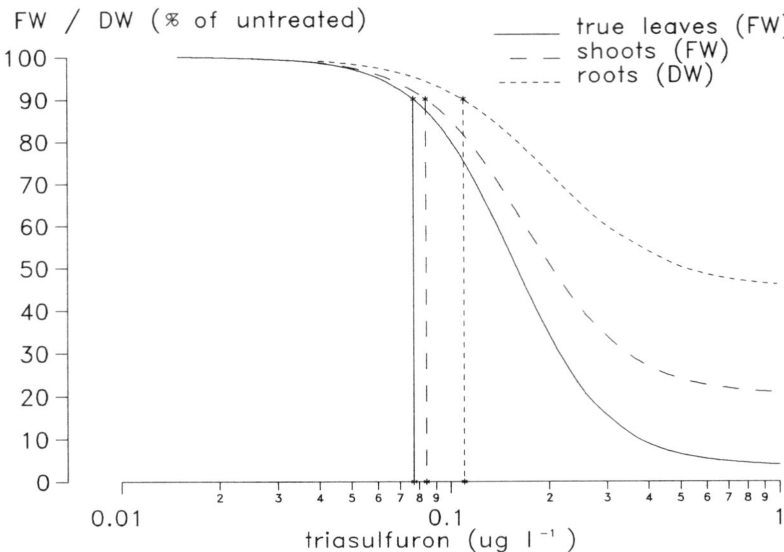

FIGURE 9. Comparison of logistic dose-response curves of different response variables and the corresponding NOELs (ED_{10}) (FW: fresh weight, DW: dry weight.) (From Günther, P., Pestemer, W., Rahman, A., and Nordmeyer, H., A technique to study the leaching behaviour of sulfonylurea herbicides in different soils, *Weed Res.*, in press.)

no incidence of tumors occurs at very low doses of carcinogenous substances, because the lifetime of exposed individuals is too short to produce a response; the higher the dose, the shorter the duration for tumor incidence will be.[73,74] This fact should also be taken into account in herbicide research.

Ecotoxicological testing methods should include experiments over a prolonged period, e.g., the vegetation period of a specific test plant, because under natural conditions a chronic exposition to very low doses over a long period is likely for persistent compounds.[45]

Bioassays for determination of herbicides in soil or water can show a higher sensitivity by prolonging the exposition period. An increase in sensitivity of duckweed (*Lemna minor*) to diquat and paraquat was found by prolonging the exposition period; the limit of detection was decreased by a factor of 10.000 when prolonging the test period from 3 to 72 h.[75] Turbak et al.[68] stated that an experiment over a longer period will affect more physiological processes than a short time assay, and thus be more sensitive.

A long term bioassay method using oats (*Avena sativa*) for testing xenobiotics has been developed by the authors.[27] The results of an assay with atrazine are shown in Figure 10.[52] Up to the second week no differences in fresh weight were observed. From the third week onwards, marked effects were found. Logistic curves, fitted to the data for each week, resulted in an increase of the NOEL (ED_5) during the first 6 weeks from 0.002 to 0.011 mg l^{-1}. On the other hand, the ED_{50} values decreased during this time from 0.27 to 0.03 mg l^{-1}. Thereafter changes were much smaller. This may indicate, that plants were slightly damaged during the first stages of growth but recovered, even though they were still exposed to low doses of atrazine. Consequently, NOELs obtained in short time assays may be too low if extrapolated to the whole vegetation period and therefore bear an inherent safety margin. Similar results were obtained using various environmental chemicals.[27] Further experimentation will prove whether a NOEL can be defined as a dose that damages the plants so much that recovery is precluded.

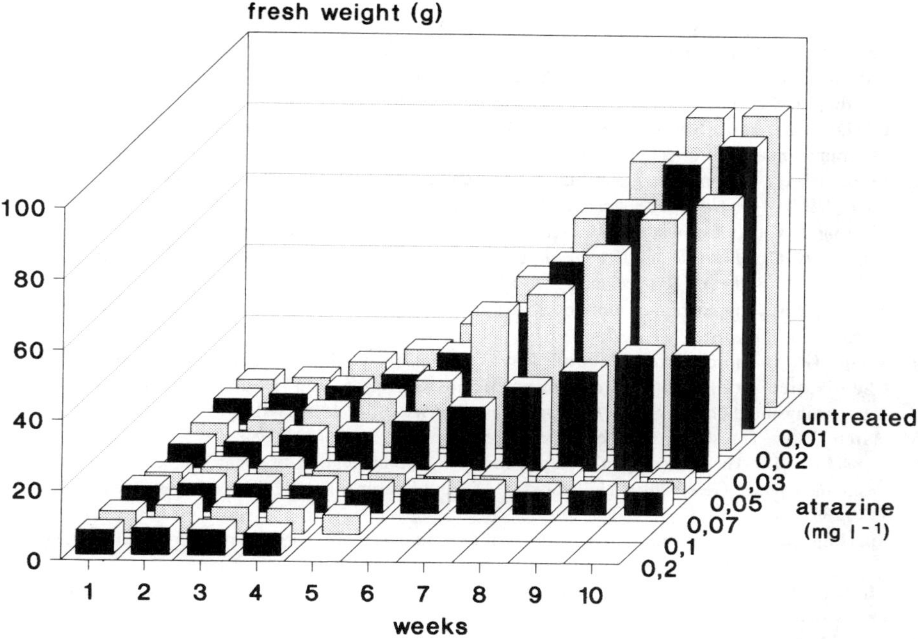

FIGURE 10. Effect of atrazine on the fresh weight production of oats in a long-term assay in hydroponic culture. (From Günther, P., Biotests mit höheren Pflanzen zur Untersuchung und Bewertung des Verhaltens von Sulfonylharnstoff-Herbiziden und anderen Xenobiotika im Böden, Dissertation, Universität Hannover, 1991.)

III. CONCLUSION

The problems with NOEL are becoming important in herbicide research because of a rising demand for assessment of side effects and potential ecotoxicological risk. This is particularly true for new herbicides with high biological activity and selectivity such as the sulfonylureas. A NOEL value, however, is always closely related to the experimental conditions and must be interpreted as such. Uncritical use of statistics can lead to unrealistic NOEL values. Therefore, statistical as well as biological relevance must be considered when assessing the results.

REFERENCES

1. **Blackman, G. E.,** Studies in the principles of phytotoxicity. I. The assessment of relative toxicity, *J. Exp. Bot.,* 3, 1, 1952.
2. **Sampford, M. R.,** Studies in the principles of phytotoxicity. II: Experimental designs and techniques of statistical analysis in the assessment of phytotoxicity. *J. Exp. Bot.,* 3, 28, 1952.
3. **Streibig, J. C.,** Models for curve-fitting herbicide dose response data, *Acta Agric. Scand.,* 30, 59, 1980.
4. **Brain, P. and Cousens, R.,** An equation to describe dose responses where there is stimulation of growth at low doses, *Weed Res.,* 29, 93, 1989.
5. **Günther, P., Rahman, A., and Pestemer, W.,** Quantitative bioassays for determining residues and availability to plants of sulfonylurea herbicides, *Weed Res.,* 29, 141, 1989.
6. **Wiedman, S. J. and Appleby, A. P.,** Plant growth stimulation by sublethal concentrations of herbicides. *Weed Res.,* 12, 65, 1972.
7. **Stebbing, A. R. D.,** Hormesis — the stimulation of growth by low levels of inhibitors, *Sci. Total Environ.,* 22, 213, 1982.

8. **Breeze, V. G.,** Methods to investigate sub-lethal effects of herbicides on plant species, in Field Methods for the Study of Environmental Effects of Pesticides, Monogr. No. 40, Greaves, M. P., Greig-Smith, P. W., and Smith, B. D., Eds., British Crop Protection Council, 1988, 255.
9. **Streibig, J. C.,** Herbicide Bioassay, *Weed Res.,* 28, 479, 1988.
10. **Obrigawitch, T. T., Strek, H. J., and Lichtner, F. T.,** Evaluation of laboratory and field techniques for determination of a no-effect level of activity for sulfonylurea herbicides, *Aspects Appl. Biol.,* 21, 131, 1989.
11. **Pestemer, W.,** Methodenvergleich zur Bestimmung der Pflan-zenverfügbarkeit von Bodenherbiziden, *Ber. Fach. Herbologie Univ. Hohenheim,* 24, 85, 1983.
12. **Streibig, J. C.,** Assessment of herbicidal effect in relation to mode of action, in *Weed Control on Vine and Soft Fruits,* Cavalloro, R. and Robinson, D. W., Eds. (Proceedings, EC Expert's Group Meeting, Dublin, 1985), A. A. Balkema, Rotterdam, 1987, 133.
13. **Streibig, J. C.,** Joint action of root-absorbed mixtures of herbicides in *Sinapis alba* L. and barley (*Hordeum vulgare* L.), *Weed Res.,* 27, 337, 1987.
14. **Welp, G., Brümmer, G. W., and Rave, G.,** Dosis-Wirkungs-Beziehungen zur Erfassung von Chemikalienwirkungen auf die mikrobielle Aktivität von Böden: I. Kurvenverläufe und Auswertungsmöglichkeiten, *Z. Pflanzenernähr. Bodenk.,* 154, 159, 1991.
15. **Hatch, T. F.,** Thresholds, do they exist?, *Arch. Environ. Health,* 22, 687, 1971.
16. **Henschler, D.,** The toxicologist's responsibility in the evaluation of plant protectants, in *Advances in Pesticide Science,* Geissbühler, H., Brooks, G. T., and Kearney, P. C., Eds., Pergamon Press, Oxford, U.K., 1979, 39.
17. **Sperling, K.-R.,** Lethal and sublethal effects of cadmium on marine organisms — a critical discussion about "safety levels", *Ecotoxicol. Environ. Saf.,* 7, 521, 1983.
18. **Elias, P.,** Toxikologische Beurteilung von Fremdstoffen in Nahrungsmitteln, *Mitt. Biol. Bundesanst. Land Forstwirtsch. Berl-Dahlem,* 232, 59, 1986.
19. **Chan, P. K., O'Hara, G. P., and Hayes, A. W.,** Principles and methods for acute and subchronic toxicity, in *Principles and Methods of Toxicology,* Hayes, A. W., Ed., Raven Press, New York, 1982, 1.
20. **Gaylor, D. W.,** Statistical interpretation of toxicity data, in *Toxic Substances and Human Risk — Principles of Data Interpretation,* Tardiff, R. G. and Rodricks, J. V., Eds., Plenum Press, New York, 1987, 77.
21. **Zbinden, G.,** The no-effect level, an old bone of contention in toxicology, *Arch. Toxicol.,* 43, 3, 1979.
22. **Brown, K. G. and Erdreich, L. S.,** Statistical uncertainty in the no-observed-adverse-effect level, *Fundam. Appl. Toxicol.,* 13, 235, 1989.
23. **Yermakoff, J. K.,** Toxicologic units, in *Toxic Substances and Human Risk — Principles of Data Interpretation,* Tardiff, R. G. and Rodricks, J. V., Eds., Plenum Press, New York, 1987, 13.
24. **Walker, A. and Brown, P. A.,** Measurement and prediction of chlorsulfuron persistence in soil, *Bull. Environ. Contam. Toxicol.,* 30, 365, 1983.
25. **Bringmann, G. and Kühn, R.,** Grenzwerte der Schadwirkung wassergefährdender Stoffe gegen Bakterien (*Pseudomonas putida*) und Grünalgen (*Scenedesmus quadricauda*) im Zellvermehrungshemmtest, *Z. Wasser-Abwasser-Forsch.,* 10, 87, 1977.
26. **Welp, G. and Brümmer, G.,** Ermittlung des Non-Effect-Levels für Umweltchemikalien mit Hilfe eines Mikroorganismentests, in *Spezielle Berichte der Kernforschungsanlage Jülich,* Vol. 224, Methoden zur ökotoxikologischen Bewertung von Chemikalien 2, 1983.
27. **Günther, P. and Pestemer, W.,** Risk assessment for selected xenobiotics by bioassay methods with higher plants, *Environ. Manage.,* 14, 381, 1990.
28. **Schaeffer, D. J., Cox, D. K., and Deem, R. A.,** Variability of test systems used to assess ecological effects of chemicals. *Water Sci. Technol.,* 19, 39, 1987.
29. **Geng, G. and Hills, F. J.,** A procedure for determining numbers of experimental and sampling units, *Agron. J.,* 70, 441, 1978.
30. **Montgomery, D.C.,** *Design and Analysis of Experiments,* 2nd ed., John Wiley & Sons, New York, 1984.
31. **Cousens, R. and Marshall, C.,** Dangers in testing statistical hypotheses, *Ann. Appl. Biol.,* 111, 469, 1987.
32. **Duefer, B.,** Ursachen der Minderwirkung von Harnstoffderivaten gegenüber *Alopecurus* auf hochgradig verseuchten Standorten in Norddeutschland, Ph.D. thesis, University Göttingen, 1991.
33. **Breeze, V. G. and Timms, L. D.,** Some effects of low doses of the phenoxyalkonic herbicide mecoprop on the growth of oilseed rape (*Brassica napus* L.) and its relation to spray drift damage, *Weed Res.,* 26, 433, 1986.
34. **Cousens, R.,** Misinterpretation of results in weed research through inappropriate use of statistics. *Weed Res.,* 28, 281, 1988.
35. **Morse, P. M. and Thompson, B. K.,** Presentation of experimental results, *Can. J. Plant Sci.,* 61, 799, 1981.
36. **Perry, J. N.,** Multi-comparison procedures: a dissenting view, *J. Econ. Entomol.,* 79, 1149, 1986.
37. **Weil, C. S.,** Statistics vs safety factors and scientific judgement in the evaluation of safety for man, *Toxicol. Appl. Pharmacol.,* 21, 454, 1972.

38. **Dawkins, H. C.,** Multiple comparisons misused: why so frequently in response-curve studies?, *Biometrics,* 39, 789 1983.
39. **Gaylor, D. W.,** Quantitative risk analysis for quantal reproductive and developmental effects, *Environ. Health Perspect.,* 79, 243, 1989.
40. **Deutsche Forschungs Gemeinschaft (DFG),** *Beiträge zur Beurteilung der Umweltwirksamkeit chemischer Stoffe,* Harald Boldt Verlag, Boppard, 1979.
41. **Brown, C. C.,** The statistical analysis of dose-effect relationships, *SCOPE 12: Principles of Ecotoxicology,* John Wiley & Sons Ltd., Chichester, 1978, 115.
42. **Brown, C. C.,** Statistical aspects of extrapolation of dichotomous dose-response data, *J. Nat. Cancer Inst.,* 60, 101, 1978.
43. **Sielken, R. L., Jr.,** Cancer risk assessment, 3: cancer dose-response extrapolations, *Environ. Sci. Technol.,* 21, 1033, 1987.
44. **Finney, D. J.,** *Statistical Methods in Biological Assay,* 3rd ed., Griffin, London, 1978.
45. **Pestemer, W. and Auspurg, B.,** Eignung eines Testpflanzensortiments zur Risikoabschätzung von Stoffwirkungen auf höhere Pflanzen im Rahmen des Chemikaliengesetzes, *Nachrichtenbl. Dtsch. Pflanzenschutzdienstes,* 38, 120, 1986.
46. **Beckett, P. H. T. and Davis, R. D.,** Upper critical levels of toxic elements in plants, *New Phytol.,* 79, 95, 1977.
47. **Lowy, R., Albrecht, R., Pélissier, M. A. B., and Manchon, P.,** Determination of the "no-effect levels" of two pesticides, lindane and zineb, on the microsomal enzyme activities of rat liver, *Toxicol. Appl. Pharmacol.,* 42, 329, 1977.
48. **Snedecor, G. W. and Cochran, W. G.,** *Statistical Methods,* Iowa State University Press, Ames, Iowa, 1967.
49. **Hansen, K.,** Die toxikologische Beurteilung von Pestizidrückständen, *Ber. Landwirtsch.,* 50, 383, 1972.
50. **Rudolph, P. and Boje, R.,** *Ökotoxikologie — Grundlagen für die ökotoxikologische Bewertung von Umweltchemikalien nach dem Chemikaliengesetz,* Ecomed, Landsberg, München, 1986, chap. 3.
51. **Maximov, V. N., Moshinskaya, L. R. and Nosov, V. N.,** Statistical estimation of the threshold of toxic effects, *Acta Hydrobiol.,* 16, 139, 1988.
52. **Günther, P.,** Biotests mit höheren Pflanzen zur Untersuchung und Bewertung des Verhaltens von Sulfonylharnstoff-Herbiziden und anderen Xenobiotika im Böden, Dissertation, Universität Hannover, 1991.
53. **Duffus, J. H.,** Interpretation of Ecotoxicity, in *Toxic Hazard Assessment of Chemicals,* Richardson, M., Ed., Royal Chemical Society, London, 1986, chap. 10.
54. **Johnson, G. B. and Berger, R. D.,** On the status of statistics in phytopathology, *Phytopythology,* 72, 1014, 1982.
55. **Luckey, T. D.,** Insecticide hormoligosis, *J. Entomol.,* 61, 7, 1968.
56. **Stebbing, A. R. D.,** Growth hormesis: a by-product of control, *Health Phys.,* 52, 543, 1987.
57. **Pestemer, W., Stalder, L., and Eckert, B.,** Availability to plants of herbicide residues in soil. II. Data for use in vegetable crop rotations, *Weed Res.,* 20, 349, 1980.
58. **Hance, R. J., Smith, P. D., and Cotterill, E. G.,** The effect of age on the availability of linuron and simazine residues in soil, *Weed Res.,* 17, 429, 1977.
59. **Pestemer, W., Stalder, L., and Potter, C. A.,** Nachbauprognosen bei Atrazinrückständen im Boden mit Hilfe von Verfügbarkeits- und Langzeit-Biotest-Daten, *Ber. Fachg. Herbologie Univ. Hohenheim,* 24, 53, 1983.
60. **Nyffeler, A., Gerber, H.-R., Hurle, K., Pestemer, W., and Schmidt, R. R.,** Collaborative studies of dose-response curves obtained with different bioassay methods for soil-applied herbicides, *Weed Res.,* 22, 213, 1982.
61. **Pestemer, W., Bunte, D., and Günther, P.,** Persistenz: der bestimmende Faktor für das Verhalten und die Wirkung von Herbiziden im Boden, in *Proc. EWRS Symp. on Herbicidal Activity and Selectivity,* Wageningen, 1988, 281.
62. **Günther, P., Pestemer, W., Rahman, A., and Nordmeyer, H.,** A technique to study the leaching behaviour of sulfonylurea herbicides in different soils, *Weed Res.,* in press.
63. **Schweiger, G., Sellner, M., Golle, B., and Lüttge, U.,** Effects of ecotoxicological chemicals on passive plasmalemma permeability in plants, *Ecotoxicol. Environ. Saf.,* 7, 366, 1983.
64. **Lüttge, U., Sellner, M., Schnabl, H., and Zimmermann, U.,** Ökotoxikologische Bewertung von Umweltchemikalien und Entwicklung von biologischen Testsystemen aufgrund der Eigenschaften pflanzlicher Membranen, *GIT Suppl.,* 4, 36, 1984.
65. **Rodecap, K. D. and Tingey, D. T.,** Stress ethylene: a bioassay for rhizosphere-applied phytotoxicants, *Environ. Monit. Assess.,* 1, 119, 1981.
66. **Stalder, L. and Pestemer, W.,** Availability to plants of herbicide residues in soil. I. A rapid method for estimation of potentially available residues of herbicides, *Weed Res.,* 20, 341, 1980.
67. **Schmidt, C.,** Actual standard and further development of an algal fluorescence bioassay, *Ecotoxicol. Environ. Saf.,* 7, 276, 1983.

68. **Turbak, S.C., Olson, S.B., and McFeters, G.A.,** Comparison of algal assay systems for detecting waterborn herbicides and metals, *Water Res.*, 20, 91, 1986.
69. **Schuphan, I., Schärer, E., Heise, M., and Ebing, W.,** Use of laboratory model ecosystems to evaluate quantitatively the behaviour of chemicals, in *Pesticide Science and Biotechnology,* Greenhalgh, R. and Roberts, T. R., Eds., Blackwell Scientific Publications, Oxford, U.K., 1986.
70. **Huber, W.,** Vergleich zwischen "Single-Species" und "Multi-Species"-Testverfahren, *Mitt. Biol. Bundesanst Land Forstwirtsch. Berlin-Dahlem,* 234, 34, 1987.
71. **Sloof, W., van Oers, J. A. M., and de Zwart, D.,** Margins of uncertainty in ecotoxicological hazard assessment, *Environ. Toxicol. Chem.,* 5, 841, 1986.
72. **Mancini, J. L. and Plummer, A. H., Jr.,** Organism response to time variable pollutant dosages, *Proc. APCA Annu. Meeting,* 80(2), 87/33.3, 1987.
73. **Kroes, R., Van Esch, G. J., and Weiss, J. W.,** Philosophy of no effect level for chemical carcinogens, in *Proc. Int. Symp. Nitrite Meat Prod.,* 1974, 227.
74. **Hartung, R.,** Dose-response relationships, in *Toxic Substances and Human Risk - Principles of Data Interpretation,* Tardiff, R. G. and Rodricks, J. V., Eds., Plenum Press, New York, 1987, 29.
75. **O'Brien, M. C. and Prendeville, G. N.,** A rapid sensitive bioassay for the determination of paraquat and diquat in water, *Weed Res.,* 18, 301, 1978.

Chapter 9

HERBICIDE RESIDUES IN SOILS AND PLANTS AND THEIR BIOASSAY

D. V. Clay

TABLE OF CONTENTS

0-8493-6603-8/93/$0.00+$.50

I. INTRODUCTION

The earlier chapters of this book deal with the theory, development, and use of methods of assay in which the herbicide is directly available to the assay agent. Herbicide applications in the field, however, result in a very different situation. In this chapter the complex nature of herbicide distribution and availability in the environment is discussed. When an assay is used to assess the inherent biological activity of a new compound, residue considerations may be of no relevance; however, as soon as the persistence of a herbicide effect in the treated plant, or its fate in plants or soil, is being studied an understanding of the residual behavior of that compound and the factors controlling it becomes important. It is even more important that those using bioassay to investigate environmental fate and significance of herbicides are aware of the nature of residue distribution, the factors controlling availability in the system, and the possible influence of breakdown products on the result. This understanding will help to ensure that representative substrates are used and unrealistic claims as to the meaning of the results are avoided.

Some residual activity is important for most herbicides to be effective, whether they are foliar acting and susceptible to postspraying rainfall or are soil acting with the functional need to remain active for weeks or months. Problems arise when this activity persists to affect following crops or to have adverse environmental consequences. There is currently considerable effort being devoted to understanding and quantifying the problem, so that necessary remedial or preventative action can be taken. Bioassay, in its widest sense, has a part to play in this task through the assessment of the agricultural and ecological significance of residues.

In the early years of the use of residual herbicides understanding of their long term distribution and persistence in soil came from field sampling and residue measurement. Since the early 1970s modeling of persistence based on herbicide and soil properties and weather data has developed; useful prediction of the soil behavior of compounds can now often be made to help in the assessment of the environmental acceptability of potential herbicides or, for existing products, in predicting residue levels at the end of the growing season. This development has relied on field bioassay data for its validation.

A number of valuable reviews on herbicide residues have been used, particularly Hance,[1,2] Grover,[3] and Calderbank.[4]

II. DISTRIBUTION OF RESIDUES

As soon as a herbicide is applied a process of redistribution and degradation begins which is only completed when all has been transformed to nonherbicidal products. This may be relatively short in the case of simple compounds or may take months or years for more persistent products.

The possible routes of herbicide redistribution are well understood and illustrated elsewhere.[5] It is important to remember that the process is dynamic — as redistribution occurs the different consequences of volatilization, adsorption, binding, desorption, degradation, or further redistribution are possible.

A. DISTRIBUTION AT APPLICATION TIME

The amounts of herbicide reaching the target depend on wind speed at spraying the accuracy of the application equipment and its operation and the nature of the target. Willis and McDowell[6] have summarized available information on amounts reaching sprayed plants showing the large variability in deposits. Data on sprays applied to the soil also show large point to point variation in deposits.[7,8] This suggests either uneven application or that some droplet drift may be occurring as a result of turbulent fluctuations in nominally similar meteorological conditions. Walker and Brown[9] found a large difference in precision according to the equipment used; coefficients of variation for point deposits were 16% for a knapsack sprayer and 41% for a tractor-mounted boom. Measurements of drift of herbicides in typical spraying conditions have indicated losses of the order of 0.2 to 5% of the total amount applied, half of which was deposited within 10 m downwind and 10^{-5} of the applied dose carried beyond 300 m.[10-12]

B. REDISTRIBUTION AFTER APPLICATION

The processes of redistribution and degradation will start as soon as a herbicide is applied. As will be seen the relative importance of the different processes is dependent on the properties of the herbicide as well as on formulation, weather, vegetation, and soil conditions.

1. Plant Uptake

Although the purpose of herbicide use is activity on target plants the amount of herbicide actually used for this appears to be small and it may not be a major route for residue dissipation. For foliar acting herbicides <1% of that applied may reach the site of action.[13] Similarly for soil applied herbicides only a small proportion may be taken up; Hoffman and Lavy[14] found less than 5% of soil applied atrazine was absorbed by plants. With both foliar applied herbicides and strongly adsorbed soil-acting herbicides there has to be an excess herbicide present to ensure enough is absorbed to give toxicity. Once absorbed the herbicide may still not be broken down within the plant, remaining in senescent vegetation and subsequently incorporated in the soil. There is evidence for the long term persistence in vegetation of a number of herbicides including clopyralid,[15] dalapon,[16] glyphosate,[17] and picloram.[18] There is also evidence that some volatile herbicides in plants may subsequently be evaporated from leaf surfaces, e.g., dichlobenil.[19] Other herbicides may be exuded from roots into soil, e.g., dicamba,[20] glyphosate,[21] and imazapyr.[22]

2. Volatilization

For many herbicides volatilization is a significant means of loss. The comprehensive review by Taylor and Glotfelty[23] gives figures varying from 74% EPTC lost after 74 h, 90% of nonincorporated trifluralin in 120 days, 21% 2,4-D ester in 5 days, to 2.4% atrazine in 24 days. Chemical properties are important in determining volatility, although vapor pressures are not of prime importance once the herbicide is adsorbed. Volatilization occurs by both convective and diffusive flow and is affected by the pattern of water loss from the soil surface more than by wind speed or temperature,[23] losses being greater from wet soils. Herbicide formulation also affects degree of evaporative loss, lower rates being reported for wettable powder and microencapsulated formulations than for emulsifiable concentrates.[24] Soil incorporation reduces volatility losses but also increases variability of deposits.[25] Evaporation occurs from vegetation and from soil; in the case of a sprayed cereal crop that from vegetation was not dependent on soil moisture conditions.[12]

Damage to adjoining crops from herbicide vapor has been a problem particularly with some ester formulations of the phenoxy alkanoic herbicides applied to cereals,[10] but some dinitroaniline, diphenyl ether, and benzonitrile herbicides can also cause damage. Once volatilized the herbicides are clearly subject to degradative processes in the atmosphere but they may also be redeposited by rain.[23]

3. Wind Erosion

This means of loss may be important where herbicide is applied to dry cultivated soils or on highly organic or sandy soils subject to "wind blows". There are few quantitative studies but both atrazine and chlorthal-dimethyl have been found in sampled air in conditions where volatilization could be discounted.[23] Such movement has also been responsible for contamination of lake water by atrazine.[26] Granular herbicides on dry soil can also be displaced by wind.

4. Herbicide Leaching

The principles involved in herbicide leaching and data on its occurrence have been extensively reviewed.[27,28] Precipitation is the main factor involved in downward movement of herbicide after application to land whether or not it has a vegetation cover. Herbicide not absorbed or bound by vegetation will be washed from plant surfaces on to soil. Estimates of the amount of herbicide that reaches the soil range from 50% on land with a complete plant cover to 100% where cover is sparse.[12,29] Dead vegetation probably retains only a small proportion of herbicide applied to it.[30] Normally precipitation causes herbicide movement into deeper soil layers unless the infiltration capacity is exceeded when local flooding or run-off will result (see Section II.B.5).

Herbicide moves into soil in four ways:[2,31]

1. As undissolved herbicide particles
2. In solution in soil water
3. Adsorbed on soil particles or colloids
4. For volatile herbicides in the vapor phase.

Of these, transport in solution is generally the most important and occurs primarily by mass (convective) flow of water, though diffusion in soil water is clearly significant for movement within micropores. The relative importance of each mechanism will depend on herbicide properties, precipitation rates and soil factors. Similarly the depth and speed of penetration depends on these same factors as they will determine the amounts adsorbed and desorbed on each layer of soil and the amounts entering micropores and being by-passed by subsequent mass flow of water. As a result of net downward flow of water some herbicide may end up in subsoil, ground water and ultimately in river or borehole water.

Evidence for the downward movement of herbicide adsorbed on soil particles comes from studies of movement of radiolabeled material bound to particles[32] and from the recovery of paraquat residues in uncultivated soils at depths below 10 cm.[33,34] Paraquat is completely adsorbed after application to the soil surface and the only likely means of downward movement is on soil particles carried down permanent macropores by rain.

Leaching is affected by herbicide properties, soil conditions, and weather. With the increased knowledge of herbicide behavior in soils it has become apparent that environmental fate in general and leachability in particular can be broadly predicted from a knowledge of basic herbicide physicochemical properties such as melting point, vapor pressure, water solubility and partition coefficient (octanol/water).[35] Supplementing this with laboratory studies of soil adsorption, degradation rate and mobility at different soil pHs confers greater precision on predictions of leachability, particularly for ionizable chemicals such as weak acids or bases.[29] Thus as a generalization it is the weakly adsorbed herbicides (those with a low K_{ow}) that are most leached but this may be of no agronomic or environmental significance if they are rapidly degraded in top and subsoil. Herbicide dose does not appear to have a significant effect on the proportion leached down the soil profile.[36-38]

Rainfall pattern after application can profoundly affect leaching. Significant downward movement appears to occur when heavy precipitation follows within a week or so of application. In these circumstances more herbicide may be found in the subsoil or reaching ground water.[27,39,40] This has been observed with herbicides such as atrazine and simazine but may be particularly significant with weakly adsorbed herbicides.

Soil moisture content at the time of application may also be important with this type of herbicide; imazapyr has been found to be more mobile when applied to wet compared to dry soils,[41,42] although mobility was also affected by soil pH, clay type, and oxide content.

The potential for herbicide residues to reach deeper soil or ground water is affected by rainfall pattern and seasonal factors. While deep leaching can occur quickly with intense precipitation after application, normally downward movement is reduced by soil adsorption and degradation processes. To take account of this, Rao et al.[28] have suggested that, in predictions of leaching, an attenuation factor should be included as well as a mobility factor. The possibility of upward movement of leached herbicide from subsoil must also be considered; there is some evidence that weakly adsorbed herbicides such as chlorsulfuron and triasulfuron may move from deeper to shallower soil layers in spring/summer when evapotranspiration rates are high.[43] In general in temperate zones, herbicide applications in spring are less subject to leaching into subsoil than autumn applications because the net balance of water flow is upward rather than downward.[44]

Soil properties that most affect the downward movement of herbicides are organic matter content, particle size distribution (texture) and composition, pH, bulk density, and pore distribution. For strongly adsorbed herbicides, soil organic matter content is generally the most important factor, the higher the content the less the leaching. Leaching will be greater in coarse texture sand soils than in silt or clays. For some herbicides, e.g., imidazolinones, the type of clay and oxide content can affect adsorption and therefore leaching.[41,42] Leaching of weakly adsorbed acidic herbicides, e.g., sulfonylureas and imidazolinones may be much greater in high pH soils.[41,45]

With herbicides of all groups different soil physical conditions can lead to variable leaching in particular the presence or absence of fissures and permanent vertical macropores in the soil. These channels account for the increased leaching of some herbicides such as atrazine, metribuzin and simazine found under non- or minimally tilled soils compared with cultivated soils.[39,40,46,47] This difference was not found with a nonadsorbed material, bromide, which moved through tilled and nontilled soils to the same extent.[39] The presence of deep cracks in some clay soils after periods of drought could greatly increase the chances of deep leaching of subsequent herbicide applications.[27] Vertical channels in uncultivated soil are also thought to account for the presence of paraquat residues at depths of 5 to 20 cm from spray applied to the soil surface,[33,34] the herbicide being carried to these depths adsorbed on soil particles. Presence of a cover of dead vegetation or mulch material can affect downward herbicide movement directly apart from indirect effects through altered degradation rates. Jones et al.[47] found greater leaching of metribuzin under straw covered plots.

However, precise prediction on the basis of soil properties and conditions is as difficult with leaching as it is with availability and persistence since, depending on the herbicide, the effect of one factor may be overridden by another and all can be over-ridden by weather factors.[27,48]

5. Herbicide Run-Off

This subject has recently been comprehensively reviewed by Leonard.[49] It can be a major means of loss of herbicides from the target areas if these are sloping sites and rain intensity exceeds the infiltration rate of the soil. Herbicide may move in solution or adsorbed on soil particles. The time between herbicide application and heavy rainfall is critical. Rainfall within a few days of application could lead to run-off problems even with the relatively short persistence herbicides such as 2,4-D or mecoprop. Adsorbed herbicide may be at risk from movement on eroding soil for a longer period after application.

A cover of organic material such as straw generally reduces run-off problems since lateral soil and water movement is reduced and the soil surface below such cover may be less compact allowing faster water infiltration.[49,50] However, heavy rain falling immediately after application can lead to increased lateral movement of herbicide since most herbicide will still be on the organic material from which it is relatively easily removed and little may have been adsorbed on soil. Run-off problems are not associated with any particular class of compounds; instances have occurred with alachlor, atrazine, simazine, 2,4-D, and picloram among others.[49]

6. Herbicide Residues in Water

This topic is reviewed briefly because of its importance in considering the persistence and fate of herbicide residues; detailed reviews are available.[2,51-55] Evidence of herbicide presence in drainage water, water courses, and wells has been available for some years, with an estimated 0 to 2% of that applied reaching such water.[27,56,57] The route of herbicide contamination will be via both run-off and percolation water, although the time scale of percolation to underground aquifers is little known. Higher residue levels in drainage and river water can be found some weeks after a main herbicide application period and may be linked to periods of heavy rainfall.[57,58] While the presence in water of residues of herbicides such as atrazine can sometimes be clearly correlated with agricultural use, e.g., applications on maize, its presence in other areas may be linked with industrial uses.[2,57] The problems associated with the collection, analysis and interpretation of data on residues in water have been reviewed by Hance;[59] where the residues occur at levels not far above the limits of detection the accuracy of determinations needs confirmation by alternative analytical methods. A further area of concern is the relative degradation rates of pesticides in water in relation to aeration, light, sediments, etc.[60,61]

C. VARIATION IN REDISTRIBUTION

The degree of variability that may occur in deposits at application has been noted above. Some processes affecting residues, such as volatility or photolysis may not increase variation of the remaining residues but weather and localized soil factors can have a major effect sometimes reducing but generally increasing variability. More even distribution occurs in surface soil where a herbicide is applied to wet soil or light precipitation follows application and herbicide is moved away from applied particles by diffusion, mass flow, or soil movement. This can account for the uniform effectiveness of soil acting herbicides applied in different spray volume rates[62] or as granules.

Heavy precipitation after application is probably the most important source of variation particularly when it causes surface water run-off, leaving uneven deposits behind. Similarly uneven infiltration of water into soil leads to lateral variation in deposits. However, the inherent variability of soil properties such as organic matter content and bulk density between fields and sometimes between areas a few meters apart will lead to differences in downward movement and degradation rates and this may also account for much of the lateral variability encountered.[38] In practice studies of point to point pesticide residue distribution in soil weeks or months after application have shown coefficients of variation from 7 to 50% or more;[9,25,38] variability does not necessarily increase with lapse of time or sample depth[37,38] but can be affected by application method.[9]

III. AVAILABILITY AND SIGNIFICANCE OF RESIDUES

A. FACTORS AFFECTING PERSISTENCE

The dependence of disappearance rates on soil and weather conditions has made the task of quantifying the influence of individual factors difficult and that of prediction of persistence of field applications even more so. However much progress has been made with a combination of modeling techniques interacting with laboratory and field studies and the behavior of the main herbicide groups is comparatively well understood.[5,25,52]

1. Herbicide Properties

This is the main factor influencing persistence within vegetation; most herbicides are metabolized relatively rapidly in plants so that residues do not have long term effects. This is obviously essential in those applied to edible crops where residue levels in produce must be minimal. There are, however, situations where longer persistence occurs[63] and effects may be observed months later particularly on perennial plants, e.g., foliar acting herbicides such as paraquat applied to the senescing foliage of bulb plants affecting the next years shoots and glyphosate or phenoxyalkanoic herbicides applied to fruit tree foliage in summer giving damage symptoms on the following years growth.[64,65] Herbicide metabolites can also sometimes accumulate in plants and cause symptoms, e.g., the dichlobenil soil breakdown product 2,6-dichlorobenzamide causes distinct leaf margin chlorosis in some woody plants.[36]

2. Soil Properties

In soil the main factor determining persistence is the degree of adsorption of the herbicide by organic matter or soil colloids and this is a function of the physical and chemical properties of the herbicide and the origin of the soil components.[4,66] The stronger the adsorption the less the availability of the herbicide for degradative processes (see below). In the field it is impossible to be precise about the relative importance of the different soil factors since many are fluctuating according to season and also interact with one another. The main factors involved are organic matter, pH, particle size, moisture, aeration, and the presence of amendments such as fertilizers and organic material. Because much breakdown is effected by microflora it is mainly the influence of soil factors on the dynamics of these populations that affects dissipation rates and this can well be specific for a particular herbicide/soil combination.[25]

As a generalization, with strongly adsorbed herbicides, persistence of phytotoxic residues is less in soils with higher organic matter content.[25] With weakly adsorbed herbicides there appears to be no consistent effect of organic matter levels on persistence and residual phytotoxicity.[41,67] Long term residual phytotoxicity has been found to be greater on more organic soils, but this may be linked with anaerobic soil conditions.[15]

Leaching is less in higher organic matter content soils leaving more herbicide in the surface soil much of which is reversibly or irreversibly adsorbed. In this zone microbiological activity is greater due to higher aeration, temperature, and nutrient levels, and this generally leads to higher dissipation rates. Soil moisture levels will be higher with increased organic matter thus favoring dissipation. By contrast all these factors, apart from moisture may be lower in subsoil leading to reduced breakdown rates.[68,69] In many soils the mineral component may not be involved in adsorption since it is surrounded by organic material.[4] Subsequently desorbed herbicide may move to clay or oxide components where it is preferentially bound.[4,41] Soil pH can influence persistence although contradictory results on its effects are frequently reported.[25] There is evidence with sulfonylurea and imidazolinone herbicides that degradation rates can be lower in soils of higher pH.[42,45] There may be an optimum pH for the particular organisms involved in degradation of any one herbicide.[70]

Soil particle size or texture sometimes has a direct effect on persistence. Imidazolinone herbicides can be more persistent on clay soils with greater adsorption on certain clay minerals and oxides.[42,71] However it is often difficult to separate effects of particle size from associated differences in soil organic matter and moisture content and aeration. Greater persistence of herbicides sometimes reported in light sandy soils compared to silt and clay soils[25] was probably due to these associated factors. Soil cultural conditions may affect persistence, residues being less with lower aggregate density,[72] cultivation,[73] and addition of organic matter.[25,74] Higher inorganic nitrogen levels in soil may result in enhanced herbicide persistence.[25] Presence of a crop can cause, but does not always lead to, reduced persistence of herbicide compared to that in uncropped land;[25,74] loss from herbicide uptake by crop roots may be counterbalanced by the reduced degradation rate in the drier soil below the crop.

3. Climatic Factors

Weather can have an overriding effect on persistence through its effect on soil moisture and temperature. Incubation studies of herbicide incorporated in soils of different moisture contents and at different temperatures invariably show greater persistence in dry soil and at low temperatures.[5] In the field most evidence on the effect of moisture on breakdown rates comes from comparison of regional or year to year differences. Breakdown is less following applications in drier areas than wetter[75] and at the end of dry seasons compared with wet.[76] The effect of temperature has also been shown in field studies; persistence of atrazine and chlorfenac was found to be greater in cooler areas,[7] and propyzamide breakdown was retarded in cooler winter months.[77] However, the effect of temperature can be overridden by moisture deficits; in temperate regions residue levels are consistently found to be higher after hot dry summers.

4. Adaptation

With most herbicides studied, repeated annual applications to the same plots has not lead either to a build-up of residue or to reduced levels of residues with time.[25,74,78] However, there is now much evidence to show that with repeated applications of certain herbicides to the same soil there is a build-up of microorganisms adapted to breakdown the herbicide resulting in more rapid degradation than in freshly treated soil. Although first observed with the phenoxyalkanoic herbicides it has also been found with amide and thiocarbamate herbicides,[2,78,79] although not all members of a group may be affected. Accelerated breakdown of a soil-acting herbicide may significantly reduce its effectiveness and require use of alternative products. On the positive side there may be the possibility of applying adapted microorganisms to soil to speed the degradation of unwanted residues left from overdoses or accidental contamination.[2]

B. PREDICTION OF PERSISTENCE

The ability to predict the degree of persistence is important for herbicide producers, users and those concerned with environmental safety. Improved prediction helps manufacturers develop only compounds with desirable agronomic and environmental properties and advisers to decide safe cropping sequences. Although knowledge of the factors affecting persistence has developed over the years from laboratory and field experiments their relative importance in the field has often only been appreciated following crop damage incidents where all the factors leading to damage can be analyzed. The whole subject is slowly being put on a more quantitative basis with the development of mathematical models for predicting herbicide dissipation, leaching, and persistence and consequences for following crops.[25,80-83] Such models can combine all available information on herbicides, plant responses, soil properties and weather patterns but have been dependent for validation and often amendment on data from field experiments and usage. They do now offer the possibility for screening out new herbicides with unfavorable properties and thus greatly reduce the possibility of adverse effects from new products.[84] While models can only approximate the effects of important variables, refinement will continue as new information on the complex interactions between herbicide, soil, weather patterns, and plants becomes available.

C. AVAILABILITY OF RESIDUES

Many of the factors affecting availability of herbicide residues to plants, soil fauna and microorganisms have been outlined above as those affecting persistence (Section III.A). Thus availability of uniformly distributed residues will be largely governed by adsorption/desorption processes which in turn are influenced by herbicide and soil properties particularly organic matter content. Mechanisms of soil adsorption of herbicides have been discussed by Hance.[66] Good correlation between adsorption, soil organic matter content, and plant response in pot experiments has been found, providing the herbicide is uniformly incorporated in the soil.[85] Exceptionally, where the organic fraction surrounds the mineral component, as in organic sandy soils, a relatively large amount of herbicide is inactivated in spite of total organic matter content being low.[86] In the field correlation between organic matter content and plant response is not always close because of the effects of weather on distribution through the soil and the influence of local variations in properties such as pH.[66] The time taken for applied herbicide to get into solution, move to adsorption sites and, when solution concentrations decrease, be desorbed is also an important factor.[87] Where plant growth on different field sites is being used to measure availability it is important to remember that for herbicides such as photosynthesis inhibitors response is greatly effected by growing conditions particularly evapotranspiration rates. Very close correlation between adsorption and availability to plants is of course shown by very strongly adsorbed herbicides such as paraquat and diquat where none is available to plants on mineral soils until all adsorption sites are saturated.[88,89] Soil moisture at spraying can affect availability not only by affecting herbicide movement but by a direct effect on adsorption; Subagyo[42] found that imazapyr was less available for plant uptake when applied to dry compared to wet soil on which it was excluded from adsorption sites by water.

The age of the residue in the soil may affect its availability for breakdown. There is good evidence that with more persistent herbicides there can be an apparent reduction in the rate of breakdown in the second or third year following either a high dose application or the last of repeated annual treatments, e.g., with bromacil,[90] chlorthiamid and dichlobenil,[36] diuron, and simazine.[37,91] This could be evidence of reduced availability of aged residues for dissipation processes or for the release of bound residues not previously extracted by the solvents used or available to bioassay plants (see below). However, Stalder and Pestemer[92] found no evidence of an effect of age of residues on availability in tests of residual phytotoxicity with methazole, oxyfluorfen, and simazine; in their work, however, a water extract was bioassayed not the soil itself. Hance et al.[38] also did not find an effect of age of residue on availability of linuron and simazine.

D. BOUND RESIDUES

A proportion of some pesticides reaching the soil may be strongly bound to the organic or clay mineral fraction and be unavailable to plants or animals or to the organic solvents normally used for extraction. The phenomenon was first reported by Bailey and White;[93] subsequent information has been reviewed by Kaufman et al.[94] and Calderbank.[4] Quantitative study of the amount of a parent herbicide and its different metabolites bound in this way is difficult, because the solvents needed to remove them from the soil often break down the compounds. This is also true with studies with radiolabeled herbicides where total activity may be quantified but identification of which compounds are present may not be possible.

Both parent pesticide or degradation products may be bound, either by adsorption or chemical reactions. There is good evidence that chemical reactions with organic matter lead to binding of chloroaniline residues derived from anilide, phenylcarbamate and urea herbicides and binding of chlorophenol from phenoxyalkanoic herbicides.[4] The process of binding may take time and it is possible for compounds to move from weaker to stronger adsorption sites, e.g., with paraquat.[88] Calderbank[4] concluded that although most of the initial binding of basic herbicides is by adsorption to organic matter some of the stable bound herbicide will end up in the clay fraction.

Bound residues may persist indefinitely in field soil though there is evidence of subsequent degradation by microorganisms. Paraquat is regarded as strongly bound to mineral soils but slow degradation is reported. Hance[95] found that 15% of radiolabeled paraquat was lost from field soil in 6 months; results from a long term field study suggested a half-life for paraquat of about 6 years.[95]

How much herbicide is present in soil in this form? Calderbank[4] has summarized much previous work which shows that with a range of herbicides bound residues may constitute 10 to 90% of the amount applied but only some of this was identified as the parent herbicide.

With any one chemical group results can vary widely. Capriel et al.[96] applied radiolabeled atrazine to soil and found 50% of the activity, including parent compound, still present after 9 years. Bound residue formation has also been indicated for prometryne in an organic soil[97] and simazine.[98] However, Hance[95] applied labeled simazine to field soil containing aged simazine residues and to previously untreated soil and found all the radioactive residue was extractable with methanol after 4 months. Differences of this sort may be explained on the basis of variation in the amount and composition of the organic matter in the different soils.

E. SIGNIFICANCE OF RESIDUES

1. Agricultural Significance

Users of soil acting herbicides require soil persistence of the products for long term weed control but no carry over of significant amounts to damage subsequent crops. In practice residues from newly introduced herbicides have caused problems because of the impossibility of appreciating and allowing for all the seasonal and soil differences that can affect dissipation rates in drawing up recommendations for recropping of treated land. Numerous examples of such problems can be quoted. Atrazine residues from applications in maize in the U.S.[99] gave damage on following crops where dry seasons led to higher end of season residues. Some chlorsulfuron applications to cereals in the U.K. caused damage to following sugar beet crops;[100] the slower than expected dissipation rate of chlorsulfuron in poorly drained, high pH soils and subsoil appeared to cause this problem. Imazethapyr residues from applications in peas have damaged subsequent crops.[101] Where persistence is suspected the problem can be overcome by replanting only tolerant crops; residues remaining near the soil surface can also be reduced by cultural measures such as ploughing to reduce concentrations.[102,103] Field experience as well as experimentation can lead to the compilation of a considerable amount of information on residue levels and recropping response and this can give useful warning of risk situations (see Table 9.2, Riley and Eagle[103]).

There is no evidence that repeated use of herbicides at recommended rates has any direct adverse effects on the crops treated. Long term studies of monocultures receiving the same herbicides have shown no build-up of residues or effects on crop production and soil fertility.[78,104,105] Herbicide treatment of rotational crops has also had no effect on productivity.[106] Indirect effects from herbicide use such as increased compaction in minimum tillage systems[107] or acidification of soil surface and reduction in organic matter levels in plantation crops[104,108] are well documented; however, these are effects of the cultural system and the reduction in vegetation cover rather than an effect of herbicides per se.

Problems to following crops from residues from repeated annual applications in perennial crops have occurred; however, the importance of an adequate time interval between the final application of persistent herbicides such as bromacil and dichlobenil and recropping is well understood and allowed for in product recommendations. Build up of damaging residue levels from repeated application of persistent herbicides such as simazine does not appear to occur in temperate regions.[36,74,91] Where orchards with rows treated repeatedly with recommended doses of residual herbicide have been grubbed and replanted with annual crops damage to these crops has generally been caused by deteriorated soil structure rather than herbicide residues.[108]

Crops may fail shortly after the application of residual herbicides, due to drought, pest and disease attack, or winter cold. In this situation immediate recropping is difficult because of the high level of residual phytotoxicity. Where this problem occurs frequently in major crops, guidelines have been worked out for safe replanting with the same or alternative crops.[102,109] Van Himme et al.[110] have carried out extensive studies to categorize the relative tolerance of a wide range of sown crops to residues of 20 residual herbicides commonly used in arable and horticultural crops in Northern Europe. Such information together with local knowledge of soil properties, weather after spraying, and actual herbicide residue levels can all aid decisions on safe recropping.

Problems from the persistence of phytotoxic residues of herbicides in vegetation have been rare considering the scale of herbicide use and are mainly found with growth-regulating herbicides. Clopyralid has been found to persist in sugar beet foliage fed to cattle and, through manure, affect subsequent potato crops.[15] Contamination of potato tubers with clopyralid and glyphosate from accidentally treated haulm also caused severe long term foliage symptoms.[15,111] Damage has been caused to glasshouse crops grown on bales made from cereal straw previously treated with TBA.[112]

2. Environmental Significance

It is clear from the information reviewed above that a proportion of applied herbicide will not reach or will be lost from its target and that some may persist in treated soil after the harvesting of the crop it has protected. In general phytotoxic amounts reported are relatively small and there are processes at work which lead to further degradation or inactivation. There is little evidence of adverse effects of herbicides at residual levels on soil microorganisms and soil fauna.[1] There will, of course, be the same indirect effects on these organisms from reduction of vegetation cover as indicated in the agricultural situation above.[104]

The environmental importance of bound residues discussed above is difficult to assess. A major problem, with many pesticides, is quantifying the amount, because of the impossibility, with present analytical techniques, of separating parent pesticide and degradation products. There is no evidence from studies to date that damaging quantities are released into the environment by weather or cultural changes. While binding of residues may become stronger with age there would also appear to be a slow release of some residues by mineralization processes; these are then dissipated through the normal processes.[4] This scenario lends support to the view expressed by Kearney[113] that the binding of residues and their subsequent slow release and breakdown forms the safest method of disposal of pesticides in the environment. The amounts of pesticide bound in this way are small compared with the total quantity of complex organic molecules derived from plant and animal life and similarly incorporated into the organo-clay fraction of the soil all the time.[4]

The amount of information on ground water and aquifer contamination by small concentrations of herbicides is increasing and should make it possible to identify any herbicide/use situations which lead to unacceptable levels in ground water.[2] There is of course a need for more information on residue levels from herbicide uses in all parts of the world, particularly where there is intensive use, to ensure that off-target contamination is minimized and does not adversely affect human health or terrestrial and aquatic ecosystems.

Methods of herbicide residue measurement are now very sensitive and should ensure that, where resources are available, environmental contamination can be adequately monitored. There are problems, however, where residue levels are near the limit of detection whether by chemical or biological methods in that precision becomes much less and spurious results can be obtained. Thus, in connection with measurements of pesticides in ground water, Hance[59] has suggested that confirmation of a result is advisable from several methods; this practice would seem wise in other residue situations where sequential sampling is not possible.

IV. RESIDUE ASSESSMENT METHODS

Laboratory and glasshouse bioassay methods were employed extensively in the early years of herbicide use in agriculture both to identify causes of crop damage and to quantify residue amounts. As the availability of instrumental methods has developed, with their advantages of sensitivity, speed and efficiency, the use of bioassays for measuring residues has been restricted. However, there will be a continuing need for their use for residue identification, evaluation of effects of new herbicides and field residue assessment. There are a number of reviews available on principles and methods.[114-118]

A. HERBICIDE IDENTIFICATION

Bioassays have proved useful for suggesting or confirming which herbicides may have caused crop damage from product contamination, spray drift, or soil or straw contamination. Symptoms on test plants can be matched with those treated with the suspect herbicide.[119] However, effects are not normally specific to a particular herbicide or herbicide group so only the general mode of action may be indicated. Illustrations of herbicide effects can be useful in aiding identification,[120,121] but it must always be remembered that symptoms may differ according to growing conditions and growth stage.[109] In addition other causes such as mineral imbalance, virus infection or toxic products from other plant material may cause similar symptoms.[109] Where contamination on organic material such as straw is suspected dilution of the material with untreated soil or sand is necessary to obtain reasonable growth.[122]

B. HERBICIDE MEASUREMENT

1. Soil Sampling and Storage

The problems of variability in residue distribution have been discussed above and this must be allowed for in the method of sampling, sample distribution and number. The object of the assessment will determine the correct method — whether the range of concentrations in high and low crop damage areas is being sought or an average value for a field. A sample number of 12 per plot has been suggested for replicated trials[123] and larger numbers for assessing whole field residue levels.[9] With stratified samples extreme care must be taken to avoid contamination from downward movement of herbicide on soil particles particularly where there is a high herbicide concentration near the surface. Removing large diameter intact cores, inverting and resampling taking a narrow diameter core can overcome this problem as can digging trenches across plots and sampling horizontally at required depths.[37] With most assay methods soil is air dried and passed through a 3 mm diameter sieve before use to aid preparation of the assay. If there is delay in starting the assay, soil must be deep frozen to avoid herbicide dissipation and changes in pH during storage which may affect response.[124] Untreated soil for control treatments in the assay and for preparing standards must be prepared identically. There can be a problem in obtaining appropriate control soil particularly when assaying field soil for suspected damage where the whole field has been sprayed. The nearest source of equivalent untreated soil must be used but the influence of a different organic matter content, pH, and nutrient level on response must be considered. In field experiments studying persistence in uncropped land regular treatment of control plots with a contact herbicide can prevent weed growth affecting soil properties.[86]

2. Bioassay Methods

Methods available for the measurement of residues have been reviewed, e.g., Horowitz[114] for some foliar-applied herbicides and Horowitz,[115] Hurle,[124] Clay,[116] and Lavy and Santelmann;[117] the latter describe practical details on methods. They include techniques suitable for assaying dry or moist soil samples. Laboratory methods involve measurements of effects on root or shoot growth, leaf discs, or O_2 evolution by unicellular algae. Whole plant tests are carried out in controlled environment or glasshouse and often involve dilution of treated soil with untreated to establish 50% effect levels for comparison with a reference series.

The laboratory and glasshouse methods enable measurement of biologically important residues for most herbicides although sensitivity and accuracy may vary considerably. Examples of some problems are

1. The reduced sensitivity of assays of photosynthesis inhibitor herbicides in low transpiration conditions
2. Availability, distribution, and dissipation of herbicide in the soil during the assay period may be different for freshly applied and aged residues

3. Poor standardization and control of the growing conditions in whole plant assays on the assumption that they require little expertise

The need for adequate standards is emphasised by Hurle[124] and borne out by the large variation in results given by collaborative multisite tests.[125,126] Inadequate mixing of herbicide in soil samples and shortcomings in watering plants are some possible sources of error. Factors affecting variability were reviewed by Lavy and Santelmann.[117]

Field assays have been a valuable means of obtaining information on the toxicity of residues to following crops, especially when linked with measurement of residue amounts at the time of planting. They must, however, be carried out at an appropriate time of year since seasonal effects on toxicity can be large;[117] preplanting cultivations must also be appropriate since different operations such as ploughing can affect residue distribution and phytotoxicity.[73,127] Planting a range of crops provides useful information on relative tolerance.[36,110]

3. Interpretation of Measurements

Once information on residue amounts has been obtained by bioassay or instrumental methods there remains the task of assessing its significance for the current or a following crop. This has been done in numerous cropping regions by developing a data bank of residue measurements and effects on crops from experiments and cases of damage, which, as it expands, enables satisfactory recommendations to be made for recropping.[102,103] An alternative approach was proposed by Stalder and Pestemer[92] that involves measuring the concentration of water extractable herbicide in the soil. The effect of that concentration of herbicide on prospective crop species is obtained by comparison with results of dose/response tests in water culture so the likelihood of damage to each species as a following crop can be predicted. Pestemer et al.[128] found this gave satisfactory prediction for the response of field grown crops to residues of photosynthesis inhibitor herbicides in 26 out of 30 cases. Subsequent work with atrazine, methabenzthiazuron, and pendimethalin showed some overprediction of damage but no cases of damage without its prediction.[129] With chlorsulfuron, Duffy[130] reported good correlation between toxicity of soil water extracts in a lentil root assay and residual phytotoxicity. However, Hance[95] used the method with a wider range of herbicide groups and found poor prediction of residual phytotoxicity; generally, crop damage was overpredicted, but with the relatively insoluble and volatile herbicide trifluralin, the test often underpredicted damage. There appeared to be a big variability in the amounts of herbicide extracted by water at different dates, often not well correlated with solvent extractions done on identical samples. Investigation of the factors that may cause this variability might help refine the technique.

C. COMPARISON OF BIOASSAY AND INSTRUMENTAL METHODS

The historical pattern of soil herbicide residue measurement has already been described with bioassay, often being used initially but declining as the more efficient or sensitive instrumental methods are developed. Direct comparisons of the two methods have been made and many of these were reviewed by Hurle.[124] Thus for 2,4-D, chlorotoluron, and ametryn, Eberle and Gerber[131] found good agreement between chemical methods and bioassay using *Chlorella pyrenoidosa* in water extracts of soil and whole plants grown in soil. Stalder and Pestemer[92] also measured availability of herbicides in soil and obtained good agreement between results from a whole plant bioassay and a chromatographic method (Figure 1). However, there are many instances of anomalous results, generally with bioassays recording lower residue levels; this may sometimes be due to poor technique but may also result from the variation in herbicide availability to plants in soil discussed above. Modern instrumental techniques generally have much lower limits of detection than quantitative assays but may not measure total residual phytotoxicity. It also must be emphasized that while bioassay

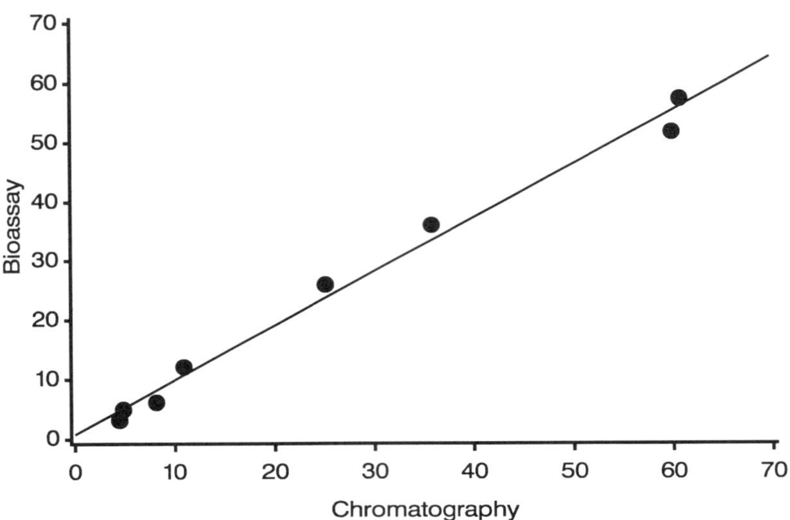

FIGURE 1. Relationship between percentage availability of eight photosynthesis-inhibitor herbicides in soil determined by bioassay and gas chromatography. (Redrawn from Stalder, L. and Pestemer, W., *Weed Res.*, 20, 341, 1980.)

methods, based on measuring a 50% growth response in comparison with a standard, should remain equally precise at all residue levels; with chemical methods precision is regarded as much poorer near the limit of detection.[59] With both instrumental and bioassay methods the assumption that the availability of the residue under study will be the same as that of the freshly applied herbicide used to determine recovery values in an instrumental method or an ED_{50} value in a bioassay may not always be justified.[131]

V. CONCLUSIONS

While herbicide residue measurement and identification are now largely done using instrumental methods, bioassay techniques remain of use in situations where instrumental methods are unavailable or for new herbicides for which adequate chemical methods have not been developed. Thus, initially, analysis of residues of some of the highly active sulfonylurea herbicides was more satisfactory using a root growth bioassay, compared with chemical methods.[118] Bioassays will continue to be of importance for the investigation of crop damage and measuring total residual phytotoxicity in soil that includes the herbicide and its breakdown products or where mixtures of residual herbicides are present. There will continue to be a need for using field bioassays to collect information on residual toxicity. This information, combined with determinations of actual residues present, may be used for building up regional data banks on safe residue levels for recropping as well as for validation of predictive models of herbicide persistence and phytotoxicity.

An aspect that is beginning to receive attention, because of increased concern about side effects of herbicides in the countryside and the local loss of plant species, is the effect of herbicide drift onto noncrop plants in land adjacent to sprayed fields.[132] There is little information on herbicide effects on these species; this is a topic to which properly planned and executed studies using whole plant bioassay can make an important contribution.

REFERENCES

1. **Hance, R. J.,** *Interactions between Herbicides and the Soil,* Academic Press, London, 1980.
2. **Hance, R. J.,** Herbicide behavior in the soil, with particular reference to the potential for ground water contamination, in *Herbicides,* Hutson, D. H. and Roberts, T. R., Eds., John Wiley & Sons Ltd., 1987, 223.
3. **Grover, R., Ed.,** *Environmental Chemistry of Herbicides, 1,* CRC Press, Boca Raton, FL, 1988.
4. **Calderbank, A.,** The occurrence and significance of bound residues in soil, *Rev. Environ. Contam. Toxicol.,* 108, 71, 1989.
5. **Nash, R. G.,** Dissipation from soil, in *Environmental Chemistry of Herbicides, 1,* Grover, R., Ed., CRC Press, Boca Raton, FL, 1988, 131.
6. **Willis, G. H. and McDowell, L. L.,** Pesticide persistence on foliage, *Rev. Environ. Contam. and Toxicol.,* 100, 23, 1987.
7. **Harris, C. I., Wookon, E. A., and Hummer, B. E.,** Dissipation of herbicides at three soil depths, *Weed Sci.,* 17, 27, 1969.
8. **Fryer, J. D. and Kirkland, K.,** Field experiments to investigate long-term effects of repeated applications of MCPA, tri-allate, simazine and linuron, report after six years, *Weed Res.,* 10, 133, 1970.
9. **Walker, A. and Brown, P.,** Spatial variability in herbicide degradation rates and residues in soil, *Crop Protection,* 2, 17, 1983.
10. **Elliott, J. G. and Wilson, B. J., Eds.,** The influence of weather on the efficiency and safety of pesticide application, *BCPC Occasional Pub. No. 3,* 1983, 135.
11. **Maybank, J., Yoshida, K., and Grover, R.,** Spray drift from agricultural pesticide applications, *J. Air Pollut. Control Assoc.,* 28, 1009, 1978.
12. **Grover, R., Shewchuk, S. R., Cessna, A. J., Smith, A. E., and Hunter, J. H.,** Fate of 2,4-D iso-octyl ester after application to a wheat field, *J. Environ. Qual.,* 14, 203, 1985.
13. **Caseley J. C. and Walker, A.,** Herbicide entry and transport, in *Weed Control Handbook: Principles,* 8th ed., Hance R. J. and Holly, K., Eds., Blackwell Scientific Publications, Oxford, 1990, 183.
14. **Hoffman, D. W. and Lavy, T. L.,** Plant competition for atrazine, *Weed Sci.,* 26, 94, 1978.
15. **Eagle, D. J.,** Agrochemical damage to UK crops, *Pestic. Outlook,* 1(2), 14, 1990.
16. **Foy, C. I.,** The chlorinated aliphatic acids, in *Herbicides: Chemistry, Degradation and Mode of Action,* 2nd ed., Kearney, P. C. and Kaufman, D. D. Eds, Marcel Dekker, New York, 1975, 399.
17. **Gottrup, O., O'Sullivan, P. A., Schraa, R. J., and Vanden Born, W. H.,** Uptake, translocation, metabolism and selectivity of glyphosate, *Weed Res.,* 16, 197, 1976.
18. **Sassman, J., Pienta, R., Jacobs, M., and Cioffi, J.,** *Pesticide background statements, 1, herbicides,* USDA Handbook No. 633, 1984, P-23.
19. **Verloop, A.,** Fate of dichlobenil in plants and soils in relation to it's biological activity, *Residue Rev.,* 43, 55, 1972.
20. **Frear, D. S.,** The benzoic acid herbicides, in *Herbicides, Chemistry, Degradation and Mode of Action,* Kearney, P. C. and Kaufman, D. D., Eds., Marcel Dekker, New York, 1976, 514.
21. **Coupland, D. and Lutman, P. J. W.,** Investigations into the movement of glyphosate from treated to adjacent untreated plants, *Ann. Appl. Biol.,* 101, 315, 1982.
22. **Townson, J. K.,** Influence of formulation and application variables in relation to the performance of glyphosate and imazapyr for control of *Imperata cylindrica* (L.) Raeuschel, Ph.D. Thesis, University of Reading, U.K., 1990.
23. **Taylor, A. W. and Glotfelty, D. E.,** Evaporation from soils and crops, in *Environmental Chemistry of Herbicides, 1,* Grover, R., Ed., CRC, Boca Raton, FL, 1988, 90.
24. **Turner, B. C., Glotfelty, D. E., Taylor, A. W., and Watson, D. R.,** Volatilization of microencapsulated and conventionally applied chlorpropham in the field, *Agron. J.,* 70, 933, 1978.
25. **Hurle, K. and Walker, A.,** Persistence and its prediction, in *Interactions Between Herbicides and the Soil,* Hance, R. J., Ed., Academic Press, London, 1980, 349.
26. **Wu, T. L.,** Atrazine residues in estuarine water and the aerial deposition of atrazine into Rhode River, Maryland, *Water Air Soil Pollut.,* 15, 123, 1981.
27. **Leistra, M.,** Transport in solution, in *Interactions between Herbicides and the Soil,* Hance, R. J., Ed., Academic Press, London, 1980, 31.
28. **Rao, P. S. C., Jessup, R. E., and Davidson, J. M.,** Mass flow and dispersion, in *Environmental Chemistry of Herbicides,* Grover, R., Ed., CRC, Boca Raton, FL, 1988, 21.
29. **Riley, D.,** Using soil residue data to assess the environmental safety of pesticides, BCPC Monogr. No. 4 Pesticides in Soils and Water, 1991, 11.
30. **Banks, P. A. and Robinson, E. L.,** Soil reception and activity of acetochlor, alachlor and metolachlor as affected by wheat (*Triticum aestivum*) straw and irrigation, *Weed Sci.,* 34, 607, 1986.

31. **Hutson, J. L. and Wagenet, R. J.,** Predicting the fate of herbicides in the soil environment, in *Proc. Brighton Crop Protection Conf. — Weeds,* 1989, 1111.
32. **Mercer, E. R. and Hill, D.,** Movement of simazine adsorbed on to soil particles, *Report ARC Letcombe Laboratory,* 63, 1974.
33. **Fryer, J. D., Hance, R. J., and Ludwig, J. W.,** Long-term persistence of paraquat in a sandy loam soil, *Weed Res.,* 15, 189, 1975.
34. **Khan, S. U., Marriage, P. B., and Saidak, W. J.,** Residues of paraquat in an orchard soil, *Can. J. Soil Sci.,* 55, 73, 1975.
35. **Briggs, G. G.,** Theoretical and experimental relationships between soil adsorption, octanol-water partition coefficients, water solubilities, bioconcentration factors and the parachlor, *J. Agric. Food Chem.,* 29, 1050, 1981.
36. **Clay, D. V. and McKone, C. E.,** Effects of repeated applications at high rates of chlorthiamid and dichlobenil to blackcurrants, *Pestic. Sci.,* 10, 429, 1979.
37. **Clay, D. V. and Stott, K. G.,** The persistence and penetration of large doses of simazine in uncropped soil, *Weed Res.,* 13, 42, 1973.
38. **Hance, R. J., Smith, P. D., Byast, T. H., and Cotterill, E. G.,** Variability in the persistence and movement in the field of abnormally high rates of simazine and linuron: some wider implications, in *Proc. Br. Crop Protection Conf.— Weeds,* 1976, 643.
39. **Steenhuis, T. S., Staubitz, W., Andreini, M. S., Surface, J., Richard, T. L., Paulsen, R., Pickering, N. B., Hagerman, J. R., and Geokring, L. D.,** Preferential movement of pesticides and tracers in agricultural soils, *J. Irrigation Drainage Eng.,* 116, 50, 1987.
40. **Hall, J. K., Murray, M. R., and Hartwig, N. L.,** Herbicide leaching and distribution in tilled and untilled soil, *J. Env. Qual.,* 18, 439, 1989.
41. **Wehtje, G., Dickens, R., Wilcut, J. W., and Hajek, B. F.,** Sorption and mobility of sulfometuron and imazapyr in five Alabama soils, *Weed Sci.,* 35, 858, 1987.
42. **Subagyo, T.,** Studies on imazapyr, a herbicide for minimum tillage purposes in areas infested with Lalang (*Imperata cylindrica* (L.) Raeuschel), Ph.D. thesis, University of Reading, U.K., 1989.
43. **Mahnuken, G. E. and Weber, J. B.,** Capillary movement of triasulfuron and chlorsulfuron in Rion sandy loam soil, in *Proc. 41st Anniversary Meeting Soil Weed Sci. Soc.,* 1988, 332.
44. **Briggs, G. G.,** Factors affecting the uptake of soil-applied chemicals by plants and other organisms, BCPC Monogr. No. 27, Soils and Crop Protection Chemicals, 1984, 35.
45. **Brown, H. M. and Kearney, P. C.,** Plant biochemistry, environmental properties, and global impact of the sulfonylurea herbicides, in *Synthesis and Chemistry of Agrochemicals II,* ACS Symp. Ser. 443, 1991, 32.
46. **Mansell, R. S., Wheeler, W. B., and Calvert, D. V.,** Leaching losses of two nutrients and a herbicide from two sandy soils during transient drainage, *Soil Sci.,* 130, 140, 1980.
47. **Jones, R. E., Jr., Banks, P. A., and Radcliffe, D. E.,** Alachlor and metribuzin movement and dissipation in a soil profile as influenced by soil surface conditions, *Weed Sci.,* 38, 589, 1990.
48. **Hance, R. J., Hocombe, S. D., and Holroyd, J.,** The phytotoxicity of some herbicides in field and pot experiments in relation to soil properties, *Weed Res.,* 8, 136, 1968.
49. **Leonard, R. A.,** Herbicides in surface waters, in *Environmental Chemistry of Herbicides 1,* Grover, R., Ed., CRC, Boca Raton, FL, 1988, 45.
50. **Atkinson, D. and Allen, J. G.,** Observations on the influence of orchard soil management on simazine movement, *Weed Res.,* 16, 305, 1976.
51. **IUPAC,** IUPAC report on pesticides (23), potential contamination of ground water by pesticides, *Pure Appl. Chem.,* 10, 1419, 1987.
52. **Jury, W .A., Winer, A. M., Spencer, W. F., and Focht, D. D.,** Transport and transformations of organic chemicals in the soil-air-water ecosystem, *Rev. Environ. Contam. Toxicol.,* 99, 120, 1987.
53. **Leistra, M.,** Behaviour and significance of pesticide residues in ground water, *Aspects Appl. Biol.,* 17 *Environmental Aspects of Applied Biology,* 223, 1988.
54. **Ritter, W. F.,** Pesticide contamination of ground water in the United States, *J. Env. Sci. Health, Part B. Pestic. Food Contam. Agric. Wastes,* 25, 1, 1990.
55. **Cohen, S. Z., Eiden, C., and Lorber, M. N.,** Monitoring ground water for pesticides, in *Evaluation of Pesticides in Ground Water,* Garner, W. Y., Honeycutt, R. C., and Nigg, H. N., Eds., ACS Symp. Series No. 315, American Chemical Society, Washington D.C., 1986, 170.
56. **Jarczyck, H. J.,** Untersuchungen zum Sicherverhalten von Pflanzenschutzmitteln in einen Monolith-Lysimeter-Anlage, *Landwirtsch. Forsch.,* 40, 273, 1984.
57. **Clark, L., Gomme J., and Hennings, S.,** Pesticides in a chalk catchment: inputs and aquatic residues, *Pestic. Sci.,* 32, 15, 1991.
58. **Hurle, K., Giessl, H., and Kirchhoff, J.,** *Uber das Vorkommen einiger ausefafhlter Pflanzenschutzmittel im Grundwasser,* Springer-Verlag, Stuttgart, 169, 1987.

59. **Hance, R. J.,** Accuracy and precision in pesticide analysis with reference to the E.C. water quality directive, *Pestic. Outlook,* 1(1), 23, 1989.
60. **Bacci, E., Renzoni, A., Gaggi, C., Calamari, D., Franchi, A., Vighi, M., and Severi, A.,** Models, field studies, laboratory experiments: an integrated approach to evaluate the environmental fate of atrazine (s-triazine herbicide), *Agric., Ecosystems and Environ.,* 37, 513, 1989.
61. **Wood, M., Harold, J., Johnson, J., and Hance, R. J.,** The potential for atrazine degradation in aquifer sediments, in *Pesticides in Soils and Water,* BCPC Monogr. No. 47, 1991, 175.
62. **Addala, M. S. A., Hance, R. J., and Drennan, D. S. H.,** Effects of application method on the performance of soil-applied herbicides. II. Field studies, *Weed Res.,* 24, 105, 1983.
63. **Newton, M., Roberts, F., Allen, A., Kelpsas, B., White, D., and Boyd, P.,** Deposition and dissipation of three herbicides in foliage, litter and soil of brushfields of Southwest Oregon, *J. Agric. Food Chem.,* 38, 574, 1990.
64. **Grossbard, E. and Atkinson, D.,** *The Herbicide Glyphosate,* Butterworths, London, 1985, 301.
65. **Clay, D. V. and Ivens, G. W.,** The response of apple, pears and plums to shoot applications of growth-regulator herbicides, in Proc. 9th Br. Weed Control Conf., 1968, 939.
66. **Hance, R. J.,** Adsorption and bioavailability, in *Environmental Chemistry of Herbicides, 1,* Grover, R., Ed., CRC, Boca Raton, FL, 1988, 1.
67. **Lavy, T. L., Roeth, F. W., and Fenster, C. R.,** Degradation of 2,4-D and atrazine at three soil depths in the field, *J. Env. Qual.,* 2, 132, 1973.
68. **Walker, A., Cotterill, E. G., and Welch, S. J.,** Adsorption and degradation of chlorsulfuron and metsulfuron in soils from different depths, *Weed Res.,* 3, 281, 1989.
69. **Upstone, M. E.,** Effects of metsulfuron-methyl on following crops of sugar beet and potatoes, in *Pesticides in Soils and Water,* BCPC Monogr. No. 47, 1991, 123.
70. **Corbin, F. T. and Upchurch, R. P.,** Influence of pH on detoxification of herbicides in soil, *Weeds,* 15, 370, 1967.
71. **Barnes, C. J., Goetz, A. J., and Lavy, T. L.,** Effects of imazaquin residues on cotton, *Weed Sci.,* 37, 820, 1989.
72. **Hazelden, J.,** Soil properties affecting the carryover of a herbicide, in *Pesticides in Soils and Water,* BCPC Monogr. No. 47, 1991, 93.
73. **Hance, R. J., Smith, P. D., Cotterill, E. G., and Reid, D. C.,** Herbicide persistence: effects of plant cover, previous history of the soil and cultivation, *Meded. Fac. Landbouwwet. Rijksuniv. Gent,* 43(2), 1127, 1978.
74. **Clay, D. V.,** Residues of simazine in soil following repeated applications to fruit, ornamental and forestry crops, *ADAS Expl Hort.,* 30, 46, 1978.
75. **Burnside, O. C., Fenster, C. R., Wicks, G. A., and Drew, J. V.,** Effect of soil and climate on herbicide dissipation, *Weed Sci.,* 17, 241, 1969.
76. **Holly, K. and Roberts, H. A.,** Persistence of phytotoxic residues of triazine herbicides in the soil, *Weed Res.,* 3, 1, 1963.
77. **Walker A.,** Persistence of pronamide in soil, *Pestic. Sci.,* 1, 237, 1970.
78. **Fryer, J. D., Smith, P. D., and Hance, R. J.,** Field experiments to investigate long-term effects of repeated applications of MCPA, tri-allate, simazine, and linuron, *Weed Res.,* 20, 103, 1980.
79. **Racke, K. D. and Coates, J. R., Eds.,** Enhanced biodegradation of pesticides in the environment, Am. Chem. Soc. Symp. Series, 426, 302, 1990.
80. **Hamaker, J. W.,** Mathematical prediction of cumilative levels of pesticides in soil, in *Organic Pesticides in the Environment, Advances in Chemistry Series No. 60,* Rusen, A. A. and Kraybill, H. F., Eds., American Chemical Society, Washington D.C., 1966, 122.
81. **Walker, A.,** A simulation model for prediction of herbicide persistence, *J. Environ. Qual.,* 3, 396, 1974.
82. **Briggs, G. G.,** Degradation in soils, Persistence of Insecticides and Herbicides, BCPC Monogr. No. 17, 1976, 41.
83. **Blair, A. M., Martin, T. D., Walker, A., and Welch, S. J.,** Measurement and prediction of chlorsulfuron persistence in soil following autumn and spring application, in *Proc. Brighton Crop Protection Conf. — Weeds,* 3, 1989, 1121.
84. **Tooby T. E. and Marsden, P.K.,** Interpretation of environmental fate and behaviour data for regulatory purposes, Pesticides in Soils and Water: Current Perspectives, BCPC Monogr. No. 47, 1991, 3.
85. **Walker, A.,** Activity and selectivity in the field, in *Interactions between Herbicides and the Soil,* Hance, R. J., Ed., Academic Press, London, 1980, 203.
86. **Clay, D. V. and Davison, J. G.,** The persistence of simazine in a range of soils in selected areas of the United Kingdom, in Proc. 10th Br. Weed Control Conf., 1970, 821.
87. **Hamaker, J. W. and Thompson, J. M.,** Adsoption, in *Organic Chemicals in the Soil Environment, 1,* Goring, C. A. and Hamaker, J. W., Eds., Marcel Dekker, New York, 1972, 49.
88. **Damanakis, M., Drennan, D. S. H., Fryer, J. D., and Holly, K.,** The toxicity of paraquat to a range of species following uptake by the roots, *Weed Res.,* 10, 278, 1970.

89. **Riley, D., Tucher. B. V., and Wilkinson, W.,** Biological unavailability of bound paraquat residues in soil, in *Bound and Conjugated Pesticide Residues,* Kaufman, D. D., Still, G. G., Paulson, G. D., and Bandal, S. K., Eds., ACS Symposium Series 29, 1976, 301.
90. **Wiseman, J. S. and Lawson, H. M.,** Effects on subsequent cereal crops of residual herbicides used in raspberry experiments, in Proc. 10th Br. Weed Control Conf., 1970, 268.
91. **Marriage, P. B., Saidak, W. J., and von Stryk, F. G.,** Residues of atrazine, simazine, linuron and diuron after repeated applications in a peach orchard, *Weed Res.,* 15, 373, 1975.
92. **Stalder, L. and Pestemer, W.,** Availability to plants of herbicide residues in soil. I. A rapid method for estimating potentially available residues of herbicides, *Weed Res.,* 20, 341, 1980.
93. **Bailey, G. W. and White, J. L.,** Review of adsorption and movement of organic pesticides by soil colloids with implications concerning pesticide bioactivity, *J. Agric. Food Chem.,* 12, 324, 1964.
94. **Kaufman, D. D., Still, G. G., Paulson, G. D., and Bandal, S. K., Eds.,** *Bound and Conjugated Pesticide Residues,* A.C.S. Symp. Series 29, Washington, D.C., 1976.
95. **Hance, R. J.,** Extractability and biological effects of some herbicides in the soil, in *Quantification, Nature and Bioavailability of Bound ^{14}C-Pesticide Residues in Soil, Plants and Food,* Food and Agricultural Organization/I.A.E.A., Vienna, 1986, 1.
96. **Capriel, P., Haisch, A., and Kahn, S. U.,** Distribution and nature of bound (non-extractable) residues of atrazine in a mineral soil nine years after herbicide application, *J. Agric. Food Chem.,* 33, 561, 1985.
97. **Kahn, S. U.,** Distribution and characteristics of bound residues of prometryn in an organic soil, *J. Agric. Food Chem.,* 30, 175, 1982.
98. **Kloskowski, R. and Fuhr, F.,** Formation of bound residues of (ring-^{14}C) simazine in a Parabraunerde (alfisol) and their availability to maize, in *Radiotracer Studies of Bound Pesticide Residues in Soil, Plants and Food (Rep. Res. Co- ordination Meeting Neuherberg),* I.A.E.A., Vienna, 1984, 133.
99. **Bucholz, K. P.,** Factors influencing oat injury from triazine residues in soil, *Weeds,* 13, 362, 1965.
100. **Nicholls, P. H., Evans, A. A., and Walker, A.,** The behaviour of chlorsulfuron and metsulfuron in soils in relation to incidents of damage to sugar beet, in *Proc. Brighton Crop Protection Conf. — Weeds,* 1987, 549.
101. **Vencill, W. K., Wilson, H. P., Hines, T. E., and Hatzios, K. K.,** Common lambsquarters (*C. album*) and rotational crop response to imazethapyr in pea (*Pisum sativum*) and snapbean (*Phaseolus vulgaris*), *Weed Tech.,* 4 (1), 39, 1990.
102. **Jan, P., Cochet, J. C., and Ador, J.,** Possibilites de semis, au printemps, apres une cereale d'automne prealablement desherbee, *C. R. 10e Conf. COLUMA,* 1, 315, 1979.
103. **Riley, D. and Eagle, D.,** Herbicides in soil and water, in *Weed Control Handbook: Principles,* 8th ed., Hance, R. and Holly, K., Eds., Blackwell Scientific Publications, Oxford, 1990, 243.
104. **Clay, D. V. and Davison, J. G.,** The effect of 8 years of different soil management treatments in raspberries on weed seed and worm populations, soil microflora and on chemical and physical properties of the soil, *Proc. Br. Crop Protection Conf.—Weeds,* 1976, 249.
105. **Fryer, J. D., Ludwig, J. L., Smith, P. D., and Hance, R. J.,** Tests of soil fertility following repeated applications of MCPA, triallate, simazine, and linuron, *Weed Res.,* 20, 111, 1980.
106. **Moffatt, J. R.,** Cultivation weedkiller experiment, Rothamsted, 1961–72, *Rep. Rothamsted Exp. Sta. 1974,* 2, 155, 1975.
107. **Ellis, F. B., Elliott, J. G., Barnes, B. T., and Howse, K. R.,** Comparison of direct drilling, reduced cultivation and ploughing on the growth of cereals. II. Spring barley on a sandy loam soil: soil physical conditions and root growth, *J. Agric. Sci., Cambridge,* 89, 631, 1977.
108. **White, G. C. and Atkinson, D.,** Orchard soil management: an appraisal, *Aspects Appl. Biol.,* 8, *Weed Control in Fruit Crops,* 159, 1984.
109. **Caverly, D. J.,** Advisory problems with residual herbicides, in *Proc. Brighton Crop Protection Conf. — Weeds,* 1987, 601.
110. **Van Himme, M., Bulke, R., and Calus, A.,** Possibilities of growing replacing spring crops after failure of a winter cereal or winter rape crop treated with persistent herbicides, *Meded. Fac. Landbouwwet. Rijksuniver., Gent,* 54(2a), 1989, 289.
111. **Lawson H. M. and Wiseman, J.,** Contamination of seed potato crops, *Rep. Scottish Crop Res. Inst. 1986,* 147, 1987.
112. **Caverly, D. J.,** Significance of residues to following crops, in *Pesticide Residues,* MAFF Reference Book 347, HMSO, London, 1983, 112.
113. **Kearney, P. C.,** Summary of soil bound residues discussion session, in *Bound and Conjugated Pesticide Residues,* Kaufman, D. D., Still, G. G., Paulson, G. D., and Bandal, S. K., Eds., A.C.S. Symposium Series 29, 1976, 378.
114. **Horowitz, M.,** Bioassay techniques for foliar acting herbicides, *Residue Rev.,* 61, 113, 1976.
115. **Horowitz, M.,** Application of bioassay techniques to herbicide investigations, *Weed Res.,* 16, 209, 1976.

116. **Clay, D. V.,** Biological assay methods for herbicide residues, in *Pesticide Residues,* MAFF Reference Book 347, HMSO, London, 1983, 153.
117. **Lavy, T. L. and Santelmann, P. W.,** Herbicide bioassay as a research tool, in *Research Methods in Weed Science, 3rd ed.,* Camper, N. D., Ed., S. Weed Science Society, Champaign, IL, 1986, 201.
118. **Strek, H. J., Burkhart, D. C., Strachan, S. D., Peter, C. J., and Ruggiero, M.,** Use of bioassays to characterize the risk of injury to follow crops by sulfonylurea herbicides, in *Proc. Brighton Crop Protection Conf. — Weeds,* 1989, 245.
119. **Way, J. M.,** The effects of sub-lethal doses of MCPA on the morphology and yield of vegetable crops. II. Cabbage, cauliflower and brussels sprouts, *Weed Res.,* 3, 11, 1963.
120. **Eagle, D. J. and Caverly, D. J.,** *Diagnosis of Herbicide Damage to Crops,* Ref. Book 221, HMSO, London, 1981.
121. **Derr, J. F. and Appleton, B. L.,** *Herbicide Damage to Trees and Shrubs: A Pictorial Guide to Symptom Diagnosis,* BlueCrab, Virginia Beach, 1988.
122. **Dumas, E., Vallee, J. C., Martin, C., Vansuyt, G., and Vernoy, R.,** *Weed Res.,* 17, 173, 1977.
123. **Taylor, A. W., Freeman, H. P., and Edwards, W. M.,** Sample variability and the measurement of dieldrin content of a soil in the field, *J. Agric. Food Chem.,* 19, 832, 1971.
124. **Hurle, K.,** Biotests for the detection of herbicides in soil, in *Crop Protection Agents — their Biological Evaluation,* McFarlane, N. R., Ed., Academic Press, London, 1977, 285.
125. **Santelmann, P. W., Weber, J. B., and Wiese, A. F.,** A study of soil bioassay technique using prometryne, *Weed Sci.,* 19, 170, 1971.
126. **Nyffeler, A., Gerber, H.-R., Hurle, K., Pestemer, W., and Schmidt, R. R.,** Collaborative studies of dose-response curves obtained with different bioassay methods for soil-applied herbicides, *Weed Res.,* 22, 213, 1982
127. **Eagle, D. J.,** Herbicide persistence, symptoms of damage and crop susceptibility, *National Pesticide Unit Report,* MAFF SS/PU/15, Cambridge, 1979.
128. **Pestemer, W., Stalder, L., and Eckert, B.,** Availability to plants of herbicide residues in soil II. Data for use in vegetable crop rotations, *Weed Res.,* 20, 349, 1980.
129. **Gottesburen, B., Pestemer, W., Wang, K., Wischnewsky, M. B., and Zhao, J.,** Concept, structure and validation of the expert system herbasys (Herbicide Advisory System) for selection, prognosis on persistence and effects on succeeding crops, in Pesticides in Soil and Water, BCPC Monogr. No. 47, 1991, 129.
130. **Duffy, M. J.,** The characterization of herbicide persistence, *Pesticides in Soil and Water,* BCPC Monogr. No. 47, 1991, 85.
131. **Eberle, D. O. and Gerber, H. R.,** Comparitive studies of instrumental and bioassay methods for the analysis of herbicide residues, *Arch. Environ. Contam. Toxicol.,* 4, 101, 1976.
132. **Marrs, R. H., Frost, A. J., and Plant, R.,** Effects of herbicide drift on higher plants, in *Environmental Impact of Herbicide Drift, Interim Report,* Davis, B. N. K., Ed., NERC, ITE, Monks Wood Exp. Stn., 1991, 4.

Chapter 10

PLANT TISSUE CULTURE AND HERBICIDE RESEARCH

Gayle Davidonis

TABLE OF CONTENTS

0-8493-6603-8/93/$0.00+$.50

I. INTRODUCTION

Plant tissue culture can be defined as the culture of plants, plant parts, tissues, organs, cells, or protoplasts *in vitro*. Pioneering research in plant tissue culture established conditions under which cell division would take place and investigated the nutritional and hormonal requirements of cultures. Subsequently, plant tissue cultures were used to elucidate primary and secondary metabolic pathways in plant cells and the processes of differentiation and organogenesis. Recently plant tissue culture has been used in agrochemical research focusing on the use of plant tissue cultures for the screening of herbicides and growth regulators, studies of the metabolism and the modes of action of herbicides, and the development of herbicide-resistant cultures.[1-5]

All herbicides eventually affect plant growth by inhibiting growth through biochemical and/or biophysical alteration of the plant. In the case of herbicides, whose primary action is the disruption of photosynthesis, growth inhibition is a secondary effect. The two components of plant growth are cell division and cell expansion. Generally, cell divisions are localized in meristems but can occur in nonmeristematic tissue. Apical meristems are found at the tips of shoots and roots, and lateral and intercalary meristems occur above the shoot nodes and in leaf sheaths. Cell expansion can produce approximately isodiametric cells or be directed longitudinally or laterally. This review compares the responses to herbicides of plant cells *in vitro* and *in planta*.

II. CHARACTERISTICS OF PLANT TISSUE CULTURES

A. INITIATION OF PLANT TISSUE CULTURES

Plant tissue cultures (callus) can be derived from almost any piece of plant tissue (explant) free of microorganisms. Vegetative tissue and seeds are surface-sterilized by treatment in ethyl alcohol and/or sodium hypochlorite and rinsed in sterile distilled water. Vegetative tissue explants are placed on agar-solidified culture media. Seeds can be germinated under sterile conditions and tissue explants transferred to culture media.

Callus tissue is composed of a mass of actively dividing unorganized cells. Some callus tissues are hard in texture and others break easily into small fragments (friable callus). Most suspension cultures are obtained by transfer of friable callus pieces to agitated liquid media. Suspension cultures consist of single cells and clumps of cells. Cell clumping complicates the rapid measurement of cell mass, number, and volume needed in bioassay applications. Mechanical, enzymatic, and chemical methods have been employed to break up cell clumps, but Kubek and Shuler[6] found no techniques that gave sustained disaggregation without affecting the biochemistry of the culture.

The growth curve of a cell suspension culture contains an initial lag period occurring prior to cell division followed by an exponential rise in cell number. Gradually, the cell division rate drops, and cells enter a stationary or nondividing stage. Growth of callus tissue is not easily separated into lag, log, and stationary stages, although the growth curve of callus tissue has the same general shape as that of suspension cultures.

B. PHOTOAUTOTROPHIC, PHOTOMIXOTROPHIC, AND HETEROTROPHIC CULTURES

Some important factors involved in the growth of callus and suspension cultures are plant species, initial inoculum size, physiological age, frequency of subculture, external culture conditions (light, temperature, culture vessel), and composition of the culture medium. By strict definition, organisms that require both organic and inorganic substances for the normal completion of their life cycle are heterotrophic; those that need to be supplied only with inorganic substances are called autotrophic. These basic definitions are altered to encompass callus and cell suspension cultures. Cultures grown in the light in the absence of added sugar and possessing a functional photosynthetic system are called photoautotrophic. Those grown in the light in the presence of added sugar and possessing a functional photosynthetic system are called photomixotrophic. Those grown in the light or dark in the presence of added sugar and not possessing a functional photosynthetic system are called heterotrophic.

Photoautotrophic suspension cultures have well developed chloroplasts with a chemical composition and photosynthetic activity similar to mesophyll cells *in planta*.[7,8] Pigment production, photosynthesis, and respiration in photoautotrophic soybean (*Glycine max* L. Merr.) and cotton (*Gossypium hirsutum* L.) suspension cultures were similar to those found in young expanding leaves.[9,10] Photoautotrophic cell lines are grown in the light without organic carbon sources in a CO_2-enriched environment, since low ribulose 1,5-bisphosphate carboxylase activity and high respiration rates lower photosynthetic efficiency of photoautotrophic cells at ambient CO_2.[10] Cultures are initiated on agar medium (containing 3% sucrose as a carbohydrate source) from green explants and grown in the dark to promote callus production. The callus is then placed in liquid medium and agitated under low intensity light to produce suspension cultures, which are transferred to a higher light intensity (200 $\mu E\ m^{-2}\ s^{-1}$). The sucrose concentration is lowered to 1% followed by subculturing into medium without sucrose.

Photoautotrophic cultures usually have a slower growth rate than other cultures. After a two week (photomixotrophic cells) or three week (photoautotrophic and heterotrophic cells) culture period, the average increase in fresh weight of photoautrophic, photomixotrophic, and heterotrophic cells at harvest was 1.0, 5.5, and 5.0 g respectively.[7] High phosphoenolpyruvate carboxylase activity was a characteristic feature of photosynthetically active photoautotrophic or photomixotrophic cultures. The most promotive conditions to increase its activity are photomixotrophic nutrition and a high rate of cell division during exponential growth of the cultures.[11] Photomixotrophic and photoautotrophic cultures have practically the same CO_2 fixation capacity on a per unit chlorophyll basis, but photoautotrophic cell suspensions have a high chlorophyll concentration.[11] Meristematic cells and cells cultured under heterotrophic conditions are relatively undifferentiated and nonphotosynthetic.

III. HERBICIDE PRESCREENS

Gressel[2] compared the costs of herbicide screens using several plant species with that of herbicide prescreens using plant tissue cultures and concluded that screening *in vitro* is not prohibitively expensive. The most important consideration in designing a herbicide prescreen is that an active herbicide not be missed. In some cases phytotoxic activity may be found *in vitro* without any effect being observed on whole plants.[12] Prescreens can be used to test many different compounds for herbicide activity or they can be tailored for evaluation of structure-activity relationships. Since herbicides can be both tissue and species specific, a number of factors need to be considered in developing a prescreen.

A. DESIGNING A PRESCREEN

Greenhouse screens use more than one species and this should also be true with prescreens *in vitro*. Haloxyfop, a herbicide used to control grasses in broadleaf vegetable crops was added to soybean and corn (*Zea mays* L.) suspension cultures, the calculated 50% lethal dose (LD_{50}) was 47 times higher for soybean cells than for corn cells.[13] A comparison of concentrations that inhibited cell growth by 50% (I_{50}) revealed that Canadian thistle (*Cirsium arvense* L. Scop.) is 30 times more sensitive to chlorsulfuron than leafy spurge (*Euphorbia esula* L.).[14]

Another important factor to consider is the tissue type used in a prescreen. The growth of photoautotrophic, photomixotrophic, and heterotrophic cultures was examined using herbicides that were thought to have different primary modes of action.[7] Herbicides that

TABLE 1
Effects of Herbicides on the Growth of Tobacco Suspension Cultures

Herbicide (1 μM)	Inhibition of cell growth (%) Photoautotrophic cultures	Photomixotrophic cultures	Heterotrophic cultures
Atrazine	90	70	0
Diuron	100	40	0
Propanil	100	40	0
Paraquat	100	20	20
Nitrofen	80	80	0
Dinoseb	100	70	100
2,4-D	30	0	0

Adapted from Sato et al.[7]

inhibited photosynthetic processes (atrazine, diuron, propanil, paraquat, nitrofen) reduced the growth of photoautotrophic cell suspension cultures to a greater degree than photomixotrophic or heterotrophic cell suspension cultures (Table 1). Equal concentrations of herbicides that had modes of action other than photosynthesis inhibition reduced the growth of all types of cells to the same degree. Photoautotrophic cell suspension cultures showed the responses most similar to those of seedlings assayed with the same herbicides.[7] Napropamide treated white tomato (*Lycopersicon esculentum* Mill) callus tissue showed greater growth inhibition than green callus tissue and thus compared favorably with the observation that green tomato shoots were less sensitive to the herbicide than root tissue.[8]

Some herbicides act at specific stages in the cell cycle, and herbicides that affect cell division cannot be detected if they are added to the culture medium during the stationary stage of growth. The cell cycle is divided into four stages: a pre-DNA synthesis period (G_1), a DNA synthesis period (S), a post-DNA synthesis period (G_2) and mitosis. Interphase consists of G_1, S, and G_2. Mitosis is further divided into prophase, metaphase, anaphase, and telophase. Common techniques used in evaluating herbicide effects on the cell cycle are preparation of a root tip squash and meristem tissue fixation, embedding, sectioning, and staining.[16,17] Preliminary evaluation of minimal herbicide concentration that inhibits growth is necessary since long treatment periods or high concentrations may be toxic. Callus tissue can also be fixed, embedded, sectioned, and stained. Suspension cultures do not require embedding.

After the tissue is prepared the cells can be catagorized. Cells in mitosis that cannot be catagorized into one of the normal mitotic stages are classed as aberrant. In some species it is difficult to decipher between stages. A useful measurement is mitotic index:

$$MI = \frac{\text{total number of nuclei, prophase to telophase}}{\text{total number of nuclei}} \times 100$$

Rapidly growing tissue cultures show a mitotic index (MI) of 3 to 6%. A herbicide that has a site of action during G_1, S, or G_2 (interphase) and causes a reduction in MI is classified as causing inhibition of mitotic entry.[18]

The growth of soybean suspension cultures monitored by changes in fresh and dry weight was inhibited by chlorsulfuron.[19] Ray[20] reported that chlorsulfuron inhibited the synthesis of the branched chain amino acids valine, leucine, and isoleucine in plants. Adding valine reversed the growth inhibition in both roots and suspension cultures.[19,21] Chlorsulfuron decreased the MI in root meristem tissue from 6.4 to 0.9%.[19] A number of carbamate herbicides disrupt the mitotic sequence by altering the function of microtubules that are

TABLE 2
Cell Elongation in Heterotrophic Suspension Cultures of Parsley

Compound	Concentration (μ*M*)	Cell length (μm)
Untreated		40.7 ± 1.0
Ethephon	30	63.8 ± 0.7
ACC	30	62.1 ± 0.7
GA_3	1	59.9 ± 0.5

Adapted from Grossmann[4] with permission.

involved in cell plate formation at telophase. Propham, chloropropham, and carbetamide blocked cell division, causing increases in cell diameter.[22] Asulam had little effect on cell expansion in *Acer* and *Apium* cultures but inhibited growth at high concentrations.[22,23]

Nitroanilines disrupt the mitotic sequence by preventing the proper alignment of chromosomes at metaphase.[18] The effect of trifluralin on fresh weight gain of tobacco (*Nicotiana tabacum* L.) and carrot (*Daucus carota* L.) callus tissue was concentration dependent. The inhibition of growth was more pronounced in tobacco callus tissue.[24]

Disruption of normal cell enlargement results in reduced growth, even if cell division is not affected. Oat coleoptiles and pea hypocotyls have been used in assay systems to study cell enlargement in the absence of cell division.[25] Diallate and triallate inhibited cell division at high concentrations and cell expansion at low concentrations in shoot tissue.[26] The inhibition of growth caused by the application of alachlor and metolachlor has been attributed to an inhibition of both cell division and cell expansion.[27]

Inhibition of cell enlargement in cell suspension cultures may be a sensitive parameter for prescreening herbicides or growth regulators. Any cell suspension can be used to study the effect of herbicides on cell division, but relatively few suspensions have been used to study the disruption of cell enlargement. Grossmann[4] has shown a modest increase in the cell size of parsley (*Petroselinum crispum* Mill.) cells during the culture period. Cell cultures sensitive to growth regulators may hold a key to understanding cell enlargement *in vitro*. Cell enlargement in suspension cultures is promoted by the ethylene-producing compounds, ethephon, and aminocyclopropane-carboxylic acid (ACC).[4,28] A 50% increase in cell length was obtained with the addition of ethephon, ACC, or gibberellic acid (GA_3) to parsley suspension cultures (Table 2). However most suspension cultures do not respond to GA_3 treatment.[29] Gibberellic acid promoted cell expansion in auxin independent suspension cultures of spinach (*Spinacia oleracea* L.) and auxin dependent rose (*Rosa* spp.) suspension cultures.[29] The rose cells elongated, the spinach cells remained isodiametric.

Cotton fibers are derived from epidermal cells on the surface of ovules. These single celled fibers can reach lengths over 20 mm. Callus cultures have been derived from preanthesis cotton ovules. When these ovule-derived cotton cells were placed into liquid medium containing GA_3 and agitated, they elongated to lengths of 1 mm; cotton suspension cultures derived from hypocotyl explants did not elongate.[30] Both auxin independent and auxin dependent suspension cultures derived from ovules were sensitive to GA_3.[31] Auxin or cytokinin, applied alone or in combination with GA_3, did not promote elongation above that observed with GA_3 alone (Table 3). Cotton tissue culture system may be useful as a prescreen method to identify herbicides and growth regulator compounds that inhibit cell elongation.

B. GROWTH MEASUREMENTS

Some herbicides kill rapidly and others inhibit growth only after several days. With time herbicides that inhibit growth can be detoxified and growth of the culture resumes. Therefore, the time of evaluation of herbicide effects should also be considered in the development of

TABLE 3
Cell Elongation in Auxin-Dependent Cotton Cells

Compounds	Numbers of cells (frequency) Cell length (mm) <0.5	0.5–1.0	>1.0
Untreated	79	21	0[a]
5.7 μ*M* GA_3	33	53	14
0.22 μ*M* 2,4-D	73	24	3[a]
1.2 μ*M* isopentyladenine (2ip)	93	7	0[a]
5.7 μ*M* GA_3 + 0.22 μ*M* 2,4-D	35	53	12
5.7 μ*M* GA_3 + 1.2 μ*M* 2ip	35	55	10

Adapted from Davidonis.[31]

[a] Frequency distributions significantly different from 5.7 μ*M* GA_3 at $p = 0.05$.

a prescreen. Growth is one of the most convenient physiological parameters to measure, but it is neither simple nor rapid. Most techniques for measuring growth have their advantages and disadvantages. Cell counting is one of the best growth parameters, but it is time consuming. Unless the culture consists of single cells or very tiny clumps, cell counting requires the separation of clumped cells by pectinase or chromic acid treatment.[32] Under special circumstances the rapid electronic Coulter cell counter can be used.[33] Changes in cell size during the culture period can be determined by microscopic examination. The growth of cells in culture (the sum of cell division and cell expansion) is measured by other parameters such as dry weight, fresh weight, packed cell volume (PCV), settled cell volume (SCV), and turbidimetric measurements. Packed cell volume is determined by centrifuging an aliquot and recording the cell volume. Settled cell volume is a variant of PCV in which the volume is measured after sedimentation under gravity. Cell suspension cultures not composed of small aggregates can be sonicated and the optical density of the fragments measured.[34] Simple weight determinations require large sample sizes. An increase in fresh weight, PCV, and SCV may, in addition to measuring growth, reflect a change in the water content of cells.

Turbidimetric measurements can be taken in a nondestructive way using calibrated side arm flasks.[35] Turbidity was shown to vary proportionally with cell number and dry weight. This method cannot be used with extremely heterogenous cultures composed of various sized clumps and single cells, or when the addition of the test substances causes aggregation. Other nondestructive methods used are variations of PCV and SCV methods.

Relative growth rate of cell suspensions in flasks can be determined. The flasks are tilted backwards at a certain angle in a fixed stand, the cells are allowed to settle, and the length of the chord formed by the surface of the packed cells across the base of the flask is measured on the outside of each flask. This method requires a calibration curve.[36] Aliquots of cell suspensions can also be pipetted onto filter paper discs placed on agar medium. Under sterile conditions, the discs and adhering cells are removed periodically, weighed, and replaced. This method provides an immediate measurement of fresh weight.[37]

An indirect biophysical measure of growth involves measurement of the electrical conductivity of the medium. Hahbrock and Kuhlen[38] showed that a decrease in conductivity during the culture period was due to an uptake of inorganic ions by the cells. A decrease in conductivity was inversely correlated with changes in PCV and cell number.[38,39] Assessments of various growth parameters in suspension cultures were made, and it was concluded that dry weight, SCV, and conductivity were the most useful methods.[40]

Large scale experiments using flasks require considerable space, and numerous researchers have developed microtest systems. Suspension cultures were grown in tubes or microtiter dishes on a shaker.[40,41] Because it is difficult to guarantee uniform illumination in these microtest systems, a stationary method was developed by Gaba et al.,[42] who described a method in which thin layers of cell suspensions were pipetted into petri plates. At the end of the culture period, cells were washed into centrifuge tubes and the PCV determined.

In plant tissue cultures phytotoxicity has been correlated with metabolic activities, membrane permeability, and cessation of cytoplasmic streaming. Most herbicides do not directly inhibit protein biosynthesis, however, inhibition of other metabolic pathways will lead indirectly to a decrease in the rate of protein biosynthesis. Herbicides were more inhibitory to leucine incorporation into protein than were nonphytotoxic compounds. Herbicides that disrupt photosynthesis were poor inhibitors of leucine incorporation. Thiocarbamate herbicides were very phytotoxic to seedlings, but they were more effective inhibitors of suspension cell leucine incorporation. Five out of seven thiocarbamates that were inactive on whole plants inhibited suspension cell leucine incorporation.[43] Cell viability can be determined by staining with Evan's blue or neutral red. Dead cells treated with Evan's blue will absorb the stain but will remain colorless when treated with neutral red. Living cells accumulate neutral red in the vacuole. Zilkah and Gressel[28] tested several stains that penetrate cells with defective membranes and found most of them unsatisfactory. They concluded that positive vital staining indicated that the cells were not yet dead.

IV. HERBICIDE METABOLISM

Herbicide metabolism has been reported to be qualitatively similar between plant tissue cultures and intact plants.[1,5,44] Meristems are sensitive sites of action for diclofop-methyl. Therefore, absorption, translocation, and metabolism may be expected to influence its activity and selectivity. However, long distance transport of the herbicide is extremely limited. The parent compound and its metabolites are retained in the tissue where initial absorption occurs.[45] Considering these features diclofop-methyl is an excellent choice of herbicide to compare metabolism *in vitro* and *in planta*. Both *Avena sativa* and *Avena fatua* are susceptible to diclofop-methyl under field conditions. The metabolic products of diclofop-methyl metabolism appear to be the same in intact plants and in suspension cultures of *A. sativa* and callus suspended in liquid cultures of *A. fatua*.[46,47] The formation of the ester conjugate is a reversible reaction occurring *in vitro* and *in planta* in susceptible species. The formation of the phenolic conjugate constitutes a major detoxification pathway in resistant species.[48]

The resistance of cotton to chlorotoluron has been attributed to extensive *N*-dealkylation of the parent herbicide. In susceptible ryegrass (*Lolium multiflorum* Lam.) *N*-dealkylation does not progress beyond the formation of *N*-monodealkylated metabolites.[49] Rapid uptake of chlorotoluron in cotton suspension cultures was followed by rapid detoxification, whereas uptake in ryegrass suspension cultures was slow; this was attributed to an ineffective detoxification mechanism. These results were consistent with greenhouse trials using cotton and ryegrass. In agreement with chlorotoluron metabolism *in planta,* cotton suspension cultures metabolized chlortoluron by complete *N*-dealkylation. The cotton suspension cultures showed a greater capacity for hydroxylation of the *N*-demethylated metabolite. Although some small qualitative differences existed the metabolism of chlorotoluron in ryegrass plants and ryegrass suspension cultures was similar and occurred mainly by *N*-monodealkylation coupled with ring methyl oxidation.[49,50] In wheat (*Triticum aestivum* L.) the level of free *N*-monodimethyl derivative was higher in suspension cultures than in wheat plants.[51] This led to the conclusion that the first *N*-demethylation was probably more important in wheat suspension cultures than in plants. Diphenamid is converted to a number of metabolites. Glucose conjugation is an important component of diphenamid metabolism in soybean plants where these conjugates accounted for 46 to 48% of the metabolites. Glucosides in soybean suspension culture account for only 6% of the metabolites.[52]

Numerous nutrient media have been formulated for plant tissue culture. These media may influence herbicide metabolism. In one study cells were grown in four different nutrient media that differed in organic nutrients and growth regulator concentrations.[53] The rate of chlorotoluron metabolism in cotton cells was related to the rate of the growth of these cells in four different media. Variation in the rate of chlorotoluron metabolism in corn suspension cultures was not related to growth rate in the different media.

Metabolic capabilities of callus and suspension cultures may change with time. Just as herbicides may act at specific times during cell growth, herbicides may be metabolized at different rates during the culture period. The mitotic index of soybean root callus cultures peaked at two weeks during a 9-week culture period. The metabolism of 2,4-dichlorophenoxyacetic acid (2,4-D) was determined in 3 and 9-week-old soybean root callus tissue after a 48 h treatment period. The internal concentration of 2,4-D in 3-week-old callus tissue was higher than in 9-week-old cultures, but the level of amino acid conjugates was higher in 9-week-old cultures. It was concluded that the metabolism found in 9-week-old callus tissue closely paralleled the metabolism of 2,4-D in differentiated root cultures.[54] Suspension cell cultures were treated with diphenamid during early log, log, and stationary stages of growth. All cultures formed the same metabolites but log stage and stationary stage cells produced larger amounts of metabolites than did early log phase cells.[52] Metabolic pathways may change during long term culture of plant cells. In a suspension originating from newly isolated lettuce (*Lactuca sativa* L.), callus chlorotoluron was metabolized by sequential-*N*-demethylation and by oxidation of the 4-methylphenyl group. A cell suspension derived from 5-year-old lettuce callus tissue showed a reduction in the ability to form the ring hydroxymethyl metabolite of chlorotoluron.[55]

Cuticle penetration and translocation of herbicides in whole plants are not significant factors in plant tissue culture-herbicide interactions. Herbicides can move across cell membranes by diffusion or active transport, and cell suspension cultures offer an opportunity to examine herbicide uptake. Bentazon, a photosynthetic inhibitor, must penetrate the plasma and thylakoid membranes to reach its site of action.[56] Cell suspensions of susceptible species do not metabolize bentazon.[57] Characteristics of herbicide accumulation can be investigated, such as binding to cellular constituents, energy dependence, and ion trapping. There is evidence that bentazon is absorbed across membranes via diffusion and that bentazon accumulation in plant cells was via an energy dependent, ion trapping mechanism that resulted in the cytoplasmic accumulation of bentazon.[58]

The presence of herbicide metabolites in the culture medium may be due to either the efflux of the metabolites from intact cells or the release of metabolites from dead cells. The auxin-like herbicide, 2,4-D, was readily metabolized in soybean tissue cultures.[54,59] The glycosides of hydroxylated 2,4-D and the amino acid conjugates are retained in the tissue while 2,4-D efflux was rapid. Efflux data from tissue, incubated in 20% dimethylsulfoxide, which disrupts vacuolar membranes, indicated that 2,4-D metabolites were located in the vacuole.[60] Further, Schmitt and Sandermann[61] isolated vacuoles from 2,4-D treated cultures and reported that the glycoside conjugates were localized in vacuoles.

V. CONCLUSIONS

Plant tissue cultures are being used as model systems in herbicide research. Plant tissue culture has the following advantages: aseptic conditions, control of nutritional and hormonal parameters, uniform environmental conditions, ease of extraction and purification of metabolites, more efficient use of radioactively labeled compounds, elimination of the cuticular barrier, and very little intercellular transport.[1-4] An examination of the characteristics of cells in culture has shown that cultures can mimic leaf mesophyll cells or serve as nonphotosynthetic equivalents of meristematic cells. Plant tissue cultures can be used in the development of herbicide prescreens, and screening assays can be used to identify phytotoxic compounds or evaluate structure/activity relationships. Plant tissue culture systems are also useful in studies of herbicide uptake and metabolism, and they can be used to localize metabolites in the cells. In all applications, however, factors such as media composition, growth stage of culture and age of culture, must be considered.

REFERENCES

1. **Mumma, R. O. and Davidonis, G. H.,** Plant tissue culture and pesticide metabolism, in *Progress in Pesticide Biochemistry and Toxicology,* Vol. 3, Hutson, D. H. and Roberts, T. R., Eds., John Wiley & Sons, Ltd., New York, 1983, 255.
2. **Gressel, J.,** *In vitro* plant cultures for herbicide screening, in *Biotechnology in Agricultural Chemistry,* ACS Symposium Series, No. 334, LeBaron, H., Mumma, R. O., Honeycutt, R. C., and Deusing, J. H., Eds., American Chemical Society, Washington D.C., 1987, 41.
3. **Gressel, J.,** Plant tissue culture systems for screening of plant growth regulators: hormones, herbicides, and natural phytotoxins, *Adv. Cell Cultures,* 3, 93, 1984.
4. **Grossmann, K.,** Plant cell suspensions for screening and studying the mode of action of plant growth retardants, *Adv. Cell Cultures,* 6, 89, 1988.
5. **Camper, N. D. and McDonald, S. K.,** Tissue and cell cultures as model systems in herbicide research, *Rev. Weed Sci.,* 4, 169, 1989.
6. **Kubek, D. J. and Shuler, M. L.,** On the generality of methods to obtain single-cell plant suspension cultures, *Can. J. Bot.,* 56, 2521, 1978.
7. **Sato, F., Takeda, S., and Yamada, Y.,** A comparision of effects of several herbicides on photoautotrophic, photomixotrophic and heterotrophic cultured tobacco cells and seedlings, *Plant Cell Rep.,* 6, 401, 1987.
8. **Husemann, W., Fischer, K., Mittelbach, I., Hubner, S., Richter, G., and Barz, W.,** Photoautotrophic plant cell cultures for studies on primary and secondary metabolism, in *Primary and Secondary Metabolism of Plant Cell Cultures 2,* Kurz, W. G. W., Ed., Springer, Berlin, 1989, 35.
9. **Rogers, S. M. D., Ogren, W. L., and Widholm, J. M.,** Photosynthetic characteristics of a photoautotrophic cell suspension culture of soybean, *Plant. Physiol.,* 84, 1451, 1987.
10. **Blair, L. C., Chastain, C. J., and Widholm, J. M.,** Initiation and characterization of a cotton (*Gossypium hirsutum* L.) photoautotrophic cell suspension culture, *Plant Cell Rep.,* 7, 266, 1988.
11. **Neumann, K. H. and Bender, L.,** Photosynthesis in cell and tissue culture systems, in *Plant Tissue and Cell Culture,* Green, C. E., Somers, D. A., Hackett, W. P., and Biesboer, D. D., Eds., Alan R. Liss, Inc., New York, 1987, 151.
12. **Bastide, J., Carre, P., Gomez, F., and Greiner, A.,** Metabolism of phenyl-3H-indoles by cell suspension of *Acer pseudoplatanus* and their phytotoxicity, *Pest. Sci.,* 27, 33, 1989.
13. **Cho, H-Y., Widholm, J. M., and Slife, F. W.,** Effects of haloxyfop on corn (*Zea mays*) and soybean (*Glycine max*) cell suspension cultures, *Weed Sci.,* 34, 496, 1986.
14. **Swisher, B. A. and Weimer, M. R.,** Comparative detoxification of chlorsulfuron in leaf disks and cell cultures of two perennial weeds, *Weed Sci.,* 34, 507, 1986.
15. **Zilkah, S., Bocion, P. F., and Gressel, J.,** Target tissue for napropamide inhibition: effects on green and white callus cultures and seedlings, *Weed Sci.,* 26, 711, 1978.
16. **Van't Hof, J.,** Experimental procedures for measuring cell population kinetic parameters in plant root meristems, in *Methods in Cell Physiology,* Vol. 3, Prescott, D., Ed., Academic Press, New York, 1968, 95.
17. **Jensen, W. A.,** *Botanical Histochemistry,* W. H. Freemann and Co., San Francisco, 1962, 55.
18. **Hess, F. D.,** Herbicide effects on the cell cycle of meristematic plant cells, *Rev. Weed Sci.,* 3, 183, 1987.
19. **Scheel, D. and Casida, J. E.,** Sulfonylurea herbicides: growth inhibition in soybean cell suspension cultures and in bacteria correlated with block in biosynthesis of valine, leucine or isoleucine, *Pestic. Biochem. Physiol.,* 23, 398, 1985.
20. **Ray, T. B.,** Site of action of chlorsulfuron: inhibition of valine and isoleucine biosynthesis in plants, *Plant. Physiol.,* 75, 827, 1984.
21. **Ray, T. B.,** Mode of action of chlorsulfuron: the lack of direct inhibition of plant DNA synthesis, *Pestic. Biochem. Physiol.,* 18, 262, 1982.
22. **Nurit, F., Gomes de Melo, E., Ravanel, P., and Tissut, M.,** Specific inhibition of mitosis in cell suspension cultures by a *N*-phenylcarbamate series, *Pestic. Biochem. Physiol.,* 35, 203, 1989.

23. **Watts, M. J. and Collin, H. A.,** The effect of asulam on the growth of tissue cultures of celery, *Weed Res.,* 19, 33, 1979.
24. **Camper, N. D., Ahmed, F. A., and Figliola, S.,** Growth effects, uptake and metabolism of trifluralin in tissue cultures, *J. Environ. Sci. Health,* B24, 291, 1989.
25. **Yopp, J. H., Aung, L. H., and Steffens, G. L.,** *Bioassays and Other Special Techniques for Plant Hormones and Plant Growth Regulators,* Plant Growth Reg. Soc. Am., Lake Alfred, FL, 1986.
26. **Banting, J. D.,** Effect of diallate and triallate on wild oat and wheat cells, *Weed Sci.,* 18, 80, 1970.
27. **Deal, L. and Hess, F. D.,** An analysis of the growth inhibitory characteristics of alachlor and metalachlor, *Weed Sci.,* 28, 168, 1979.
28. **Zilkah, S. and Gressel, J.,** The estimation of cell death in suspension cultures evoked by phytotoxic compounds: differences among techniques, *Plant Sci. Lett.,* 42, 305, 1978.
29. **Fry, S. C. and Street, H. E.,** Gibberellin-sensitive suspension cultures, *Plant Physiol,* 65, 472, 1980.
30. **Davidonis, G. H.,** Fiber development in preanthesis cotton ovules, *Physiol. Plant.,* 75, 290, 1989.
31. **Davidonis, G. H.,** Gibberellic acid-induced cell elongation in cotton suspension cultures, *J. Plant Growth Reg.,* 9, 243, 1990.
32. **Street, H. E.,** Cell suspension cultures-techniques, in *Plant Tissue and Cell Culture,* Bot. Monogr. Vol. 11, Street, H. E., Ed., Blackwell Scientific Publications, Oxford, 1977, 59.
33. **Kubek, D. J. and Shuler, M. L.,** Electronic measurement of plant cell number and size in suspension culture, *J. Expt. Bot.,* 29, 511, 1978.
34. **Kubek, D. J. and Shuler, M. L.,** A rapid quantitative method to measure growth of plant cell suspension cultures, *Can. J. Bot.,* 56, 2340, 1978.
35. **Sung, Z. R.,** Turbidimetric measurement of plant cell culture growth, *Plant Physiol.,* 57, 460, 1976.
36. **Gilissen, L. J. W., Hanisch ten Cate, C. H., and Keen, B.,** A rapid method of determining growth characteristics of plant cell populations in batch suspension culture, *Plant Cell Rep.,* 2, 232, 1983.
37. **Horsch, R. B. and King, J.**, Measurement of cultured plant cell growth on filter paper discs, *Can. J. Bot.,* 58, 2402, 1980.
38. **Hahlbrock, K. and Kuhlen, E.,** Relationship between growth of parsley and soybean cells in suspension cultures and changes in the conductivity of the culture medium, *Planta,* 108, 271, 1972.
39. **Grossmann, K. and Jung, J.,** A new micromethod for testing plant growth retardants in cell suspension cultures, *Plant Cell Rep.,* 3, 156, 1984.
40. **Davis, D. G., Stolzenberg, R. L., and Dusky, J. A.,** A comparison of various growth parameters of cell suspension cultures to determine phytotoxicity of xenobiotics, *Weed Sci.,* 32, 235, 1984.
41. **Thiemann, J., Nieswandt, A., and Barz, W.,** A microtest system for the serial assay of phytotoxic compounds using photoautotrophic cell suspension cultures of *Chenopodium rubrum,* *Plant Cell Rep.,* 8, 399, 1989.
42. **Gaba, V., Cohen, N., Shaaltiel, Y., Ben-Amotz-A., and Gressel, J.,** Light-requiring acifluorfen action in the absence of bulk photosynthetic pigments, *Pestic. Biochem. Physiol.,* 31, 1, 1988.
43. **Egli, M. A., Lou, D., White, K. R., and Howard, J. A.,** Effects of herbicides and herbicide analogs on [^{14}C] leucine incorporation by suspension-cultured *Solanum nigrum* cells, *Pestic. Biochem. Physiol.,* 24, 112, 1985.
44. **Swisher, B. A.,** Use of plant cell cultures in pesticide metabolism studies, in *Biotechnology in Agricultural Chemistry,* ACS symposium Series, No. 334, Le Baron, H., Mumma, R. O., Honeycutt, R. D., and Deusing, J. H., Eds., American Chemical Society, Washington, D.C., 1987, 18.
45. **Shimabukuro, R. H.,** Selectivity and mode of action of the post emergence herbicide diclofop-methyl, *Plant Growth Reg. Soc. Am. Q.,* 18, 37, 1990.
46. **Dusky, J. A., Davis, D. G., and Shimabukuro, R. H.,** Metabolism of diclofop-methyl in cell cultures of *Avena sativa* and *Avena fatua,* *Physiol. Plant.,* 54, 490, 1982.
47. **Shimabukuro, R. H., Walsh, W. C., and Hoerauf, R. A.,** Metabolism and selectivity of diclofop-methyl in wild oat and wheat, *J. Agric. Food Chem.,* 27, 625, 1979.
48. **Dusky, J. A., Davis, D. G., and Shimabukuro, R. H.,** Metabolism of diclofop-methyl (Methyl-2[2′, 4′-dichlorophenoxy) phenoxy] propanoate) in cell suspensions of diploid wheat (*Triticum monococcum*), *Plant Physiol.,* 49, 51, 1980.
49. **Owen, W. J. and Donzel, B.,** Oxidative degradation of chlortoluron, propiconazole, and metalaxyl in suspension cultures of various crop plants, *Pestic. Biochem. Physiol.,* 26, 75, 1986.
50. **Ryan, P. J., Cross, D., Owen, W. J., and Laanio, T. L.,** The metabolism of chlortoluron, diuron and GGA 43057 in tolerant and susceptible plants, *Pestic. Biochem. Physiol.,* 16, 213, 1981.
51. **Canivenc, M.-C., Cagnac, B., Cabanne, F., and Scalla, R.,** Induced changes in chlorotoluron metabolism in wheat cell suspension cultures, *Plant Physiol. Biochem.,* 27, 193, 1989.
52. **Davis, D. G., Hodgson, R. H., Dusbabek, K. E., and Hoffer, B. L.,** The metabolism of the herbicide diphenamid (*N-N*-dimethyl-2-2-diphenyl-acetamide) in cell suspensions of soybean, *Physiol. Plant.,* 44, 87, 1978.

53. **Cole, D. J. and Owen, W. J.,** Influence of composition of nutrient medium on the metabolism of chlortoluron in plant-cell suspension cultures, *Pestic. Sci.,* 19, 67, 1987.
54. **Davidonis, G. H., Hamilton, R. H., and Mumma, R. O.,** Metabolism of 2,4-dichlorophenoxyacetic acid in soybean root callus and differentiated soybean root cultures as a function of concentration and tissue age, *Plant Physiol.,* 62, 80, 1978.
55. **Cole, D. J. and Owen, W. T.,** Metabolism of chlortoluron in cell suspensions of *Lactuca sativa*: a qualitative change with age of culture, *Phytochem.,* 27, 1709, 1988.
56. **Retzlaff, G., Hilton, J. L., and St. John, J. B.,** Inhibition of photosynthesis by bentazon in intact plants and isolated cells in relation to the pH, *Z. Naturforsh.,* 34c, 944, 1979.
57. **Sterling, T. M. and Balke, N. E.,** Differential bentazon metabolism and retention of bentazon metabolites by plant cell cultures, *Pestic. Biochem. Physiol.,* 34, 39, 1989.
58. **Sterling, T. M., Balke, N. E., and Silverman, D. S.,** Uptake and accumulation of the herbicide bentazon by cultured plant cells, *Plant Physiol.,* 92, 1121, 1990.
59. **Davidonis, G. H., Hamilton, R. H., and Mumma, R. O.,** Metabolism of 2,4-dichlorophenoxyacetic acid (2,4-D) in soybean root callus: evidence for the conversion of metabolites to free 2,4-D, *Plant Physiol.,* 66, 537, 1980.
60. **Davidonis, G. H., Hamilton, R. H., and Mumma, R. O.,** Evidence for compartmentalization of conjugates of 2,4-dichlorophenoxyacetic acid in soybean callus tissue, *Plant Physiol.,* 70, 939, 1982.
61. **Schmitt, R. and Sandermann, H.,** Specific localization of ß-D-glucoside conjugates of 2,4-dichlorophenoxyacetic acid in soybean vacuoles, *Z. Naturforsch.,* 37c, 772, 1982.

Chapter 11

IMMUNOLOGICAL ASSAYS

Sigmund Josef Huber and Peter Ulvskov

TABLE OF CONTENTS

0-8493-6603-8/93/$0.00+$.50

I. INTRODUCTION

Human food, originating from cultivated plants and animals, is exposed to pesticides and is thus a target of pesticide contamination. For the last 50 years herbicides have been widely used to control unwanted vegetation as a matter of routine in modern crop production. Consequently, herbicides and their degradation products are found in plants, soil, water, humans, and animals. For this reason suitable analytical methods are imperative for effective environmental monitoring. Many physiochemical methods, such as HPLC and GC, often combined with mass spectroscopy, have been developed for residue analysis of herbicides and their metabolites. These methods, however, are time consuming and require laboratory equipment of high technical standard.

An alternative to the physiochemical methods is the immunological methods. The first radioimmunoassays for the determination of the herbicide, paraquat, in body fluids were published by Levitt[1,2] in the late 1970s. This test system was developed for medical purposes, because paraquat was in some instances used in suicide attempts. The radioimmunoassay was a useful diagnostic aid, because the then known analytical methods were too time consuming to be of practical use in clinical practice.

Briefly, immunoassays are based upon the principle that antibodies towards herbicides can be prepared in animals. These antibodies can recognize the chemical configuration of the herbicide and attach with exquisite specificity to the herbicide molecule.

The first published general immunoassay was a radioimmunoassay for the determination of insulin.[3] Other immunoassays using an enzyme rather than radioactivity as the label were also developed quite early.[4,5] All these immunoassays, however, came from the medical research and were designed for diagnostic purposes in endocrinology and toxicology.

A radioimmunoassay to screen for the herbicides 2,4-D and 2,4,5-T in surface water was developed for environment monitoring in the early 1980s.[6] At this time, one of the first enzyme-linked immunosorbent assays for the determination of paraquat had already been published by Niewola et al.[7]

The development of immunoassays for determination of herbicides depends on the access to specific antibodies against the herbicidal substances.

Usually substances with molecular weight exceeding 5000 Da act as antigens, but for an immunogenic activity their individual structure is also important. In general, herbicides are compounds of a low molecular weight and they are therefore not antigenic per se, but so-called haptens. Haptens are substances which can be recognized and bound by specific antibodies, but they do not evoke an immunological response in animals.

Herbicides as haptens can only evoke an immunological response if they are bound to high molecular weight molecules, e.g., proteins. For this reason they must be coupled covalently to the macromolecular carrier prior to immunization. Antibody specificity is determined by the design of the hapten-carrier conjugate in a predictable manner. Making the optimal choice of coupling chemistry is therefore of paramount importance in the development of immunoassays against low molecular weight substances. Binding of the analyte, the herbicide, to the antibody does not result in any chemical reaction which can be monitored and used as a quantitative signal in the assay. To overcome this difficulty a tracer of the herbicide is included in the assay and the competition of the herbicide molecules in the sample with the tracer for a limited number of binding sites on the antibodies is the basis of quantification of the analyte. For the development of enzyme immunoassays (EIA or ELISA: enzyme linked immunosorbent assay) the selected coupling method should be so mild that the enzyme to which the herbicide molecules are coupled retains most of its activity following conjugation. If this is not possible, either a radioimmunoassay (RIA) should be considered where, as the name implies, the tracer is radiolabeled, or an indirect enzyme immunoassay where the antibody rather than the

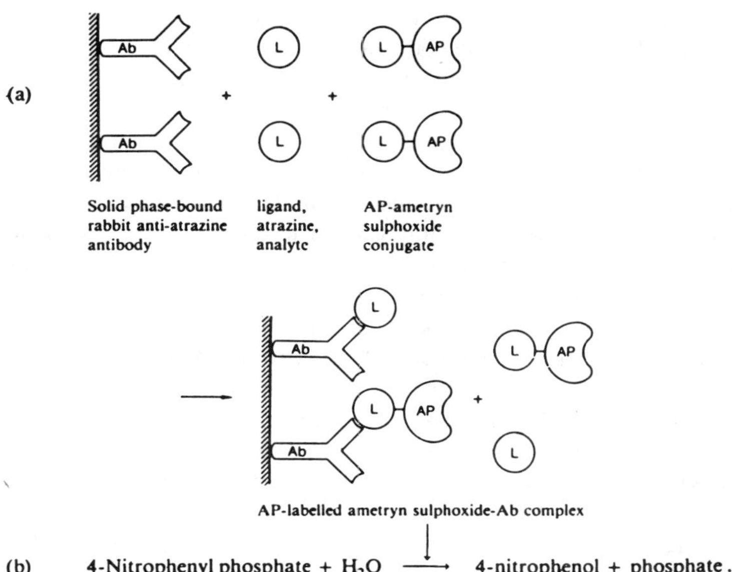

FIGURE 1. Principle of a solid-phase enzyme immunoassay (according to Huber and Hock[17]).

herbicide is enzyme labeled (either the primary antiherbicide antibody or a secondary antibody, see Section III.A.5).

Irrespective of the type of tracer used in the assay, a method is required for the separation of tracer bound to the antibodies and free tracer molecules that are displaced from the antibodies by either sample molecules or known standards of "cold" herbicide. Due to the difference in size between the antibodies and the herbicide molecules, physical methods of separation can be used, e.g., equilibrium dialysis or ammoniumsulfate precipitation. In solid phase assays this step is bypassed by taking advantage of the ability of polystyrene (and a few other synthetic polymers) to bind proteins unspecifically. The adsorption to the solid phase depends on time, buffer system (pH, ionic strength), temperature, and concentration of the coating substance.[8] Antibodies are incubated with the solid phase, e.g., a polystyrene surface that is simply washed free of unbound antibodies and a mixture is added of a known amount of labeled herbicide and either known standards for the calibration curve or the unknowns were measured (see Figure 1). The standards or samples are then allowed to compete with the tracer for the binding sites at the surface. Following a washing step the quantity of bound tracer can be measured by scintillation counting, if the labeled herbicide is radioactive, or by adding the substrate of the enzyme used for the herbicide tracer. Due to the competitive nature of the immunoassays the measured radioactivity or enzyme activity is inversely related to the amount of herbicide in the samples and standards. Methods of data analysis will be dealt with in Section III.A.3. For the solid phase small beads, which may be magnetic in order to facilitate washing, can be used, but by far the most widely used system for assays with an enzymic label is the 96-well microtiter plate which fits all ELISA-readers.[7,9,10] The sensitivity of immunoassays depend on a number of parameters: the coupling chemistry used for the preparation of the herbicide antigen, the antiserum, or the selection criteria for monoclonal antibodies, to name a few. Detection limits as low as 0.01 pmol in 100 μl are not uncommon. A reasonable quantitative precision is usually attained at around 10 times this value, i.e., 0.1 pmol/100 μl, in simple enzyme immunoassays. If this sensitivity is insufficient reagents for ELISA signal amplification are commercially available, e.g., the streptavidin-biotinylated peroxidase complexes.

Specific antiherbicide antibodies may also be used for the direct demonstration of herbicide conjugated in the plant tissue. Immunocytological localization of low molecular weight substances is, however, hampered by the diffusible nature of the antigen, i.e., the herbicide, to be localized. A fixative must be diffused into the tissue prior to sectioning for microscopy. The free herbicide molecules are normally as diffusible as the fixative; consequently they will move inside the tissue until fixation is effective. If the fixative immobilizes the herbicide by a method which does not prevent the recognition by the antibodies, the fixed herbicide molecules can be visualized with the antibodies. The results can be interpreted with more certainty when the plant tissue itself immobilizes the herbicide. Atrazine, for example, is bound to glutathione-*S*-transferase in the plant tissue. Experiments by Huber and Sautter[11] clearly demonstrate that conjugated atrazine can be localized in atrazine-treated leaf disc pieces of corn using highly specific antibodies for recognition of atrazine. Conjugated atrazine recognized by antiatrazine antibodies can then be visualized by fluorescein labeled secondary antibodies (secondary antibodies are raised against antibodies from the species used for the primary antiherbicide antibodies) as described in Section III.C.

Immunoassays and immunofluorescence localization experiments are based upon antibodies that can recognize and strongly bind substances that are indentical to or structurally related to the original antigens. Herbicide specific antibodies can be developed and used in combination with an appropriate indicator system to identify specific herbicide at quantitative levels approaching 1 ppb or less; or conjugated herbicide molecules in cryosections of plant organs.

Immunological techniques are thus able to extend our knowledge of the fate of herbicides in organisms and in the environment.

II. PRODUCTION OF HERBICIDE BINDING ANTISERA

A. PREPARATION OF IMMUNOGEN

Because of low molecular weight, herbicides are not immunogenic themselves. Consequently, the preparation of antiherbicide antibodies requires a suitable immunogen, consisting of the herbicide as hapten and an immunogenic carrier, e.g., protein. During the reaction steps necessary for a covalent binding reaction between protein and herbicide suitable coupling groups of herbicide and/or protein must be activated or otherwise changed. It is important that, except for the coupling group, side chains or other characteristic structures of the herbicide are unaffected. A few general rules apply to the choice of coupling method:

1. The introduced bond should be chemically stable and should not be cleaved by enzymes in the animal (ester linkages for example).
2. The antibody specificity to the position of the hapten chosen for coupling is lost due to the linkage between hapten and carrier. A certain shielding of adjacent positions can be observed and may result in reduced specificity at those positions.
3. Hydrophilic substituents on the herbicide are more important for antibody recognition than hydrophobic ones.

The circulation of free hapten molecules in the blood stream of the immunized animal inhibits production of antibodies against the hapten. Hapten-carrier conjugates are therefore prepared with stable bond between carrier and herbicide and the conjugates are always dialyzed extensively prior to immunization. The importance of rule number 2 is inversely proportional to the size of the herbicide: for very small molecules every feature is close enough to the coupling position to be affected by the bond to the carrier.

The reagent used for coupling of the ligand to the carrier may either do so directly or it may introduce a spacer molecule between carrier and ligand. The use of a spacer molecule may affect antibody specificity towards the free herbicide. The specificity of antibodies against a bulky hapten attached directly to the carrier molecule may be influenced by the steric hindrance encountered by the antibody producing cells when binding to the hapten-carrier conjugate. Judicious use af spacer molecules, usually from 5 to 10 carbon atoms long, can therefore lead to the production of antibodies with improved recognition of the free hapten.

Rule number 2 can also be used advantageously. If two herbicides differ in only one substituent, an immunoassay can be developed with equal sensitivity to both substances, provided that the covalent bond to the carrier is established via this substituent. This allows the development of herbicide class specific antibodies. An example is described in Section III.A.5.b.

Rule number 3 is often neglected. The hydrophilic substituents are generally more readily accessible to chemical modification and coupling and, therefore, chosen even though antibodies of suboptimal quality are raised in the animal. Especially for very small molecules it is better, although more involved, to prepare a derivative of the herbicide with an extra substituent to which coupling can occur. The derivative should have the same structural determinants except for the group or side chain introduced for the coupling reaction. An example of this, including experimental evidence for the importance of rule 2 and 3, has been provided by Marcussen et al. for the plant hormone indole-3-acetic acid (MW 175).[12]

For most coupling reactions, lysine-rich carrier proteins are preferred. Haptens can be coupled to the ε-amino groups of the lysine residues. Coupling reagents formining stable amide linkages are often used. The most frequently used immunogenic carrier proteins are bovine serum albumin (BSA), human serum albumin (HSA), rabbit serum albumin (RSA), ovalbumin, thyroglobulin, key hole limpet hemocyanin (KLH), or fibrinogen. The commonly used BSA provides 59 reactive ε-amino groups of lysine on its surface, but only one ε-amino and one sulfhydryl group.[13]

The stability of the immunogenic herbicide protein conjugate in the animal organism, the accessibility of the characteristic hapten determinants for the immunological system, as well as the binding ratio (moles of hapten per moles of carrier protein) play a major role for a successful immunization.[14] An optimal binding ratio (epitop density) should be between 8 and 25 haptens per carrier protein to obtain high antibody titers.[13]

Coupling methods for the preparation of immunogenic conjugates with carrier proteins will not receive extensive coverage. Additional examples can be found in a review by van Regenmortel et al.[15] Commercial suppliers of coupling reagents should also be consulted.[16]

1. Immunogenic *s*-Triazine-Protein Conjugates

In case of atrazine the analogous *s*-triazine ametryn, in which the chlorine is substituted by a methylthio group, can be used as a model compound for preparation of an immunogenic conjugate producing antiatrazine antibodies.[17] Ametryn is converted to the corresponding sulfoxide derivative by oxidizing the methylthio group with *m*-chloroperbenzoic acid in chloroform.[18] This does not affect the ethylamino and the propylamino group. The sulfoxide derivative is separated and purified by preparative thin layer chromatography, developed in a chloroform/acetone mixture.[17] The eluted sulfoxide derivative, evaporated to dryness, is activated for reaction with a suitable immunogenic carrier protein, e.g., BSA or keyhole limpet haemocyanin. The binding reaction of sulfoxide-groups with freely accessible NH_2^- and SH-groups of a carrier protein is based upon a nucleophilic substitution.[17,18] The ametryn sulfoxide-haemocyanin conjugate (Figure 2) can be stored in the dry state at 4°C for subsequent use. The methylthio *s*-triazines themselves, e.g., terbutryn are coupled directly to the carrier protein.[19] In this case a binding ratio is approximately 20 terbutryn molecules per molecule BSA. However, these antibodies cannot effectively distinguish between a chlorinated *s*-triazine herbicide and its corresponding methylthio derivative.

To prepare antibodies that recognize all side groups of atrazine, three different immunogenic conjugates are necessary.[20] The first is the ametryn sulfoxide-protein conjugate described above. The second is a simazin-protein conjugate prepared by coupling through the two

carrier protein + [sulfoxide derivative of ametryn] →

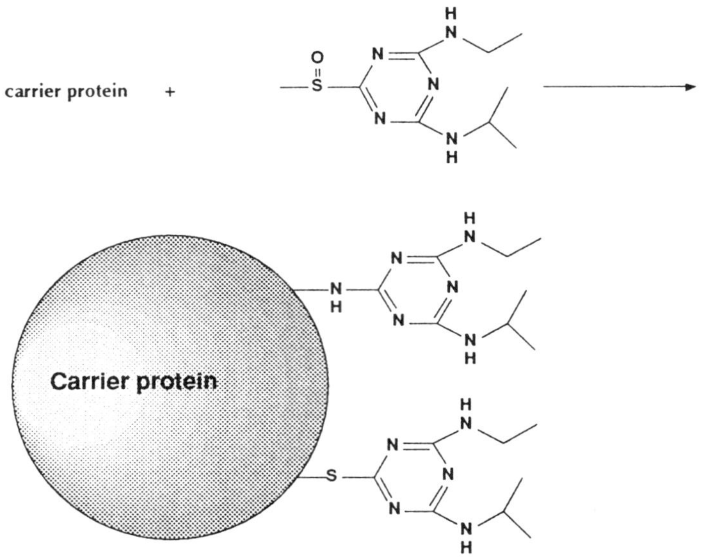

Carrier protein

FIGURE 2. Immunogenic ametryn-hemocyanin conjugate using the sulfoxide derivative of ametryn as the starting material. (From Huber, S. J. and Hock, B., Atrazine in water, in *Methods of Enzymatic Analysis, Vol. XII, Drugs and Pesticides,* 3rd ed., Bergmeyer, H. U., Bergmeyer, J., and Grabl, M., Eds., VCH Verlagsgesellschaft, Weinheim, 1986, 438. With permission.)

ethylamino groups of simazine. Finally, the third conjugate consists of propazine coupled by one of the two isopropylamino groups to the carrier protein using the carbodiimide method.[21] The three conjugates show the characteristic side chains of atrazine, which are exposed to the immunological system.

Another suitable derivative for synthesis of an immunogenic atrazine-protein conjugate, containing 6-amino caproic acid as spacer molecule, has been proposed.[22] The hapten, selected for conjugation to protein, is 2-chloro-4(isopropylamino)-6-(amino caproic acid)-*s*-triazine. The chlorine at the 2 position and one of the amino groups at either the 4 or 6 position are exposed to the immunological system. While the chlorine should remain exposed at the 2 position, the hapten may conjugate to a lysine-rich protein at either the amino ethyl group in the 4 or the amino isopropyl group in the 6 position.[22]

2. Immunogenic Chlorophenoxyacetic Acid-Protein Conjugates

The acetic acid moiety of chlorophenoxyacetic acid herbicides, e.g., 2,4-D and 2,4,5-T, is an accessible reactive coupling group for immunogen synthesis. Direct coupling of the acetic acid moiety of 2,4-D to the NH_2^- residue of a protein is possible, using the mixed anhydride method.[23,24] In this case, a binding ratio of 10 to 15 hapten molecules (2,4-D) per mol carrier protein could be obtained.[24] A better method leaving the hydrophilic and immunologically easily recognizable acetic acid group free, is based on a derivative of 2,4-D, namely 2,4-dichloro-5-amino-phenoxyacetic acid.[6] Aromatic amino groups can be activated with sodium nitrite at low temperature. The resulting diazonium salt reacts with BSA in an alkaline solution (the reaction mechanism is analogous to the one used with chlorsulfuron in Figure 3). The binding ratio of the dialyzed conjugate is in the range of 5 mol of ring-linked 2,4-D per mol BSA.[6]

3. Immunogenic Bipyridilium Herbicide-Protein Conjugates

For antibody preparation against the bipyridilium herbicide paraquat, a model compound, 6-(1,1′methyl-4,4′-bipyridilium) hexanoic acid, coupled to BSA using carbodiimides can

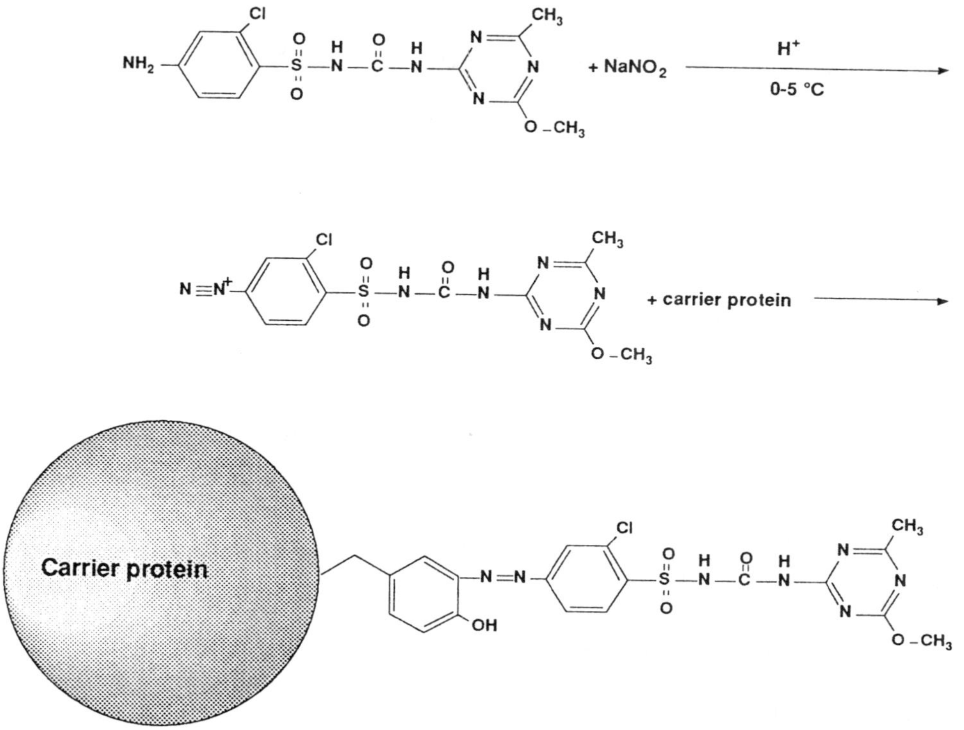

FIGURE 3. Coupling of chlorsulfuron to a protein carrier using diazotation of an aromatic amino group.

serve as immunogen.[2,7,25] By this method, the entire paraquat molecule with an additional spacer substance (hexanoic acid) is bound to the protein carrier and then exposed to the immunological system of the animal.

In another coupling reaction, a paraquat acid derivative is used for reacting with BSA.[26] The conjugation ratio of this highly immunogenic conjugate is about 20 to 25 mol of hapten per mol BSA.

4. Immunogenic Conjugates of Other Herbicide Classes

To prepare an immunogenic conjugate for trifluralin, a member of the dinitroaniline family is used as model compound.[27] This model compound can be conjugated with a lysine rich protein (BSA) using the mixed anhydride method.[23] Approximately 20 hapten units were bound to each protein molecule.

The sulfonylurea herbicide, chlorsulfuron, can be coupled by diazotation and coupling as described for 2,4-D above using a derivative of chlorosulfuron carrying an aromatic amino group for diazotation and coupling to aromatic amino acid residues of keyhole limpet hemocyanin (KLH) or BSA.[28]

5. The Use of Activated Esters

Amide bond formation between hapten and carrier are often used for establishing a stable bond between hapten and carrier. The formation of intramolecular amide bonds between lysine residues of one carrier molecule and glutamic or aspartic acid residues on another carrier molecule cannot be avoided when carbodiimide reagents are used. Depending on the conditions, this cross-linking of carrier molecules may be the favored reaction rather than the desired coupling of the hapten to the carrier. This problem is solved by the use of activated esters, which provide the experimenter with full control over the degree of carrier substitution

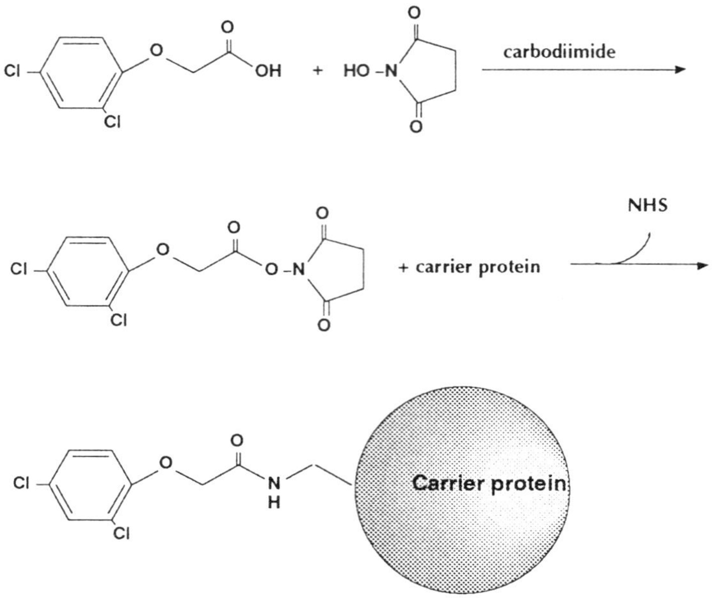

FIGURE 4. Coupling of 2,4-D to a protein carrier using an activated ester of 2,4-D.

with hapten molecules. Activated esters of herbicides are often so stable that they can be isolated, crystallized, characterized by spectroscopic methods, and stored as ready to use reagents.

The most commonly used activated ester is the succinimide ester which has been used for the preparation of a carboxyl linked 2,4-D conjugate.[29] Succinimide esters are prepared by using carbodiimide and *N*-hydroxy succinimide (NHS). Figure 4 shows preparation of the ester and the subsequent coupling procedure.

The activated esters can also be used for preparation of an active carrier, a universal reagent for coupling haptens carrying a primary amino group. The primary amino groups already present in the carrier have to be protected before carrier activation. Thereafter the carboxyl groups of glutamic and aspartic acid residues are activated with *N*-hydroxy-sulfosuccinimide, a water soluble derivative of NHS. The technique has been developed for the preparation of a cyclic AMP ELISA.[30] The method is generally applicable to haptens with primary amino groups; it should be recommended especially for peptides and other haptens carrying more than one amino group or amino groups as well as carboxyl groups where polymerization during coupling may be difficult to prevent.[31] Carrier activation can also be accomplished using *N*-succimidyl bromoacetate.[32] Bromoacetylated carrier protein will couple to ligands containing an SH-group (Figure 5).

6. Bifunctional Reagents

Compounds such as *N*-succinimidyl bromoacetate are called bifunctional reagents. These reagents are conveniently used as a method of introducing a spacer molecule between hapten and carrier. The bifunctional reagent may be viewed as a spacer with an activated ester, or other activated group, at either end. Usually the two esters are different and react under different conditions, so that coupling to carrier and hapten can be carried out sequentially with minimal loss of functional groups due to polymerization.

A very versatile heterobifunctional reagent is SPDP (*N*-succinimidyl 3-(2-propyldithio) propionate). SPDP couples selectively to a primary amino group through a succinimide ester in one end and subsequently to a sulfhydryl group with the 2 pyridyl dithio structure in the

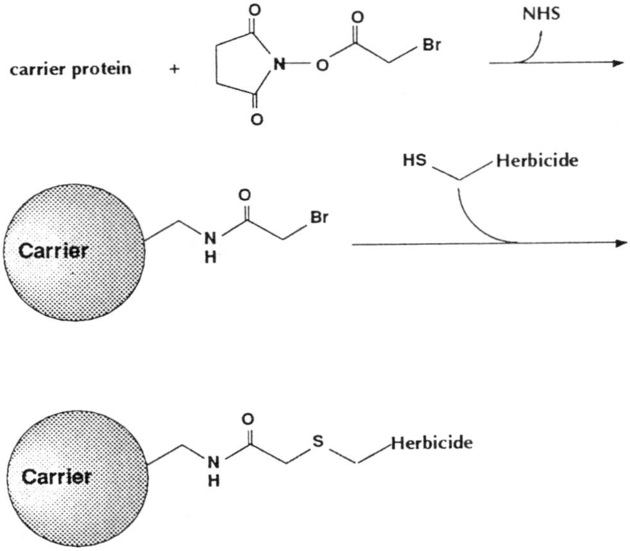

FIGURE 5. Activation of the carrier protein with *N*-succinimidyl bromoacetate and subsequent coupling with a haptenic sulfhydryl group.

other end, see Figure 6. If neither the carrier nor the herbicide contain an accessible SH-group SPDP can be used for the thiolation of either ligand or carrier with the introduction of a spacer molecule at the same time. SPDP can thus be used alone or in combination with *N*-succimidyl bromoacetate mentioned above.

B. IMMUNIZATION PROCEDURE

In the immunized animal each cell strain responds to one of the epitopes of the injected antigen. Antisera are consequently termed polyclonal. When monoclonal antibodies are prepared a single cell strain producing only one antibody is isolated. Monoclonal antibodies thus recognize a single determinant of the antigen. In 1975 Köhler and Milstein succeeded for the first time to produce monoclonal antibodies by fusion of antibody-producing mouse spleen cells with mouse myeloma cells.[33] The resulting hybridoma cells are able to divide and continually produce antibodies. Specific antibody-producing cell clones can be selected for subsequent cultivation.

1. Immunization for Preparation of Polyclonal Antisera

Rabbits, guinea pigs, rats, mice, and chickens can be immunized for preparation of polyclonal antisera.[34] In most cases, rabbits (for instance New Zealand, Belgischer Riese, Hvid Dansk Landrace) are preferred, because they are relatively tranquil, and easy to handle and to keep. Furthermore, they are easy to bleed and their relatively high body weight, compared to rats or mice, yields a blood volume of 30 to 80 ml.[2,6,7,24,26-28,35]

Many different immunization schemes are practiced. The effects of varying the immunization protocol are usually very limited. Traditionally the antigen, a few hundred micrograms, is emulsified in Freund's complete adjuvant. Freund's adjuvant is a vaccine consisting of inactivated tubercle bacilli and paraffin oil.[36] The bacillus cell walls have a strong stimulatory effect on the immune system. Repeated immunization with complete adjuvant may result in anaphylactic shock. Consequently Freund's incomplete adjuvant (the same oil without the killed bacilli) are usually employed for subsequent immunizations. In order to enhance the slow release effect of the antigen-adjuvant emulsion, immunization in peripheral skin layers,

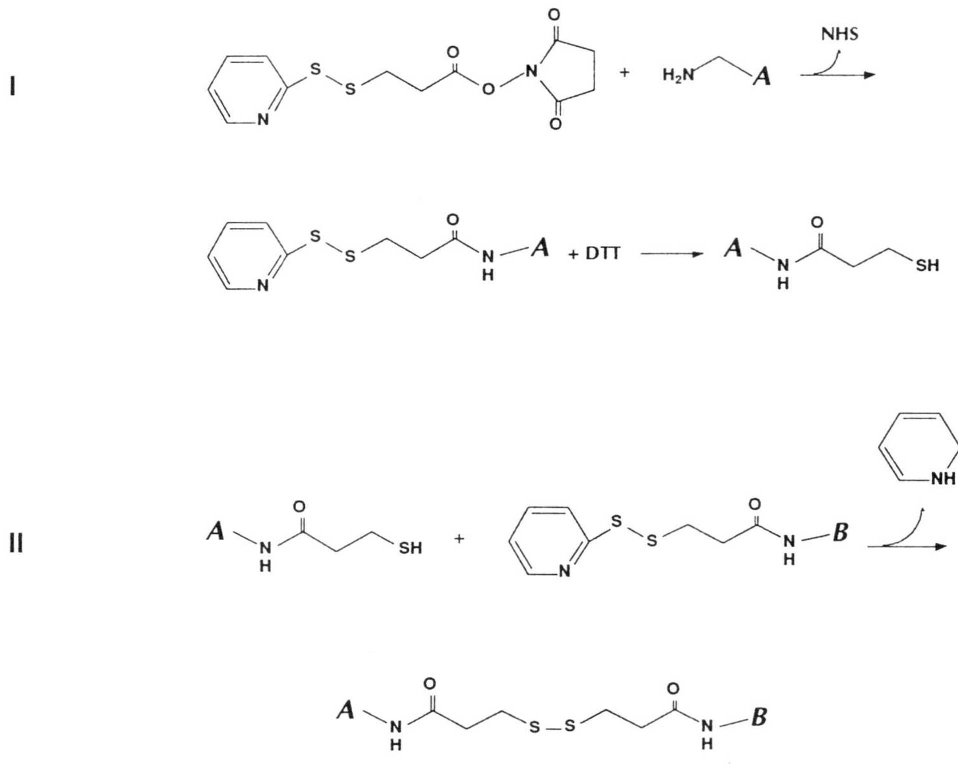

I

II

FIGURE 6. The use of the heterobifunctional reagent SPDP for thiolation (I) and for coupling of amino- and sulfhydryl groups (II). A and B may designate carrier and herbicide or vice versa.

e.g., in the food pads, was used previously. Intradermal immunization is, however, rather painful for the animal, especially if complete adjuvant is used due to the inflammatory reaction incurred by the bacilli. Other adjuvants than Freund's should be considered. The RIBI adjuvant* is a commercial adjuvant claimed to be gentler to the animal while having as strong adjuvant effect as the Freund's adjuvant.

Immunizations should be done either subcutaneously, i.e., in the fat just below the skin, or i.m. in the thigh of the rabbit. No more than 0.5 ml of emulsion should be deposited per site of injection. Injection at multiple sites have often been found to enhance the immune response. The interval between immunizations should be no shorter than 2 weeks, and as soon as the antiserum titer has stabilized it is sufficient to immunize 10 to 14 days prior to bleeding of the animal. Blood can be collected once a month. As a rabbit yields 30 ml of blood per bleeding, it is rarely necessary to bleed the rabbit more often than once every 3 months. The antibody titer and time to develop a stable immune response depends very much upon the antigen. If the hapten is only weakly antigenic, or if either the hapten or the linkage between hapten and carrier is unstable, titer is usually low and slow to develop (half a year). The plant growth regulator indole-3-acetic acid has provided experimenters with extensive experience regarding these problems.[37]

2. Immunization for Preparation of Monoclonal Antibodies

Mice are too small to allow subcutaneous or i.m. immunization to be carried out easily. Immunization is therefore done in the peritoneal cavity using adjuvant as described for rabbits,

* Immunochem Research, Inc., Montana.

but with no more than 100 μl of emulsion containing 15 to 100 μg of antigen. Three injections at least 10 days apart followed by a month rest is often used. Three days after the fourth immunization the mouse is sacrificed and the spleen is excised for cell isolation and fusion. The depot or slow release effect brought about by the adjuvant is of no relevance for this final immunization. Many experimenters prefer i.v. immunization with the antigen in PBS for the final boost. Injecting through one of the four veins accessible in the tail requires the thinnest possible canula and considerable dexterity.

In most cases, monoclonal antibodies for analytical purposes are developed in mice. There are many laboratory mouse strains that are defined by different physiological criteria.[38] For the first fusion and production of monoclonal antibodies, Köhler and Milstein used laboratory mouse strain BALB/c.[33] This mouse strain is frequently used for analytical purposes, e.g., for production of monoclonal antibodies against the herbicide paraquat.[39]

C. ANTIBODY RAISING, SEPARATION, AND CLEAN-UP PROCEDURES

Following immunization the animal responds first with a transient production of antibodies belonging to the antibody class IgM. These antibodies constitute the animals first approach to an immune response to the injected antigen. IgM antibodies are most often of relatively low affinity. Subsequent booster injection induces the animal to switch to antibodies of the IgG class and at the same time immune memory is activated and the production of IgG is not nearly as transient as that of IgM. During a prolonged immunization scheme a clonal selection among the IgG-producing cells occur favoring those cell lines making antibodies of high association constant. The association constant is a quantitative measure of the energy of the binding of the antigen to the antibody. It will be defined in Section III.A. Parallel to the clonal selection each cell attempts antibody improvement through a random scheme of gene replacement using the pre-existing antibody genes. In other words, no new antibody genes are invented in response to the antigen, but antibody structure is optimized using preformed bits and pieces. It may seem surprising that antibodies can be raised against highly unnatural molecules such as herbicides entirely depending upon pre-existing antibody genes. A number of valuable conclusions can be drawn from the understanding of this strategy:

- The immune response continues to mature over time (with repeated immunizations), and this maturation leads to an improvement from an analytical point of view. Counter examples do exist where improvement of the association constant over time simultaneously favors an unwanted cross-reaction.
- There is no way of standardizing polyclonal antisera.
- Animals with a larger antibody genome and a longer life span will eventually come up with better antisera than smaller, short lived animals with a smaller amount of genetic material allocated to antibody production.

These three points should be taken into account when deciding whether mono- or polyclonal antibodies should be raised for a given analytical purpose. Selecting a mouse derived clone producing an antibody with as high an association constant as the best clones in a rabbit antiserum is not impossible, but following the argument above, it is statistically very unlikely. If only the very best antibodies will suffice for a given analytical problem, monoclonal antibodies should not be the chosen strategy.

In addition to the desired high affinity hapten specific antibodies, polyclonal antisera contain antibodies against the carrier and also antibodies against the ligand, but with a much lower association constant than the best antibodies in the serum. The high molecular carrier can most often be chosen so that the occurrence of carrier specific antibodies becomes immaterial. Ligand specific antibodies of lower binding strength, however, cannot be easily eliminated. It should be remembered, though, that immunoassays are carried out under conditions where antibody concentration is limiting. At very low concentrations of antibody it follows from the law of mass action (the governing equation behind the association constant) that antibodies of low binding strength will not bind significant amounts of analyte or tracer and their lack of specificity is therefore irrelevant. Only the best antibodies and those which have slightly lower association constants play their part. A slightly less steep calibration curve compared to a monoclonal antibody based assay is the result.

With naturally occurring antigens monoclonal antibodies are preferred when the antigen cannot be purified to homogeneity or where the repeated isolation of antigen for a prolonged immunization scheme is too labor intensive to be practical. As the antigens used for raising herbicide specific antibodies are synthetic, these considerations normally do not apply. Monoclonal antibodies should be chosen where standardization and access to an inexhaustible source of antibodies is important. In the production of monoclonal antibodies, clonal selection is taken over by the experimenter and immune response maturation is frozen at the time of spleen excision. The cross-reactivity spectrum of the antibody can be characterized once and for all, and all assay parameters can be standardized. Furthermore, antibodies with only one association constant is particularly useful for trace enrichment and sample purification by immunoaffinity chromatography, see Section III.B.

Polyclonal antisera, produced from different antibody producing cell strains, can be purified by affinity chromatography on immobilized herbicide and thus be freed of different serum components, as well as of antibodies with no specificity towards the herbicide. Where limitied amounts of well defined antibodies are required, affinity purified antibodies provide an alternative to monoclonal antibodies, see Section III.3.b. A comparison between production and use of polyclonal and monoclonal antibodies is shown in Table 1.

1. Polyclonal Antibodies

For the preparation of the antitriazine antisera, the collected blood is allowed to coagulate.[17] The serum is separated from the coagulated part and can be used as it is.[6,24,28] If the antiserum is aliquotted and stored at –20°C it has a considerable shelf life (years) but does not tolerate repeated freeze-thaw cycles. Alternatively the immunoglobulin G (IgG) fraction of the serum can be precipitated according the method of Hurn and Chantler.[14] Following centrifugation the precipitated IgG fraction is redissolved, dialyzed, and lyophilized. The IgG fraction is stored in a dry state at 4°C. The IgG-preparations are reconditioned in the appropriate buffer prior to use.[17]

Affinity-purified antitriazine antibodies, e.g., anti-atrazine antibodies, are obtained by affinity chromatography with the hapten as ligand. For this purpose, an ametryn-conjugated amino hexylsepharose 4B-gel (Pharmacia) in a short column is used.[17] The advantage of an affinity purified antiserum is to eliminate all nonhapten specific antibodies and to enrich hapten specific antibodies with the highest affinity to the specific antigen, if these can be eluted from the column without loss of functionality.[17]

2. Monoclonal Antibodies

The preparation of monoclonal antibodies is different from that of polyclonal antibodies. In a first step, polyclonal antibodies are developed against the hapten or antigen in a suitable laboratory animal species, e.g., mice (polyclonal phase), using immunization procedures described in Section II.B.2.

Antibody producing cells isolated from an immunized animal can only survive in a culture medium for a short period of time. For this reason, they are fused with mouse myeloma cells (plasmacytoma cells). Myeloma cells are immunoglobulin producing cells of malign tumors of the immunological system, which can be cultivated indefinitely in culture media. The fusion takes place in the presence of polyethylene glycol. Sorting out the fusion products from the other cell types is accomplished using aminopterin, an inhibitor of nucleic acid biosynthesis. Mammalian cells possess a rescue pathway circumventing the aminopterin block. The alter-

TABLE 1
Comparison of Polyclonal and Monoclonal Antibody Techniques

Parameters	Polyclonal	Monoclonal
Time to develop reagent	Minimum 3 months	Minimum 6 months
Expertise required	Injection and bleeding	Injection and bleeding Cell culture skills
Special equipment	None	Cell culture supplies Laminar flow hood CO_2 incubator
Reagent characteristics		
Chemical structure	Heterogeneous	Homogeneous
Supply	Limited by lifespan of donor animal	Potentially unlimited
Affinity	Heterogeneous population of antibodies that vary in affinity for antigen Antisera has overall average affinity	Can select for clones producing high or low affinity antibody to suit intended use of reagent
Specificity	Usually high, but heterogeneous population may include antibodies that cross react with structurally related antigens	Can select for clones producing high or low affinity to suit intended use of reagent

From Bernatowicz, M. S. and Matsueda, G. R., Preparation of peptide immunogens using *N*-succinimidyl bromoacetate as a heterobifunctional crosslinking reagent, *Anal. Biochem.*, 155, 95, 1986. With permission.

native biosynthetic pathway is dependent upon hypoxanthine and thymine (alternative precursers of the DNA bases) and the enzyme hypoxanthine guanine phosphoribosyl transferase (HGPRT). If a mutant myeloma cell strain, which lacks the enzyme HGPRT, is used for the fusion, the myeloma cells can be eliminated with aminopterin while the spleen cells and the fusion products are rescued by hypoxanthine and thymine. The unfused spleen cells live but they have a finite life span and will eventually perish.The spleen cells contain the enzyme HGPRT and the binucleate fusion products, called hybridoma cells, therefore combine the vitality and infinite life span of the myeloma cells with the genes for the immunoglobulin biosynthesis and HGPRT of the spleen cells. Consequently, only the hybrids survive and can be further cultivated. Finally the hybrids are tested for production of antibody of desired specificity and cloned by limiting dilution.[33,39-41]

Antibody producing cells can be reintroduced in an animal of the same isogenic strain as the one from which the myeloma cell line originated (or, in some cases, in related strains). Usually the animals are primed with a tumor inducing substance. These include Freund's adjuvant, Pristane, and other oily substances.[41] With mice $\sim 10^8$ cells are injected into the peritoneum where they form a soft tumor. The tumor excretes relatively high concentrations of the monoclonal antibody in the so-called ascites fluid while growing. The growing tumor eventually doubles the size of the mouse and causes much suffering of the animal. This method of antibody production is therefore largely abandoned. If, however, a valuable cell line is contaminated, for example, by a mycoplasma infection, the technique of reintroduction of cells into the mouse is very useful. The immune system of the mouse will attempt to eliminate the infection and the purified cell line can be recovered from the ascites fluid and reconditioned for *in vitro* cultivation.

Large scale production of antibodies from cell cultures is possible under laboratory conditions,[41] for example, using the "connected vessel system" known as Cell Factories™.[42] Recovering antibodies from large volumes of the rather dilute cell culture supernatants can be accomplished by thiophilic interaction chromatography.[43]

III. IMMUNOASSAY SYSTEMS

A. IMMUNOASSAYS FOR QUANTITATIVE DETERMINATION OF HERBICIDES

Competitive immunoassays are the most commonly employed immunoassay systems for quantitative determination of antigens like peptides,[4,44] proteins,[3] and hapten herbicides.[1,2,6,7,11,17,19,24-26,28,35] In competitive immunoassays, free antigen or hapten molecules and labeled antigen or hapten molecules compete for a defined number of specific binding sites of antibody molecules.[45] The most common competitive immunoassay techniques are the radioimmunoassay (RIA), the direct enzyme immunoassay that is analogous to the radioimmunoassay and the indirect enzyme immunoassay where a secondary antibody carries the label rather than the hapten.

The basic principle of competitive immunoassays is

$$Ab + Ag + AgL \Leftrightarrow AbAg + AbAgL$$

where *Ab* denotes free antibody molecules, *Ag* denotes free antigen or hapten, and *AgL* denotes free labeled antigen or hapten. The free *Ag* and the labeled antigen *AgL* or hapten molecules, compete for a fixed and limited number of specific binding sites on the antibody molecules *Ab*. After an incubation period a separation step follows in which the free and antibody bound antigen molecules are separated from each other and the amount of labeled antigen or hapten is determined. At low concentrations of free antigen or hapten, more labeled antigen or hapten molecules will be bound by antibodies. From standard curves with known concentrations of antigen or hapten, unknown samples can be determined. The standard curve also provides valuable information about the antiserum or antibody in case of monoclonal antibodies. The antigen-antibody complex (*AbAg* and *AbAgL*) is in dynamic equilibrium with free antibody and free antigen. The equilibrium may be described by the dissociation constant, but as binding rather than dissociation is of prime interest with antibodies, the inverse of the dissociation constant, the association constant, is often used. Experimentally the association constant is determined by transforming a standard curve using a Scatchard plot.[46] Association constants of polyclonal antihapten antisera are normally found to be in the range from 10^7 to $10^{11}\,M^{-1}$. The association constant of a polyclonal antiserum is actually an average of many association constants spanning about two orders of magnitude. This results in a slight nonlinearity of the Scatchard plot. For corresponding monoclonal antibodies the Scatchard plots are strictly linear, since each monoclonal antibody has only one association constant. The values for monoclonal antibodies may vary as widely as the clones contributing to a polyclonal antiserum, but comparing polyclonal antisera from rabbits with corresponding hapten specific monoclonal mouse antibodies, it is often observed that the rabbit antisera have 2- to 10-fold higher association constant than the monoclonal antibodies (see also Section II.C).

1. Radioimmunoassay

a. Principle of Radioimmunoassay

For the use in radioimmunoassay labeled herbicide must be available. ^{14}C may be used as an internal standard to estimate losses during sample preparation, but the specific activity of ^{14}C is insufficient for the use as the label in a radioimmunoassay. Tritium has adequate specific activity and it is often possible to label the unmodified herbicide with tritium. Stability, however, may be a serious problem with tritium.[36] Very high specific activities may be obtained by labeling with sulfur or iodine. Due to the shorter half-life of these isotopes fresh tracer has to be prepared repeatedly. The tracer need not be identical to the herbicide. Derivatizations which modifies the herbicide significantly can be introduced with little loss of binding to the antibody if the substitution is introduced at the coupling position used for the antigen preparation, see also Section II.A. The radioimmunoassay for 2,4-D with ^{125}I-tyramine-2,4-D amide as the tracer described below exemplifies the use of a tracer which differs significantly from the herbicide, but only in the coupling position.

Free antigen or hapten molecules and radiolabeled tracer molecules, which are incubated with diluted antiserum, compete for the specific binding sites of the antibody molecules. After incubation, a separation step is introduced which selectively precipitates either the free, labeled and unlabeled hapten or the antibody-antigen complexes, i.e., the bound tracer. Precipitation can be carried out by charcoal, which unspecifically adsorbs low molecular haptens.[47] For precipitation of antibodies, *Staphylococcus* cells carrying protein A can be used.[6] Protein A in the bacterial cell wall specifically and strongly binds the Fc portion of immunoglobulins from most mammalian species. The Fc part of the immunoglobulin is the constant region not involved in antigen recognition. Thus, binding of the antibodies to protein A does not interfere with the binding of the herbicide.

Adding a precipitating agent can be avoided by attaching the antibodies to a solid phase. Two methods have already been mentioned in the introduction. Covalent linkage of the antibodies to microcrystalline cellulose or to Sepharose 4B activated with cyanogen bromide provide a solid phase reagent that can be precipitated by low G centrifugation.[26]

Either the bound or the free phase is then counted for radioactivity and the standard curve can be plotted in a number of different ways. For direct visualization of the data the bound tracer relative to total added tracer (B/T) is plotted against log dose. Alternatively B/B_0, bound tracer relative to the amount of tracer bound in the absence of competing cold herbicide or B/F, bound tracer/free tracer ratio may be used. All three methods yield sigmoid calibration curves. The steeper the standard curve the greater the signal difference for a given difference in sample dose. Therefore, precision in the estimation of unknowns is greater where the slope of the standard curve is steep, i.e., in the middle of the measuring range, and poorer in either end.

Plotting B/F against the molar concentration of bound tracer yields a straight line with the negative association constant as the slope and the concentration of functional antibody binding sites as intercept with the x-axis.This is known as the Scatchard plot[46] and is suitable for the characterization of antisera and monoclonal antibodies but impractical for calibration curves for the estimation of unknowns. Standard curves are linearized by using the logit transformation (see Chapter 3). Contrary to the Scatchard plot, which has a theoretical foundation, the logit transformation is empirical.[48] The logit tranformation is explained in connection with the herbicide-ELISA (Section III.A.3.a).

Logit transformation as well as most other transformation strategies will in theory linearize the standard curve over an infinite range. In either end experimental error takes over and limits reproducible linearization to a certain range of concentrations typical of a given immunoassay. Consequently, the measuring range of the immunoassay is defined as a range over which the calibration curve can be linearized. The detection limit is operationally defined as the lowest concentration giving a signal which exceeds the background noise by a factor of 2.5.[26] Linearization of the standard curves with logit transformation tends to conceal the poor precision at either end of the measuring range where the slope of the untransformed calibration curve is shallow (see Chapter 3).

b. Radioimmunoassays for Herbicides

Radioimmunoassay (RIA) can be used for determination of 2,4-D and 2,4,5-T in surface water.[6] River water is clarified and forced through a ^{18}C cartridge (SEP-PAK ^{18}C), which retains 2,4-D and 2,4,5-T. After eluating and drying, the residue is dissolved in a buffer for direct use in the assay. For the immunoassay a ^{125}I-labeled ligand, prepared by iodination of 2,4-dichloro-5-carboxymethoxyphenyl-4′-hydroxyphenyl-diazene with Na ^{125}I. Phase separation was accomplished with Staphylococcus cells. After centrifugation the pellet is assayed for radioactivity in a gamma counter. The measuring range obtained for 2,4-D was from 5 to 50 pmol 100 μl^{-1} in this radioimmunoassay. The trace enrichment obtained by the use of a ^{18}C cartridge lowers the detection limit to about 0.5 pmol 2,4-D 1 ml^{-1} sample water.

A different RIA method for quantitative determination of 2,4-D is published by Knopp et al.[24] Two different radiolabeled compounds are used as tracer substances. The first tracer is a radioligand prepared by the synthesis of a 2,4-D-tyramine conjugate. Formation of the amide between 2,4-D and tyramine introduces a site suitable for iodination with Na ^{125}I. The other radiotracer is monotritiated 2,4-D, selectively labeled in the 6 position of the aromatic moiety ([6-^{3}H] 2,4-D).

Rabbit serum, sample (water, plasma, or urine), tracer, and specific antisera are put into propylene reaction tubes. Following incubation the antibodies are precipitated with a buffered PEG solution and after further incubation and centrifugation the redissolved pellet is measured in a liquid scintillation counter.

Using the monotritiated tracer [6-^{3}H] 2,4-D (specific radioactivity 465 GBq $mmol^{-1}$) a highly sensitive and specific RIA can be developed within an effective dose range of 0.5 to 500 pmol 2,4-D 100 μl^{-1}. This wide range and high sensitivity cannot be obtained using the radioligand 2,4-D-tyramine, because of structural homology between the tracer molecule and the 2,4-D lysin region in the immunogen (2,4-D BSA). Consequently, the antibody binding site comprises more than the hapten, and the free 2,4-D is a relatively poor competitor for the binding sites as compared to the tyramine amide.

The binding of immunogen related substances from the sample is a problem, especially if hapten carrier conjugates are used where essential features of the herbicide have been masked by the covalent linkage to the carrier protein, as is the case in the example with 2,4-D linked through its carboxyl group described above. See also the discussion in Section II.A.

Radioimmunoassays for paraquat have been developed using both tritium and iodine as tracers.[26]

2. Enzyme Immunoassay

a. Principle of Enzyme Immunoassay

In heterogenous competitive enzyme imunoassays sample hapten or antigen compete with enzyme-labeled hapten or antigen for specific binding sites of antibody molecules.[45] A phase separation is also necessary.

In the following, the principle of an enzyme immunoassay is described for quantitative determination of atrazine.[17] The tracer substance is an ametryn-alkaline phosphatase conjugate. Highly purified enzyme preparations with very highly specific activities are preferred as marker enzymes. The enzyme alkaline phosphatase from calf intestine is commercially available and has a specific activity of 2500 units mg^{-1}.

The principle of the atrazine assay is shown in Figure 1. In a first step (coating phase), the specific antiatrazine antibodies (*Ab*) are bound to a polystyrene surface, e.g., wells of microtiter plates. Alternatively, polystyrene spheres, which are kept in vessels containing the antibody solution or the test solution, can be used.

Solid phase-bound antiatrazine antibodies (*Ab*) bind the ligand (*L*) atrazine (hapten) of the sample solution. The enzyme tracer (*L-AP* alkaline phosphatase-hapten conjugate) occupies the free antibody-binding sites. All unbound molecules are removed by washing. The specific substrate, in the present case 4-nitrophenyl phosphate, is added and the amount of bound enzyme tracer is determined. The reaction rate is determined either by measuring the increase in absorbance at 405 nm per unit time, $\Delta A/\Delta t$, or the absorbance reached after a finite time period, conveniently when $A_{405,max} \approx 1$. The maximal absorbance is obtained in the absence of competing unlabeled herbicide (B_O) and the readings are inversely related to the atrazine concentration in the sample.

The following equation gives the ratio of enzyme activities in presence of hapten, corrected for unspecific binding.[10]

$$\frac{A_B - A_{UB}}{A_{BO} - A_{UB}} = \frac{B}{B_O}$$

A_B denotes absorbance (405 nm) in the presence of a given amount of hapten (e.g., atrazine); A_{UB} denotes absorbance (405 nm) in the presence of a large excess of hapten, A_{BO} is absorbance (405 nm) in the absence of hapten molecules. This ratio multiplied with 100 gives the percentage relative binding. Plotted against dose or logarithm of dose, it will be approximately linear over some small range of doses. If the dose range is extended, the relationship becomes sigmoid with B/B_O ranging between the asymptotes 0 and 100%. Linear standard curves within a small dose range is preferred for manual evaluation of assay results.[9,38]

$$\operatorname{logit}\left(\frac{B}{B_O}\right) = \ln\left(\frac{B}{B_O}\right) \Big/ 1 - \frac{B}{B_O}$$

Logit transformed response of a symmetrically sigmoid curve gives a straight line when plotted against the log dose (see Chapter 3).[49] In many studies, however, a spline approximation method calculated on a computer is used for the fitting of standard curves and estimation of unknowns.[7,10,25,39,49]

b. Enzyme Immunoassays for Herbicides

For quantitative determination of atrazine in water samples, two different EIA systems have been developed.[17,51,52] One system is carried out in the wells of polystyrene microtiter plates coated with antiatrazine antibodies.

The coated microtiter wells are incubated with different concentrations of atrazine or with filtered fresh water, (e.g., from ponds). Enzyme tracer is then added and after washing the *p*-nitrophenylphosphate substrate is pipetted into each well and absorption is measured at 405 nm in an ELISA reader.

In another test system, polystyrene spheres are used as antibody adsorbent surfaces. The polystyrene spheres need higher coating concentrations and longer coating periods than the microtiter plates.[51,52]

The coated polystyrene spheres are incubated with sample or standard solutions containing different concentrations of atrazine. Filtered river or pond water can be assayed directly. The ratio of sample volume and polystyrene surface in microtiter wells is fixed, but can be varied with polystyrene spheres. Where samples are relatively pure but very dilute, large sample volumes (20 ml) can be assayed.[51,52] The longer mean diffusion distance for the different reagent molecules to reach the solid phase neccesitates a longer incubation time as compared to the microtiter plates.

After addition of enzyme tracer the unbound substances are removed by washing. The spheres are then individually transferred to glass tubes containing substrate solution. The absorbance is measured at 405 nm in a spectro photometer. The calculation of the enzyme tracer binding is described in Section III.A.3.a.

The microtiter plate assay yields a measuring range of 0.01 to 30 pmol atrazine per assay.

TABLE 2
Advantages and Disadvantages of the Two Different ELISA Systems

Solid-phase ELISA with polystyrene microtiter plates as antibody carrier	Solid-phase ELISA with polystyrene spheres as antibody carriers
Constant ratio between sample volume and moistened polystyrene surface in the wells	Variable ratio between sample volume and polystyrene surface
Increase of sensitivity by an increase of sample volume not possible	Increase of sensitivity by an increase of sample volume possible within a determined range
Suitable for small volume samples, (200 μl/assay)	Suitable for samples in large volume, (20 ml/assay), e.g., water samples
2 h incubations of sample are sufficient	5 h incubations of samples are necessary
Test system can be easily mechanized and automated	Test system difficult to mechanize and automate

From Huber, S. J. and Hock, B., Atrazine in water, in *Methods of Enzymatic Analysis, Vol. XII, Drugs and Pesticides*, 3rd ed., Bergmeyer, H. U., Bergmeyer, J., and Graßl, M., Eds., VCH Verlagsgesellschaft, Weinheim, 1986, 438.

The test system with affinity-purified antiserum and polystyrene spheres has a lower detection limit due to the larger assay volume. Using 20 ml of sample, 0.5 pmol atrazine l^{-1} can be detected. A comparison of the two systems is shown in Table 2 and typical calibration curves using affinity-purified atrazine antiserum and either microtiter plates or polystyrene spheres are shown in Figure 7.

The same test systems can be used to determine all *s*-triazine herbicides, e.g., atrazine, ametryn,[17] and terbutryn.[19]

3. Indirect Immunosorbent Assay

In the immunoassay systems mentioned so far, analyte and tracer molecules compete for a limited number of specific binding sites on the immunoglobulins. When a solid phase is used another strategy is possible. If an immunogen with the same hapten but a different carrier protein is available, microtiter plates may be coated with this conjugate. Sample or standard and herbicide specific antibodies are then incubated in the wells. The principle is shown in Figure 8. Antibodies will then either bind the free herbicide molecules in solution or they will bind the bound hapten molecules adhering to the wells. Antibodies that recognize the carrier in the original antigen are eliminated by the change of carrier protein. Occasionally the herbicide will bind to the microtiter plate without the use of a carrier. An assay for trifluralin has been developed based on this strategy.[27] Following a washing step, antibodies bound to free herbicide molecules are removed while antibodies bound to conjugates on the solid phase remain. These antibodies may then be detected and quantified using an enzyme-labeled secondary antibody. Secondary antibodies are antibodies which recognize antibodies of another species. Secondary antibodies are not dependent upon the specificity of the primary antibody, as significant portions of the antibody molecule are not involved in antigen recognition. Enzyme-labeled protein A can also be used as the secondary reagent.

An assay of this type has been developed for paraquat. A hapten-carrier protein conjugate of paraquat keyhole limpet haemocyanin (PQ-KLH) is passively adsorbed by polystyrene plates which are then incubated with the samples containing paraquat (PQ) and a known

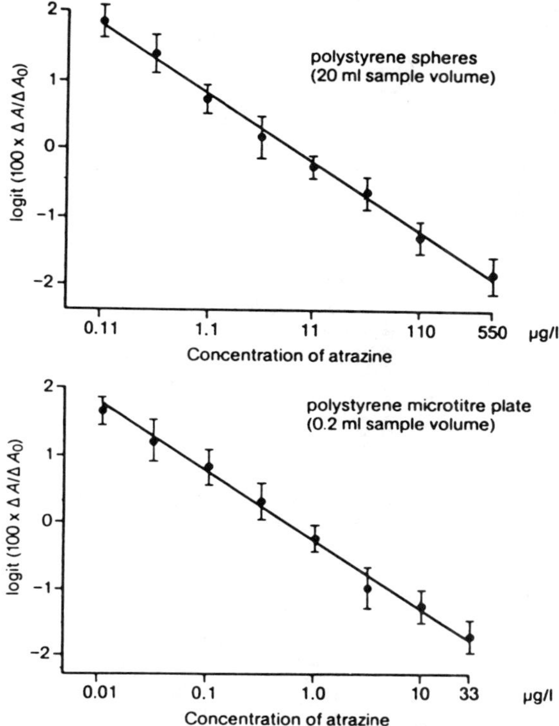

FIGURE 7. Typical linearized calibration curves for the assay of atrazine in water by the assay using polystyrene spheres or microtitre plates, respectively (according to Huber and Hock[17]).

amount of the (monoclonal) antibody (*m Ab*) specific for paraquat. The residual monoclonal antibody (*m Ab*), which has not reacted with free paraquat (PQ) in the sample, combines with PQ-KLH on the plate. After incubation for 30 min at 20°C the incubation mixture is discarded and the plates are washed. Estimation of the fixed antibody is achieved by incubating the washed plates with rabbit antimouse immunoglobulin G (*Ab*) specific for the (fixed) monoclonal antibody and labeled with horseradish peroxidase (POD). Reaction with a chromogenic substrate (1,2 phenylenediamine) determines the enzyme activity of the solid phase by increase in absorbance per unit time, $\Delta A/\Delta t$. The absorbance is measured at 450 nm. The paraquat concentration is inversely related to $\Delta A/\Delta t$.

The limit of detection for a typical extraction procedure is 0.02 mg l^{-1}, which corresponds to about 0.2 mg paraquat kg $soil^{-1}$. This lower limit of detection is satisfactory for most analyses of paraquat in soil.

Another ELISA method for paraquat is based on the same principle but instead of monoclonal antibodies, polyclonal paraquat specific antibodies, developed by immunizing rabbits with a paraquat-BSA conjugate, are used.[7] Binding of carrier specific antibodies in the anti-BSA-paraquat antiserum is eliminated through the use of a conjugate with keyhole limpet hemocyanin carrier for the coating. Concentrations of paraquat in the range 1 to 40 pmol ml^{-1} can be assayed in this ELISA system.

A similar assay for chlorsulfuron has been developed based on rabbit antiserum against the herbicide and goat anti-rabbit IgG, conjugated to alkaline phosphatase, as the secondary antibody.[28] *p*-Nitrophenylphosphate is substrate and the developed color is determined quantitatively at 405 nm with an ELISA reader. This method can detect 0.3 pmol ml^{-1}.

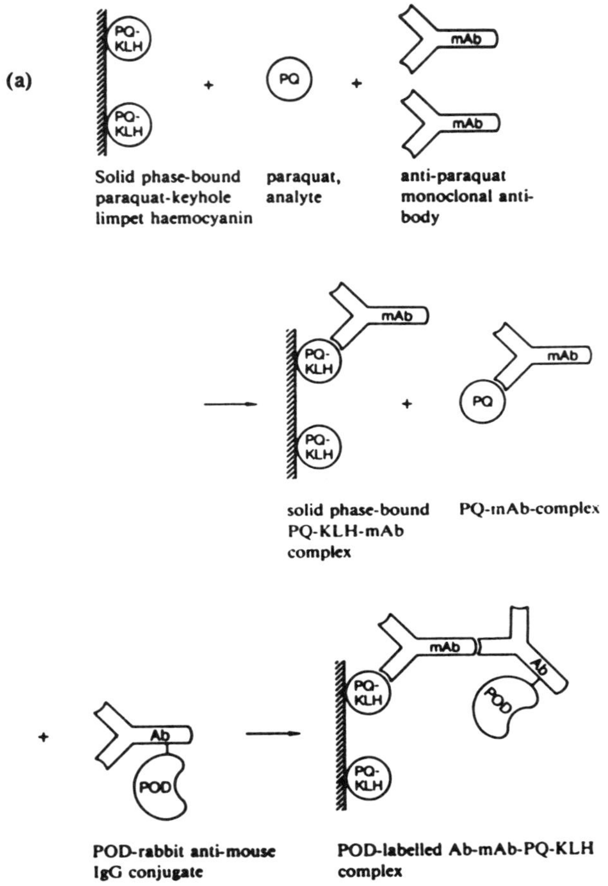

(a)

Solid phase-bound paraquat-keyhole limpet haemocyanin + paraquat, analyte + anti-paraquat monoclonal antibody

→ solid phase-bound PQ-KLH-mAb complex + PQ-mAb-complex

+ POD-rabbit anti-mouse IgG conjugate → POD-labelled Ab-mAb-PQ-KLH complex

(b) 2 1,2-Phenylenediamine + 2 H_2O_2 → 2,2′-diaminoazobenzene + 4 H_2O.

FIGURE 8. Principle of an indirect enzyme immunoassay (according to Benner and Niewola[25]).

4. Homogenous Enzyme Immunoassays — An Alternative Principle of Immunoassays

The immunoassays described above are heterogenous in the sense that at some point a phase separation is introduced. In radioimmunoassays the phase separation is often a precipitation step, whereas solid phase assays are most often employed with the enzyme labeled assays. In homogenous immunoassays no separation of the bound and the free fraction is used. The most widely used homogenous enzyme based assay is the enzyme multiplied immunoassay technique (EMIT). In this system an enzyme-herbicide conjugate is used just as in the direct immunoassays. Occasionally inhibition is observed when an antibody molecule binds to a haptenic group on the enzyme and thus sterically hinders the operation of the enzyme. In those cases phase separation is unnecessary; the relief from antibody mediated enzyme inhibition by herbicide in the sample can be assayed directly in the mixture. The assay is still competitive but the signal is proportional to the herbicide concentration.

Another enzyme based immunoassay, the enzyme modulator mediated immunoassay (EMMIA) is based upon a similar principle.[53] Enzyme modulators are compounds capable of modifying the catalytic activity of an enzyme by either inhibiting or activating enzyme activity. EMMIA is based on the ability of a ligand (e.g., herbicide) labeled enzyme modulator to modify the activity of the enzyme tracer. The modulator-ligand conjugate can compete with free ligand for a fixed and limited amount of antibodies. In absence of ligand, the modulator-ligand conjugate will combine with antibody. Bound conjugate is unable to modulate enzyme activity. As the concentration of ligand increases, it will compete more successfully with modulator-ligand conjugate for binding sites on antibodies, leaving more modulators free to complex with the enzyme tracer and thereby increasingly modulate the enzyme activity.[52] This assay also yields a signal that is proportional to the herbicide concentration in the sample.

In the heterogenous enzyme immunoassays the activity of the enzyme tracer is measured following a washing step. This ensures that the enzyme activity is always determined under optimal conditions. In the two homogenous assays the enzyme activity is assayed in the presence of the sample, hence the occurrence of alternative substrates or inhibitory substances in the sample will affect the measured enzyme activity in unpredictable ways.

Phase separation in enzyme immunoassay technique, however, is rather easy to carry out, because the adsorption of hapten-protein conjugates or glycoproteins (e.g., antibodies) to solid phase (negatively charged polystyrene or polymethacrylate surfaces) is very strong.[8,45]

Enzyme immunoassays are becoming important, because radioimmunoassays with labeled antigens or haptens involve radiation hazards. Labeled substances also show batch-to-batch variation and with a shelf life of only about 2 months, in some cases even shorter, if the radioactive decay causes destruction of the molecular structure.

Compared to the RIA technique, the reagents of enzyme immunoassay are sufficiently stable, commercially available in a standardized state (e.g., marker enzymes), not subject to special safety regulations, and produce no radioactive waste.

Microtiter plates accommodate 96 samples. Even when allowing for standards, duplicates, samples in different dilutions, etc., the potential throughput in the microtiter plate based assays is rather high.

5. Specificity of Antisera

a. Methods for Determination of Cross-Reactivities

The cross-reactivity of antisera specifies their ability to recognize and bind compounds other than their haptens or antigens, usually due to structural similarities to the original hapten or antigen.

There are some factors that influence antibody recognition of molecules. Ring systems should contain halogens for a good antibody recognition; their position within the ring system also is important as is the stability of the molecule.[54] A quantitative comparison of different cross-reacting substances is important in immunoassays.

In all described immunoassays for herbicides, the potential cross-reacting compounds are assayed at concentrations causing inhibition of 50% in the B/B_O value.[6,17,19,24,25,39]

b. Cross-Reactivities of Polyclonal Antibodies

The chemical structure of some *s*-triazine and important metabolites is shown in Figure 9. The chlorinated *s*-triazine herbicides terbuthylazine, simazine, atrazine, and propazine have the same side chains R_1 and R_2 as terbutryn, simetryn, ametryn, and prometryn except for the R_3 chlorine substituent. Cross-reactivities for some related *s*-triazine herbicides found in an atrazine assay are shown in Table 3.

Except for ametryn, which was used for the synthesis of the immunogenic conjugate, simetryn is the most effective cross-reacting triazine, followed by simazine, propazine, and prometryn. The cross-reactivities of terbutryn and terbuthylazine, however, are low. The group with mean cross-reactions only have one substituent which is identical to ametryn or atrazine, namely $-C_2H_5$ or $-CH(CH_3)_2$.[17,52]

In the cross-reaction experiments, shown in Table 3, affinity purified antisera were used. No significant difference can be found in the cross-reacting behavior between the affinity

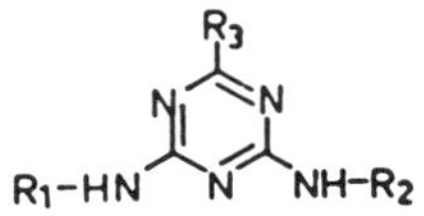

R_3 — s-triazine frame (R_1-HN, NH-R_2)

HERBICIDE:	R_1	R_2	R_3
Terbutryn	$-C_2H_5$	$-C(CH_3)_3$	$-SCH_3$
Simetryn	$-C_2H_5$	$-C_2H_5$	$-SCH_3$
Ametryn	$-C_2H_5$	$-CH(CH_3)_2$	$-SCH_3$
Prometryn	$-CH(CH_3)_2$	$-CH(CH_3)_2$	$-SCH_3$
METABOLITE:			
Hydroxyterbutryn	$-C_2H_5$	$-C(CH_3)_3$	$-OH$
N-Deethylated terbutryn	$-H$	$-C(CH_3)_3$	$-SCH_3$

FIGURE 9. Chemical structure of the *s*-triazine frame and terbutryn and related *s*-triazines (according to Huber and Hock[19]).

purified antiserum and the IgG fraction.[52] Antiserum specificity can only be manipulated efficiently through immunogen design.

Additional to the cross-reactivities of antiatrazine or antiametryn antisera, cross-reactivities of other related *s*-triazines and terbutryn metabolites with antiterbutryn antiserum are listed in Table 4. The terbutryn-BSA conjugate seems to provide an example where steric hindrance has had a major effect on antibody specificity (see Section III.A.1). Antibodies were only formed against the freely accessible antigenic determinants of coupled terbutryn, i.e., against the butylamino and the ethylamino substituent and of course against the conjugated BSA molecule. The methylthio substituent coupling site is lacking in the immunogenic conjugate. Consequently, no antibodies exist against this determinant in the IgG fraction. For this reason, the metabolite hydroxyterbutryn or the herbicide terbuthylazine, which contain the same accessible antigenic determinants as terbutryn, show high cross-reactivities. The other *s*-triazines investigated provided further information on the two antigenic determinants of terbutryn. Molecules like ametryn or atrazine, which contain an ethylamino group like terbutryn but an isopropylamino group instead of the butylamino residue, show intermediate cross-reactivities.

Simetryn and simazine have two ethylamino groups, while terbutryn has one (larger) butylamino group. Simetryn and simazine exhibit negligible cross-reactions. The effective recognition by antibodies raised against terbutryn seems to be based upon the size differences of the substituents at R_1 and R_2 (Figure 9). This corresponds with the observation that even a combination of two isopropylamino residues, as in prometryn and propazine, yield no or only negligible cross-reactivity. On the other hand, the metabolite *N*-deethylated terbutryn shows a cross-reactivity of 33%.

Compared to the R_1 and R_2 substituents, the R_3 plays a somewhat minor role as can be seen by comparison of the terbutryn/terbuthylazine/hydroxyterbutryn and ametryn/atrazine groups, which differ only in their R_3. This effect can be anticipated if the R_3 group is considered to function as a bridge between the herbicide and BSA.[19]

c. Cross-Reactivities of Monoclonal Antibodies

The specificity of monoclonal antibodies to paraquat was determined by using the bipyridilium derivatives diethyl paraquat, monoquat, diquat, and morphamquat to inhibit the

TABLE 3
Cross-Reactivities of Different *s*-Triazines in the Atrazine ELISA in Microtiter Plates Using Affinity-Purified Antibodies

s-Triazine	Apparent concentration in the atrazine ELISA ng l^{-1}	nmol l^{-1}	Relative amount as compared to atrazine %
Atrazine	550	2.50	100
Propazine	319	1.45	57
Simazine	297	1.35	55
Terbuthylazine	66	0.30	12
Ametryn	583	2.65	106
Prometryn	308	1.40	56
Simetryn	418	1.74	75
Terbutryn	77	0.35	14

Note: 0.2 ml were used per assay.

From Huber, S. J. and Hock, B., Atrazine in water, in *Methods of Enzymatic Analysis, Vol. XII, Drugs and Pesticides*, 3rd ed., Bergmeyer, H. U., Bergmeyer, J., and Graßl, M., Eds., VCH Verlagsgesellschaft, Weinheim, 1986, 438.

binding of the monoclonal antibody in the ELISA.[39] It could be shown, that the only derivative to cross-react significantly was diethyl paraquat and there was only little binding to the other derivatives.

Parallel studies were carried out in the same way by using a polyclonal antiserum, developed in rabbits. The cross-reactivity pattern, however, was similar to that observed with monoclonal antibodies.[39] The standard curve obtained with the monoclonal antibody is steeper than that of the polyclonal rabbit antiserum, and consequently, the precision of a paraquat ELISA is better with monoclonal antibodies than with polyclonal antisera. The cross-reacting behavior is, however, almost the same. There seems to be no significant difference between hapten specificity of monoclonal antibodies and polyclonal antisera in this case.

As discussed in Section II.C standardization of the immunoassay procedure is the main reason for the generation of monoclonal antibodies instead of conventional antisera.

d. Validation of Immunoassays

Since immunoassays are competitive no specific information is obtained about the analyte in the extract. The substance competing with the tracer for the available binding sites may be the anticipated herbicide, a component structurally related to it and present in trace amounts, or it may be a compound that is only poorly recognized by the antibody but present in vast excess. Finally the interference may be unspecific, for example, from residual organic solvent from an extraction step. The mere detection of a signal in an immunoassay does not distinguish between these different possibilities. Immunoassays should therefore be validated for a given type of sample subjected to a specific sample preparation procedure.

Validation of an assay may be done simply by comparison of measurements on typical samples carried out using both the immunoassay and a method which provide more information about the substance to be quantified. The preferred method is isotope dilution using GC-MS. This method requires both access to the equipment and the availability of a suitable tracer, preferably the herbicide labeled with a stable isotope. Since immunoassays are often developed where access to this equipment is difficult, other methods of assay validation should be considered. The best strategy for assay validation which does not rely on special equipment is the method called "succesive approximation". The strategy is based on the observation that nonrandom error in immunoassays is always positive — there is no such thing as a negative

TABLE 4
Cross-Reactivity of Different *s*-Triazines in the Terbutryn ELISA

s-Triazine	Apparent concentration in the terbutryn ELISA (pmol assay^{-1})	Amount (%) as compared to terbutryn (%)
Terbutryn	200	100
Simetryn	n.d.[a]	—
Ametryn	100	50
Prometryn	n.d.[a]	—
Simazine	24	12
Atrazine	110	55
Propazine	18	9
Terbuthylazine	128	64
Hydroxyterbutryn	156	78
N-deethylated terbutryn	66	33

[a] n.d. = not detectable, because below detection limit of the ELISA.

From van Regenmortel, M. H. V., Briand, J. P., Muller, S., and Plaué, S., Synthetic polypeptides as antigens, in *Laboratory Techniques in Biochemistry and Molecular Biology,* Bourdon, R. H. and Knippenberg, P. H., Eds., Elsevier, Amsterdam, 1988, chap. 3. With permission.

competition with the tracer for the antibody binding sites. Hence, interference produces overestimates. If an extract is purified and the purification step is effective, some of the interfering substance is removed and the immunoassay produces a lower estimate of the herbicide content in the partially purified extract. If yet another purification step is introduced more of the interfering substance(s) is removed and the assay produces a still lower estimate of the herbicide content. This will not proceed *ad infinitum.* Sooner or later the estimates tend to stabilize around the "true" value fluctuating only due to random error. With "successive approximation" two pieces of information are obtained: the true herbicide content of the sample and the minimal sample purification needed to produce an estimate free of interference, defined as the step of purification where the introduction of extra steps of purification does not lower the estimate any further. Sample purification is inadvertently associated with loss of the analyte. Care should be exercised in tracking the recovery of the herbicide through the purification scheme by using a suitable internal standard. The assay verification does not validate the immunoassay universally; it is only valid for the type of sample tested.

B. IMMUNOAFFINITY PURIFICATION AS A METHOD FOR TRACE ENRICHMENT

Using a reverse phase cartridge is the only example mentioned for trace enrichment of the analyte. Immobilized antibodies constitute a very useful alternative. The technology is widely used with plant hormones,[55-60] whereas examples of immunoaffinity purification of herbicides are scant at present.[61] The plant hormones are, however, quite similar to herbicides with respect to size, chemistry, and immunological properties.

While a reverse phase cartridge effectively concentrates the analyte from a relatively simple sample, e.g., pond water, immobilized antibodies retain the analyte in complex samples, e.g., leaf extracts. The bound fraction can be eluted from the antibodies and analyzed by a number of different methods. An immunoassay, preferably based on different antibodies than those of the immunoaffinity matrix, can be used. Alternatively a physicochemical method can be used. Complex extracts can very often be purified in a single step for identification using GC-MS. This strategy is the method of choice where rigorous identification of the analyte is of importance. The method may also be used prior to separation with HPLC combined with a sensitive detector (e.g., fluorescence or electrochemical detection). This method may furthermore be automated.[62] Combining immunoaffinity purification with a physicochemical method of detection is most useful where antibodies of broad specificity (usually a selected monoclonal antibody) have been selected and more than one sample component can be identified and quantified. The different components may be a family of herbicides of related structure or it may be the herbicide and its primary catabolite in the organism from which the sample was extracted.

Antibodies can be bound to agarose by stable chemical bonds under mild conditions with a high recovery of functional antibody binding sites. The matrix should be hydrophilic but with a low charge density ensuring low unspecific binding of sample impurities to the matrix. The formation of amide bonds* or secondary amine linkages** between a spacermolecule on an agarose gel and a lysine residue on the antibody molecule are popular methods. Less stable methods (cyanogenic bromide activated agarose for example) or methods requiring extreme coupling conditions (epoxide activated agarose) should not be used for this purpose. Relatively large amounts of antibody is required for the preparation of an immunoaffinity matrix with good performance (i.e., low sample loss), 10 mg specific antibody g wet gel^{-1} has been found to be adequate. The immunoaffinity matrix must not be stored frozen. With NaN_3 added as a bactericidal the immunoaffinity matrix keeps well at 4°C in a phosphate buffer. The antibodies are stabilized by the covalent bonds to the agarose matrix. Hence, immobilized antibodies have a much longer shelf life than antibodies in solution.

The analyte is applied to the column dissolved in a neutral phosphate buffer. Column volumes of 250 to 500 µl of gel will often be able to withhold several nanomoles of the hapten. Following the application of the sample, the gel is washed with buffer (10 to 20 column volumes), buffer plus a detergent or buffer plus 500 m*M* of salt; experience will tell what is required to rinse the matrix prior to elution. The final washing step is always deionized water so that a low salt eluate is obtained. The immunoaffinity matrix is eluted with 10 column volumes of methanol. The column is immediately reconditioned with water and phosphate buffer. The eluate is reduced to near drynes under a stream of N_2 or in a vacuum centrifuge and analyzed.

C. IMMUNOFLUORESCENCE LOCALIZATION OF CONJUGATED HERBICIDES IN LEAF TISSUE

1. Conjugation of *s*-Triazine Herbicides in Plants

Some plants are able to metabolize herbicides either by degrading or inactivating the compound. Differential degradation and inactivation mechanisms in plants may govern herbicide tolerance.

Tolerance to atrazine and other *s*-triazine herbicides is found in for example corn (*Zea mays*), and is important for their selective use in this crop. There are three metabolic pathways responsible for 2-chloro-*s*-triazine selectivity in plants.[63] A nonenzymatic detoxification of the herbicide to its hydroxy analogue and hydroxylation reaction catalyzed by the cyclic hydroxamate benzoxazinone in roots has been described.[64] *N*-dealkylation is another 2-chloro-*s*-triazine degradation pathway in corn. The formation of an *S*-(4-ethylamino-6-isopropylamino-2-*s*-triazino) glutathione conjugate in leaves of treated plants, catalyzed by the enzyme glutathione-*S*-transferase also inactivates atrazine.[64]

The enzyme glutathione-*S*-transferase has been purified from corn. Its molecular weight is 45 kDa and its pH optimum is between 8.0 and 8.5. Substrate specificity studies showed that atrazine was the best substrate followed by simazine and propazine.[64]

* Affi-Gel 10, Bio Rad, U.S.

** MiniLeak, Kem-En-Tek, Denmark.

Chemical methods were applied in the past to detect ^{14}C-radiolabeled atrazine and its metabolites by cation exchange chromatography and separation of the metabolites by TLC. However, methods requiring homogenization of the plant tissue cannot localize the site of atrazine conjugation in tissue.

For this reason an alternative localization method is under development using the specific recognition and binding activity of antibodies to their hapten molecules. This recognition and binding activity does not only take place against haptens (or antigens) being free movable in a solution, but also against hapten bound to polystyrene surfaces in ELISA test systems as well as being bound to biological surfaces like cellular structures.[66]

2. Immunofluorescence Labeled Conjugated Atrazine in Corn Leaf Sections

For the model studies, described in the following, corn seedlings (*Zea mays* L., Off. cultivar "Edo"), germinated at 20°C on water-soaked filter paper for 5 days, grew for about 15 days in water culture at room temperature under daylight conditions.[1]

Hand cut leaf pieces of about 5 mm length (the mid rib is removed) are incubated in buffer under slight evacuation. The buffer solution contains 100 $\mu mol\ l^{-1}$ atrazine, ametryn, or terbuthylazine, or no herbicide in untreated controls. Hand cut cross sections of 2 mm from the edge of the incubated leaf pieces are used for fixation. Fixation is carried out in formaldehyde. A series of increasing sucrose concentrations in the fixative solution is used for cryoprotection and samples are mounted on copper rods and quickly frozen in liquid nitrogen.[66] Cryosectioning is performed on a microtom at –30°C. Sections of 30 μm thickness are used for immunostaining.[11]

The incubation of cryosections for immunohistochemistry follows the procedure of Tokuyasu and Singer.[67] The entire incubation protocol is as follows: incubation of nonspecific goat IgG in buffer to reduce nonspecific background, incubation of specific antiserum, incubation of goat antiIgG coupled to fluorescein using fluoresceinisothiocyanat (FITC). In the control experiments, the specific antiserum is replaced by preimmune serum (serum collected from an animal prior to immunization). After each incubation step a washing is performed. After the last washing, the sections are viewed in a photomicroscope equipped with epifluorescence.[11] The results of different localization experiments is shown in Figure 10.

Each epifluorescence micrograph (Figure 10A,C) is accompanied by the corresponding bright field micrograph (Figure 10B,D). The red autofluorescence of the chloroplasts was so weak compared to the green FITC-fluorescence that it is negligible on the micrographs.

Cross-sections of the atrazine treated leaf discs show significant fluorescence after labeling with antiatrazine antibodies followed by FITC-conjugated secondary antibodies (Figure 10A). The fluorescence is mainly restricted to the cells of the bundle sheet. Outside this tissue, in the mesophyll cells around the bundle sheet, the FITC-fluorescence was weak. No fluorescence was observed in epidermis cells. The corresponding bright field micrograph illustrates the distribution of the various tissues in this cross section (Figure 10B). Control experiments, in which the specific antiatrazine antibodies were replaced by preimmune serum, exhibit no fluorescence (not shown).

Ametryn ($-SCH_3$ group at position 2) is a structural analogue to atrazine (–Cl group at position 2). Thus, at its surface, ametryn contains identical antigenic determinants to atrazine. Ametryn is recognized by the antiatrazine antibodies in cross-reaction studies to the same extent as atrazine (Table 3).[17,52] Cross-sections of ametryn treated leaf pieces labeled with antiatrazine antiserum followed by FITC-coupled secondary antibodies showed just the same negligible fluorescence as in the control, in which the specific antiserum was replaced by preimmune serum.

Additional controls are performed in order to exclude artifacts. First, an antigen-free leaf section (not atrazine-treated) was incubated with antiatrazine antibodies. The fluorescence pattern (not shown) was exactly the same as in the atrazine-treated leaf section incubated with

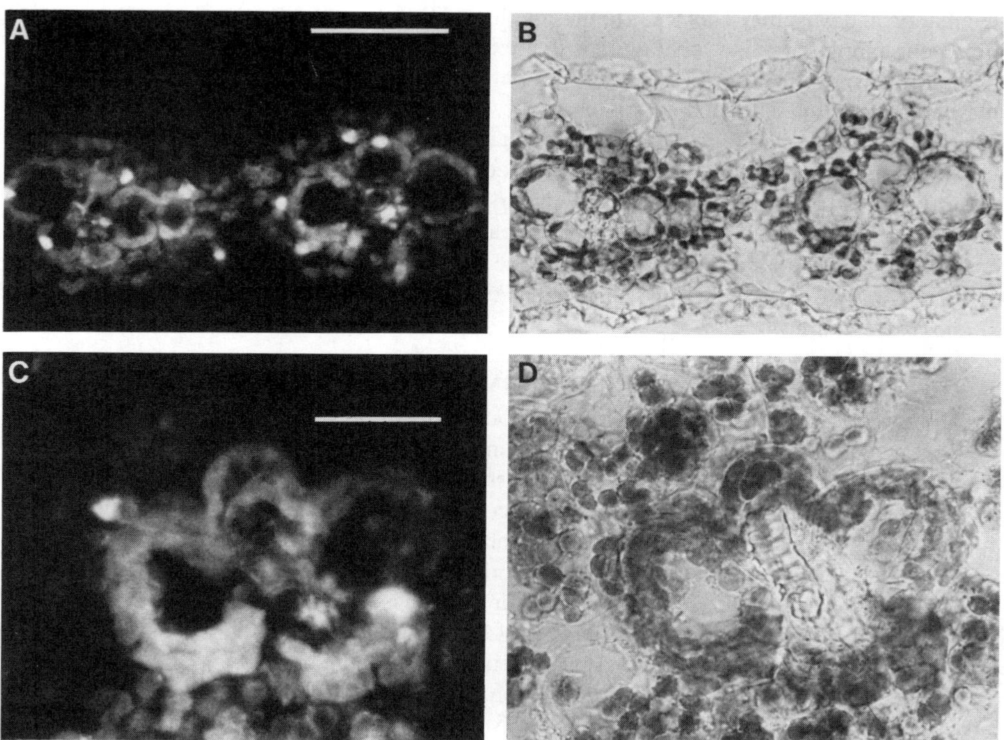

FIGURE 10. Immunofluorescence labeling of atrazine in corn leaf pieces: (A) epifluorescence micrograph after incubation of the leaf pieces with atrazine and labeling of the sections with anti-atrazine antibodies (bar represents 50 μm); (B) bright field micrograph corresponding to Figure A; (C) epifluorescence micrograph after incubation of the leaf pieces with atrazine followed by labeling with anti-atrazine antibodies at higher magnification than Figure A and B (bar represents 30 μm); (D) bright field micrograph corresponding to Figure C (according to Huber and Sautter[11]).

preimmune serum. This control excludes false positive labeling by the antiatrazine antibodies. Second, leaf pieces were treated with terbuthylazine instead of atrazine. Terbuthylazine (–Cl group at position 2) can be conjugated by glutathione-*S*-transferase,[67] but in contrast to atrazine, it does not significantly bind to antiatrazine antibodies; the cross-reactivity is only 12% (Table 3).[17,52] Terbuthylazine-treated leaf sections incubated with antiatrazine serum showed no detectable fluorescence (not shown) similar to the atrazine-treated leaf sections incubated with preimmune serum. A higher magnification leads to higher resolution in the localization of the conjugated atrazine (Figure 10C,D).

The conjugated atrazine, immobilized by formaldehyde, can be localized by immunofluorescence. Free atrazine molecules, however, are removed by washing. This can be assumed because the molecular weight of the herbicide-antibody complex is only slightly higher than that of the free antibody and, therefore, have diffusion properties similar to those of the antibodies alone.

In localization experiments with antiatrazine antibodies, no conjugate could be found in leaf sections pretreated with ametryn instead of atrazine. These findings are in agreement with specificity studies by Guddewar and Dauterman.[69] No atrazine-glutathione conjugate can be found because the enzyme has no activity against ametryn, consequently, no further conjugated intermediates can be formed. Therefore, no or only negligible fluorescence was observed in these leaf sections. This labeling pattern indicates that the ametryn-antibody conjugates as well as the unbound atrazine-antibody complexes were removed from the section during incubation and washing. This leads to the conclusion that the entire fluorescence after atrazine treatment and antiatrazine labeling was due exclusively to atrazine bound in the structures.

The intensive FITC-fluorescence in the bundle sheath cells leads to the conclusion that the herbicide is accumulated in this cell type. The bundle sheath cells are located directly around the bundle, and there is extensive contact between these cell systems. If soil-applied herbicides like atrazine are absorbed by the roots, they are translocated to the leaves by the xylem. The first leaf cells the herbicide enters are the bundle sheet cells. Therefore, these cells must have an efficient inactivation mechanism for atrazine.

Immunohistochemical methods like the one described above can increase our knowledge about the binding of herbicides in plant tissue. For subcellular localization, electron microscopy studies are required.

D. OUTLOOK FOR THE FUTURE OF IMMUNOLOGICAL ASSAYS

The development of antiherbicide antibodies opens a wide field for diagnostic and environmental monitoring studies, using immunoassays for quantitative determination as well as for physiological studies using immunohistochemical techniques in crops and weeds.

Immunohistochemical studies are able to contribute to our knowledge about the binding of herbicides in plant tissue. Glutathione conjugation and its localization might become important in connection with development of herbicide safeners, which can activate glutathione-conjugation mechanisms (see Chapter 12). In every case specific antibodies are necessary for recognition and binding of herbicide molecules or their metabolites.

The quality of immunological assays depends on binding strength and specificity of the antibodies, which in turn are dependent upon the antigen designs employed. The *in vitro* production of monoclonal antibodies[41] is also important for making standardized and fully optimized assays available to different laboratories, and thereby avoiding the development of similar assays by all laboratories engaged in herbicide analysis. Knowledge of production, handling and storage of antibodies will accumulate from many branches in the future, from medical, pharmaceutical, physiological, and environmental studies.

Enzyme based immunological assays wil probably replace radiolabeled assays, due to radioactive waste and radiation hazards, especially in routine analysis. Further improvement for enzyme based immunological assays includes current examination of new enzyme labels, production of more specific cross-linking reagents and improved techniques for coating solid surfaces with proteins and haptens. Enzyme immunoassays are suitable for routine analysis of large sample quantities.

Simple performance, rapidity in assaying, and the relatively low cost of equipment justify the use of enzyme immunoassays in drinking and ground water analysis. Most water samples can be assayed without prior extraction or clean-up procedures. Thus, immunological assays like enzyme immunoassays can contribute considerably to an improved and effective environment monitoring for herbicides or their metabolites in future.

REFERENCES

1. **Levitt, T.,** Radioimmunoassay for paraquat, *Lancet,* 13, 358, 1977.
2. **Levitt, T.,** Determinations of paraquat in clinical practice using radioimmunoassay, *Proc. Anal. Div. Chem. Soc.,* 16, 72, 1979.
3. **Berson, S. A. and Yalow, R. S.,** Recent studies on insulin-binding antibodies, *Ann. N. Y. Acad. Sci.,* 82, 338, 1959.
4. **Engvall, E. and Perlmann, P.,** Enzyme linked immunosorbent assay (ELISA). Quantitative assay of immunoglobulin G, *Immunochemistry,* 8, 871, 1971.

5. **Van Weemen, B. K. and Schuurs, A. H. W. M.,** Immunoassay using antigen enzyme conjugates, *FEBS Lett.,* 15, 232, 1971.
6. **Rinder, D. F. and Fleeker, J. R.,** A radioimmunoassay to screen for 2,4-dichlorophenoxyacetic acid and 2,4,5-trichlorophenoxyacetic acid in surface water, *Bull. Environ. Contam. Toxicol.,* 26, 375, 1981.
7. **Niewola, Z., Walsh, S. T., and Davies, G. E.,** Enzyme linked immunosorbent assay (ELISA) for paraquat, *Int. J. Immunopharmacol.,* 5, 211, 1983.
8. **Løvborg, U. L.,** *Guide to Solid Phase Immuno Assays,* AIS NUNC, Roskilde, Denmark, 1984.
9. **Huikeshoven, H., Landheer, J. E., v. Denderen, A. C., Vlasman, M., Leiker, D. L., Das, P. K., Goldring, O. L., and Pondman, K. W.,** Demonstration of dapsone in urine and serum by ELISA inhibition, *Lancet,* 4, 280, 1978.
10. **Weiler, E. W., Jourdan, P. S., and Conrad, W.,** Levels of indole-3-acetic acid in intact decapitated coleoptiles as determined by a specific and highly sensitive solid-phase enzyme immunoassay, *Planta,* 153, 561, 1981.
11. **Huber, S. J. and Sautter, C.,** Immunofluorescence localization of conjugate datrazine in leaf pieces of corn (*Zea mays*), *J. Plant Dis. Protect.,* 93, 608, 1986.
12. **Marcussen, J., Ulvskov, P., Olsen, C. E., and Rajagopal, R.,** Preparation and properties of antibodies against indoleacetic acid (IAA)-C5-BSA, a novel ring-coupled IAA-antigen, as compared to two other types of IAA specific antibodies, *Plant Physiol.,* 89, 1071, 1989.
13. **Erlanger, B. F.,** The preparation of antigen hapten-carrier conjugates: a survey, in *Methods in Enzymology,* Vol. 70 (Part A), Van Vunakis, H. and Langone J. J., Eds., Academic Press, New York, 1980, 85.
14. **Hurn, B. A. L. and Chantler, S. M.,** Production of reagent antibodies, in *Methods in Enzymology,* Vol. 70 (Part A), Van Vunakis, H. and Langone, J. J., Eds., Academic Press, New York, 1980, 104.
15. **van Regenmortel, M. H. V., Briand, J. P., Muller, S., and Plaué, S.,** Synthetic polypeptides as antigens, in *Laboratory Techniques in Biochemistry and Molecular Biology,* Bourdon, R. H. and Knippenberg, P. H., Eds., Elsevier, Amsterdam, 1988, chap. 3.
16. *Pierce General Handbook and Catalogue 1989,* 283.
17. **Huber, S. J. and Hock, B.,** Atrazine in water, in *Methods of Enzymatic Analysis, Vol. XII, Drugs and Pesticides,* 3rd ed., Bergmeyer, H. U., Bergmeyer, J., and Graßl, M., Eds., VCH Verlagsgesellschaft, Weinheim, 1986, 438.
18. **Hamboeck, H., Fischer, R. W., Di Iorio, E. E., and Winterhalter, K. H.,** The binding of s-triazine metabolites to rodent hemoglobins appears irrelevant to other species, *Mol. Pharmacol.,* 20, 579, 1981.
19. **Huber, S. J. and Hock, B.,** A solid-phase enzyme immunoassay for quantitative determination of the herbicide terbutryn, *J. Plant Dis. Protect.,* 92, 147, 1985.
20. **Pfeiffer, W., Dittel, R. H., and Mies, W.,** German Patent Application 37, 23, 726, A1, 1989.
21. **Bauminger, S. and Wilchek, M.,** The use of carbodiimides in the preparation of immunizing conjugates, in *Methods in Enzymology,* Vol. 70 (Part A), Van Vunakis, H. and Langone, J. J., Eds., Academic Press, New York, 1980, 151.
22. **Dunbar, B. D., Niswender, G. D., and Hudson, J. M.,** U.S. Patent 4,530,786, 1985.
23. **Erlanger, B. F.,** Principles and methods for the preparation of drug protein conjugates for immunological studies, *Pharmacol. Rev.,* 25, 271, 1973.
24. **Knopp, D., Nuhn, P., and Dobberkau, H. J.,** Radioimmunoassay for 2,4-dichlorophenoxyacetic acid, *Arch. Toxicol.,* 58, 27, 1985.
25. **Benner, J. P. and Niewola, Z.,** Paraquat in soil, in *Methods of Enzymatic Analysis, Vol. XII, Drugs and Pesticides,* 3rd ed., Bergmeyer, H. U., Bergmeyer, J., and Graßl, M., Eds., VCH Verlagsgesellschaft, Weinheim, 1986, 451.
26. **Fatori, D. and Hunter, W. M.,** Radioimmunoassay for serum paraquat, *Clin. Chim. Acta,* 100, 81, 1980.
27. **Dunbar, B. D., Niswender, G. D., and Hudson, J. M.,** U.S. Patent 4,780,408, 1988.
28. **Kelley, M. M., Zahnow, E. W., Petersen, C. W., and Toy, S. T.,** Chlorsulfuron determination in soil extracts by enzyme immunoassay, *J. Agric. Food Chem.,* 33, 962, 1985.
29. **Hall, J. C., Deschamps, R. J. A., and Krieg, K. K.,** Immunoassays for the detection of 2,4-D and picloram in river water and urine, *J. Agric. Food Chem.,* 37, 981, 1989.
30. **Marcussen, J., Seiden, P., van Onckelen, H., and Rajagopal, R. R.,** Monoclonal antibodies towards a stable cyclic AMP-C8 antigen, *Anal. Biochem.,* 198, PO 255-Pe 010, 1991.
31. **Marcussen, J. and Poulsen C.,** A non-destructive method for peptide bond conjugation of antigenic haptens to a diphteria toxoid carrier exemplified by two antisera specific to acetolactate synthase, *Anal. Biochem.,* 198, PO 254-Pe 009, 1991.
32. **Bernatowicz, M. S. and Matsueda, G. R.,** Preparation of peptide immunogens using *N*-succinimidyl bromoacetate as a heterobifunctional crosslinking reagent, *Anal. Biochem.,* 155, 95, 1986.
33. **Köhler, G. and Milstein, C.,** Continuous cultures of fused cells secreting antibody of predefined specificity, *Nature,* 256, 495, 1975.

34. **Campbell, D. H., Garvey, J. S., Cremer, N. E., and Sussdorf, D. H.,** *Methods in Immunology,* 2nd ed., W. A. Benjamin, Inc., New York, 1970, 7.
35. **Van Emon, J., Seiber, J., and Hammock, B.,** Application of an enzyme linked immunosorbent assay (ELISA) to determine paraquat residues in milk, beef, and potatoes, *Bull. Environ. Contam. Toxicol.,* 39, 490, 1987.
36. **Freund, J., Casals, J., and Hismer, E. P.,** Sentization and antibody formation after injection of tubercle bacilli and paraffin oil, in *Proc. Soc. Exp. Biol. Med.,* 37, 509, 1937.
37. **Pengelly, W. and Meins, F.,** A specific radioimmunoassay for nanogram quantities of the auxin indole-3-acetic acid, *Planta,* 136, 173, 1977.
38. **Green, M. C.,** *Genetic Variants and Strains of the Laboratory Mouse,* Gustav Fischer Verlag, Stuttgart, 1981, chap. 10.
39. **Niewola, Z., Hayward, C., Symington, B. A., and Robson, R. T.,** Quantitative estimation of paraquat by an enzyme linked immunosorbent assay using a monoclonal antibody, *Clin. Chim. Acta,* 148, 149, 1985.
40. **Hall, J. Ch., Deschamps, R. J. A., and Mc Dermott, M. R.,** Immunoassays to detect and quantitate herbicides in the environment, *Weed Technol.,* 4, 226, 1990.
41. **Kuhlmann, J., Kurth, W., and Ruhdel, J.,** *In vivo-* und *in vitro*-Produktion monoklonaler Antikörper im Labormässtab, *Forum Mikrobiol.,* 10, 451, 1989.
42. **Anon.,** Nunc Cell Factory Instruction, Nunc, P.O. Box 280, DK 4000, Denmark.
43. **Hutchinson, T. W. and Porath, J.,** Thiophilic adsorption of immunoglobulins — analysis of conditions optimal for selective immobilization and purification, *Anal. Biochem.,* 159, 217, 1986.
44. **Yalow, R. S. and Berson, S. A.,** Assay of plasma insulin in human subjects by immunological methods, *Nature,* 184, 1648, 1959.
45. **Blake, C. and Gould, B. J.,** Use of enzymes in immunoassay techniques, *Analyst,* 109, 533, 1984.
46. **Scatchard, G.,** The attraction of protein for small molecules and ions, *Ann. N.Y. Acad. Sci.,* 51, 660, 1949.
47. **Odell, W. D.,** Use of charcoal to separate antibody complex from free ligand in radioimmunoassay, in *Methods in Enzymology*, Vol. 70 (Part A), Van Vunakis, H. and Langone, J. J., Eds., Academic Press, New York, 1980, 274.
48. **Finney, D. J.,** *Statistical Method in Biological Assay,* 3rd ed., Charles Griffin and Company, Ltd., London, 1978, 329.
49. **Wellington, D.,** Mathematical treatments for the analysis of enzyme immunoassay data, in *Enzyme Immunoassay,* Maggio, E. T., Ed., CRC Press, Boca Raton, FL, 249, 1980.
50. **Marschner, I., Dobry, H., Erhardt, F., Landersdorfer, T., Popp, B., Ringel, C., and Scriba, P. C.,** Calculation of radioimmunoassay data by spline functions, *Ärztl. Lab.,* 20, 184, 1974.
51. **Huber, S. J. and Hock, B.,** Solid-Phase-Enzymimmunoassay zum Nachweis von Pflanzenschutzmitteln in Gewässern, *GIT Fachz. Lab.,* 29, 969, 1985.
52. **Huber, S. J.,** Improved solid-phase enzyme immunoassay systems in the ppt range for atrazine in fresh water, *Chemosphere,* 14, 1795, 1985.
53. **Ngo, T. T.,** Enzyme modulator mediated immunoassay (EMMIA), *Int. J. Biochem.,* 15, 583, 1983.
54. **Luster, M. J., Albro, P. W., Chae, K., Lawson, L. D., Corbett, J. T., and McKinney, J. D.,** Radioimmunoassay for quantitation of 2,3,7,8-tetrachlorodibenzofuran, *Anal. Chem.,* 52, 1497, 1980.
55. **Durley, R. C., Sharp, C. R., Maki, S. L., Brenner, M. L., and Carnes, M. G.,** Immunoaffinity techniques applied to the purification of gibberellins from plant extracts, *Plant Physiol.,* 90, 44, 1989.
56. **Sundberg, B., Sandberg, G., and Crozier, A.,** Purification of indole-3-acetic acid in plant extracts by immunoaffinity chromatography, *Phytochemistry,* 25, 295, 1986.
57. **Knox, J. P. and Galfre, G.,** Use of monoclonal antibodies to separate the enantiomers of abscisic acid, *Anal. Biochem.,* 155, 92, 1986.
58. **Ulvskov, P., Marcussen, J., Rajagopal, R. R., Prinsen, E., Rüdelsheim, P., and van Onckelen, H.,** Immunoaffinity purification of indole-3-acetamide using monoclonal antibodies, *Plant Cell Physiol.,* 28, 937,1987.
59. **Morris, R. O., Powel, G. K., Beaty, J. S., Durley, R. C., and Hommes, N. G.,** Cytokinin biosynthetic genes from and enzymes from *Agrobacterium tumefaciens* and other plant-associated prokaryotes, in *Plant Growth Substances,* Bopp, M., Ed., Springer Verlag, 1985, 185.
60. **Mertens, R., Stünning, M., and Weiler, E.,** Metabolism of tritiated enantiomers of abscisic acid prepared by immunoaffinity chromatography, *Naturwissenschaften,* 69, 595, 1982.
61. **Stanker, L. H., Watkins, B. E., and Vanderlaan, M.,** Environmental monitoring by immunoassay, in *Pesticide Chemistry. Advances in International Research, Development, and Legislation,* Frehse, H., Ed., VCH, Weinheim, Germany, 1991, 397.
62. **Frajam, A., De Jong, G. J., Frei, R. W., Haasnoot, W., Hamers, A. R. M., Schilt, R., and Huf, A.,** Immunoaffinity pre-column for selective on-line sample pre-treatment in high-performance liquid chromatography determination of 19-nortestosterone, *J. Chromatogr.,* 452, 419, 1988.

63. **Böger, P.,** Ein neues Resistenzprinzip gegen Herbizide, *J. Plant Dis. Protect.,* Sonderh, 9, 153, 1981.
64. **Shimabukuro, R. H., Frehar, D. S., Swanson, H. R., and Walsh, W. C.,** Glutathione conjugation, an enzymatic basis for atrazine resistance in corn, *Plant Physiol.,* 47, 10, 1971.
65. **Esser, H. O., Dupois, G., Ebert, E., Marco, G. J., and Vogel, C.,** s-Triazines, in *Herbicides,* Kearney, P. C. and Kaufman, D. D., Eds., Marcel Dekker, New York, Basel, 1975, 129.
66. **Sautter, C. and Hock, B.,** Fluorescence immuno-histochemical localization of malate dehydrogenase isoenzymes in watermelon cotyledons, *Plant Physiol.,* 70, 1162, 1982.
67. **Tokuyasu, K. T. and Singer, S. J.,** Improved procedures for immunoferritin labelling of ultrathin frozen sections, *J. Cell Biol.,* 71, 894, 1976.
68. **Lamoureux, G. L., Stafford, L. E., Shimabukuro, R. H., and Zaylskie, R. G.,** Atrazine metabolism in sorghum: catabolism of the glutathione conjugate of atrazine, *J. Agric.. Food Chem.,* 21, 1020, 1973.
69. **Guddewar, M. B. and Dauterman, W. C.,** Purification and properties of a glutathione-*S*-transferase from corn which conjugates *s*-triazine herbicides, *Phytochemistry,* 18, 735, 1979.

Chapter 12

BIORATIONAL *IN VITRO* SCREENING FOR HERBICIDE SYNERGISTS

J. Gressel, Y. Shaaltiel, A. Sharon, and Z. Amsellem

TABLE OF CONTENTS

0-8493-6603-8/93/$0.00+$.50

I. INTRODUCTION — THE PROBLEMS

There are strong pressures from various sources to reduce the amounts of herbicides used.[1] They come from environmental groups worried about the ecological impact of herbicides and their residues entering the food chain and the water supply, and from farmers wishing to decrease production costs. The uses of high rates of herbicides in monoculture with the concomitant high selection pressure has led to resistance problems that are becoming more widespread around the world.[2-4] Lowering rates, lessens selection pressure, and in many cases, delays resistance.[5,6]

Bioactively high rates, i.e., not the actual rates, but the rate used vs. the rate required to control sensitive weeds, are used for a variety of reasons: to provide enough material when coverage is not uniform, to provide residual activities, to compensate for poor penetration, and to provide control of the less sensitive weed species. The problems of coverage are agrotechnical and are not covered here. Adjuvants for better formulation, leading to less herbicide used, are discussed in Chapter 6.

The "insurance" provided by longer residual life of a herbicide must be carefully analyzed to ascertain whether it is really needed, ignoring the hard sell of the insurance salesmen. Many bioeconomic models for analyzing this need are becoming available, and most indicate that herbicide usage could be economically halved, despite the "insurance" being relatively inexpensive.[7-9]

We are left with the question of high rates to control the greatest possible spectrum of weed species, including those having less inherent sensitivity. A careful look at screening data for various herbicides shows that there are vast differences in the rates required to control various species. For example, atrazine controls many broad leaf species at 30 g ha^{-1} (some even at 10 g ha^{-1}), while 500 g ha^{-1} or more are required to control various grasses (Table 1). In maize, 1 to 2 kg ha^{-1} are typically used for the whole weed spectrum. The same is true for the typically "low rate" sulfonylureas; there are plants controlled by 30 mg ha^{-1}, yet 4 to 30 g ha^{-1} are typically used. Part of the reason for using high rates is to leave herbicide in the soil for a residual effect. Still, the rates shown in Table 1 accentuate the acute relative toxicities to various species, and this relativity carries to the field. The limits of the recommendations to use high rates have been governed only by toxicity to crops, carry over, and the economics of using competitive herbicides.

The use of biologically high rates, often two orders of magnitude greater than the rate needed to control the most sensitive species, is most extraordinary when compared to the K_i of the various herbicides at their target sites; paraquat and atrazine have K_is near 1 μM

TABLE 1
Examples of Discrepancies Between the Range of Herbicide Rates Needed to Control Plants, and the Range of K_i Values for the Target Enzymes

Herbicide	Target site	Range K_is (μM)	Range used in agriculture[14] (kg ha^{-1})	Genus	Minimum amount needed for weed control[b] (g ha^{-1})
Atrazine	Photosystem II	0.3–1	2.2–4.5	*Amaranthus, Chenopodium*	30
Paraquat	Photosystem I	0.2–3	0.6–1.1	*Bromus, Amaranthus*	10
Chlorsulfuron	Acetolactate synthase	0.007–.099[10]	0.004–0.026[a]	*Bromus, Poa, Amaranthus*[10]	0.03
Glyphosate	EPSP synthase	20–110[11]	0.3–4.5	*Xanthium*	300
Diclofop-methyl	AcetylCoA carboxylase	0.01–0.07[12]	0.8–1.4	*Echinochloa, Setaria*	100
Acifluorfen	Protoporphyrinogen-oxidase	0.03[13]	0.14–1.12	*Sinapis, Datura*	10
Alachlor	Unknown	—	1.7–9	*Echinochloa, Amaranthus*	30
2,4-D	Unknown	—	0.3–2.2	*Sinapis, Xanthium*	30

[a] Rates used in wheat.

[b] Unless referenced otherwise, these data are unpublished information from controlled herbicide screens, kindly provided by P.F. Bocion.[15]

for photosystem I and II, *in vitro*, respectively, regardless of species (Table 1). Chlorsulfuron affects acetolactate synthase in the nanomolar range *in vitro*, irrespective of species. The range of K_is among species is usually much less than one order of magnitude for a given compound. The reasons for the discrepancy between the range of K_is and the range of use rates is due to a variety of factors, including relative penetration and translocation, but the major factor involved is usually differential rates of detoxification. If one species detoxifies the herbicide (or the toxic products of herbicide action) faster than another species, more herbicide will be required to control the first species.

Many herbicides are apolar. This facilitates cuticular uptake and transport, as well as the transversing of lipid-containing membranes within cells. Additionally, the lack of water solubility can slow soil leaching. Many herbicides (e.g., diclofop-methyl) contain a lipophilic group that is enzymatically cleaved once it has penetrated into cells giving an active polar compound. Enhancing the rate of cleavage might synergize such herbicides.

Lipophilicity is necessary for inhibiting many membrane bound sites of action such as photosystem II and the pathway of carotenoid biosynthesis. The conjugation of a polar group such as an amino group, a sugar moiety, glutathione, etc, or oxidizing the herbicide to give a hydroxyl group can render the herbicide inactive just by changing polarity, as well as by changing structure that will modify its binding. Inhibiting such reactions could enhance activity. The knowledge of pathways to inhibit, and finding the right inhibitor, allow a biorational (vs. random) approach to finding synergists.

A. SYNERGISTS

One way to lower the rates of herbicide usage is to decrease the discrepancies between the narrow range of K_is and the broad range of use rates. This can be done by synergists.[16] A synergistic mixture is one that has a greater herbicidal effect than expected from the sum of each of the components used separately. The "sums of the components" are usually given as the rates of the components.

Synergistic mixtures can be of three types: (1) where one component of the mixture is a nonphytotoxic adjuvant, allowing a lower rate of herbicide to be used; (2) where both compounds are phytotoxic but synergize each other, allowing lower levels of both to be used; and (3) where neither compound is phytotoxic at the rates used, but together are phytotoxic (a "coalitive" synergy). The boundaries among the three types of synergies are not always as clear as the definitions may imply. An adjuvant may be phytotoxic at high concentrations and a herbicide nontoxic by itself at the low concentration, used in a mixture. A herbicide may act as an adjuvant, e.g., mefluidide seemed to enhance activity of bentazon and acifluorfen. This enhancement could be mimicked by a petroleum oil concentrate, a typical adjuvant.[17]

The farmer defines the sums to calculate synergism on the basis of sums of cost; the mixture must give cheaper weed control and/or greater crop yield than that achieved by the sum of the cost of the components, each giving the same level of control alone. The farmer's definition of synergism does not require that the same weed species be controlled by both components of a mixture. The classical definition based on use rates will be used herein, as prices are far less constant than the rate of herbicide that kills a weed at a specific time. Synergy here is limited to those affecting single species, not total weed control. Mixtures of herbicides that broaden the spectrum of the species controlled are discussed in Chapter 7.

B. DEDICATED ASSAYS FOR PREVENTING DETOXIFICATION

This chapter deals with synergists that prevent the detoxification of herbicides or the toxic products that they generate. To a large extent, this can be performed with a modicum of biorationale. If the pathway of detoxification is known, one can rationally devise bioassays "dedicated" to ascertain the blockage of the specific pathway. Compounds can be rationally chosen from among those that are known to specifically inhibit that general pathway, instead of randomly screening all chemicals. Such compounds can act as "leads" for further optimization prior to, or concurrently with, greenhouse and field testing.

The biorational design of such bioassay systems often requires that we ascertain the site of action, and the mode of death, as well as the detoxification pathways of the herbicide. The site of herbicide action is known in many more cases than the mode of death. For example, many herbicides including triazines block photosystem II by binding to the D_1 Q_B-binding protein of the reaction center.[18] Many researchers naively assumed that the plants die of starvation due to the blockage of photosynthesis. Plants starved by etiolation die more slowly and with different symptomology than triazine-treated plants. Light and a photosynthetic apparatus are required for triazine mediated death. Lipoxidation of membrane constituents occurs, but the initially activated species (chlorophyll, oxygen, etc.) that initiate membrane oxidation are unknown. Different herbicides directly bind to and inhibit acetylCoA carboxylase, acetolactate synthase, phytoene desaturase, EPSP synthase, protoporphyrinogen IX oxidase, etc., but it is not certain why the symptomology of the throes of death are as they are, because the precise modes of death are unknown. Knowledge allows rational design of synergistic mixtures to exacerbate the demise of weeds.

Some herbicides must be toxified to act, and they are selective in crops that lack the biochemical pathways to toxify them. Other herbicides are selectively detoxified in crops but not in weeds. The detoxification pathways in crops have been elucidated to fulfil governmental registration requirements on residues and the metabolite toxicity remaining at harvest. Consequently, industrial herbicide metabolism studies do not usually have early time points, and the identity of initial herbicide metabolites with reduced toxicity are often unknown. This initial detoxification step should be known for designing synergists. Sometimes the initial metabolite can be assumed from later metabolites. For example, if an *O*-glycoside is found, and there was no oxygen atom on the herbicide at the site of glycosylation, one must assume a primary hydroxylation.

Another major drawback of the available detoxification data is that they usually pertain to crops. We wish to synergize herbicides so that they will more effectively control weeds, not crops. Weeds need not utilize the same detoxification pathways as the crops. It will be easier to find a differential synergist, one that controls the weed and not the crop, if the pathways are dissimilar. It is an obligatory prerequisite to biorational design of synergists, to ascertain the pathway(s) to be blocked in the target weeds.

A synergist can be a member of each of the three groups cited above: another herbicide, an adjuvant, or be coalitive, but the categories can overlap. The treatment with the herbicide to be synergized can set into motion *de novo* synthesis of detoxifying enzymes and/or substrates to be conjugated with the herbicide. A second herbicide may act synergistically, because it blocks synthesis of these enzymes and/or the conjugating substrates. An adjuvant that is inactive by itself, mixed with a sublethal rate of herbicide, may have a coalitive effect.

Synergists may be advantageous for use with weeds that have evolved elevated levels of detoxification leading to herbicide resistance. The greatest use of insecticide synergists is to suppress evolved resistance by detoxification.

C. ADVANTAGES OF *IN VITRO* SYSTEMS

Well designed *in vitro* systems have many advantages over whole plant testing.[19] They can allow the assay of the system desired, often without problems of uptake and translocation.[20] They are usually highly space efficient and give more rapid answers than greenhouse or field tests. They are more amenable to highly controlled conditions, rendering them more reproducible with fewer replicates than are needed for whole plant testing. They are often axenic, decreasing the effects of microbial contamination. The personnel requirements to perform *in vitro* testing are usually far lower (per compound tested) than whole plant screening, but such testing often requires better trained personnel, even with highly automated equipment.

D. LIMITATIONS TO SYNERGISTS

Not all herbicides can be synergized in all intransigent weeds. As stated above, the detoxification pathway must exist and then be elucidated. For example, only graminae are affected by the cyclohexanedione and aryloxyphenoxy propionate inhibitors of acetyl CoA carboxylase. Even some graminae are not inhibited by this class of herbicides.[21] A synergist can be developed only if there is differential detoxification. Nongraminaceous species are resistant due to differences in the target enzyme; there is a >100-fold difference in K_i.[22-24] One can only synergize the weeds that seem to have detoxifying mechanisms to this group.[25-28] More than 80 weed species evolved a modified target site and do not bind triazines,[3] and these cannot be synergized. It should be easier to design a synergist to control the *Abutilon theophrasti* [29,30] and the *Amaranthus*[31] that evolved a biochemical mechanism to detoxify atrazine.

The cause of lack of control of a given species need not be due to detoxification or lack of target, it could be due to redundant pathways. Plants, with their nuclear and two extranuclear genomes as well as enzyme families containing discrete isozymes, alternative pathways, etc. often have more than one mechanism to achieve a given result. A herbicide may not control the weeds having the alternative mechanism. This again accentuates the need to know why the target weed is not controlled before setting out to find a synergist. In this last case, the synergist can be a compound inhibiting the alternative pathway.

A good *in vitro* system will have false positives, i.e., compounds that synergize *in vitro* but not in whole plants. This is expected. A good dedicated *in vitro* bioassay system should have no false negatives, i.e., compounds that synergize in plants but not *in vitro*. This is different from random *in vitro* screens for herbicidal action (Chapter 10). The important aspect of such a screen (compared to whole plants) is that it should allow the researcher to quickly discard negative leads and to give more accurate QSAR data for synergists, once a lead group has been pinpointed. QSAR for penetration will have to be redone once the best leads are found.

As described above, it will be hard to synergize herbicides that are not active because of the lack of a target site or redundant pathways. Herbicides that must be activated to a toxic moiety cannot be synergized if the activating enzyme system is lacking. Many herbicides are systemic and must be translocated to their site of action and may be degradated along the way. There are two ways to synergize a translocated herbicide: (1) by inhibiting detoxification and (2) by enhancing the rate of translocation. There are compounds known to decrease translocation rates (especially auxins), but there are no reports (to our knowledge) of compounds that enhance translocation. While bioassays similar to auxin translocation assays could be set up, the testing of compounds may well be an exercise in random screening, as there is no lead information.

II. SYNERGIZEABLE SYSTEMS

Cases where it can be envisaged that biorational synergism can be obtained and known synergists can be used as standards and leads are discussed below. These examples are given to whet the imagination; the reader is then invited to develop further systems. Section III describes the bioassays to find synergists.

The discussion mostly deals with inhibiting herbicide detoxification as a target for synergy. Some herbicides kill by generating toxic products, yet the plant can degrade these toxins. An equilibrium is obtained between the amount of toxic product made and its rate of degradation. More herbicide must be used to generate a more toxic product. Synergism is obtained by suppressing the degradation of the toxic product.

One case below shows that synergism need not be at the level of the plant. "Problem soils" have evolved microbial populations that rapidly degrade many herbicides.[32] Inhibiting this microbial activity with synergists preserves the activity of the herbicides.

The use of a synergist can be expected to lead to a loss of selectivity when a weed uses the same degradative pathway as the crop. Blocking this pathway could lead to the demise of crop and weed unless there are exploitable nuances of differences between them. Enhancing the spectrum of species controlled can clearly open new markets to herbicides in minimum tillage agriculture requiring preplanting contact herbicides.

A. SYNERGISTS INHIBITING GLUTATHIONE TRANSFERASE*

Many xenobiotics are degraded by glutathione transferases that initially conjugate the herbicide to glutathione (glutamyl-glycyl-cysteine) or homoglutathione (glutamyl-alanyl-cysteine).[33] "Glutathione" will be used for both tripeptides. At least three such enzymes are present in maize, each with its own herbicide substrate specificities.[34] Maize seems to have a very high activity of this enzyme system, especially the glutathione transferases conjugating atrazine and chloroacetamides.[35,36]

The genes for the various isozymes have been isolated, cloned, and sequenced.[37-39] Glutathione is irreversibly conjugated by glutathione transferases to herbicides such as atrazine. Two amino acids are eventually cleaved from the atrazine-glutathione conjugate, leaving cystinyl-atrazine. Thus, a species needs stoichiometric levels of reduced glutathione, as well as glutathione transferases for this pathway of resistance.

Glutathione transferases are also prevalent in panicoid grasses such as *Setaria* spp. and *Panicum* spp. These species do not degrade atrazine as rapidly as maize, allowing selectivity, but high rates of atrazine are required to overcome their glutathione transferases. Such high rates can increase ground and runoff water pollution, supply the high selection pressure for selecting for resistance, and leave residues precluding rotation with sensitive crops. *Abutilon theophrasti* has evolved atrazine resistance by biochemically "mimicking" maize; one gene controls an elevated glutathione transferase level.[29,30]

Various treatments to plants such as drought stress,[40] pretreatment with herbicides,[41] and compounds used as herbicide protectants,[42] can enhance the levels of glutathione and glutathione transferases in various crops. Presumably, similar treatments can enhance the levels of this system in weeds, requiring more herbicide for weed control. Synergy by inhibiting glutathione transferase conjugation was fortuitously found when tridiphane, a grass herbicide, was checked for compatibility with atrazine. Tridiphane alone had only limited activity, with no grass control after the five leaf stage. The tridiphane-atrazine mixture proved highly synergistic.[43-45] Indirect evidence showed a blockage of atrazine catabolism. Tridiphane did not directly inhibit the activity of purified glutathione transferase (Figure 1A). A glutathione transferase was found that irreversibly conjugates tridiphane with glutathione, forming a glutathione transferase inhibiting complex. This conjugate is a far more potent inhibitor of the glutathione transferase than tridiphane (Figure 1A).

Tridiphane synergizes atrazine, killing weeds in maize in the field, yet the maize is unaffected. This is surprising, as the tridiphane-glutathione conjugate inhibits the glutathione transferase that degrades atrazine in maize. This inhibition of glutathione transferase activity is less strong in maize than in *Setaria* (Figure 1B). A kinetic analysis showed that the tridiphane-glutathione conjugate is then a competitive inhibitor of glutathione, binding to glutathione transferase with a fourfold higher affinity for the enzyme from *Setaria* than the maize enzyme.[46] The *in vitro* data have been verified in whole plant experiments that also show tridiphane-glutathione conjugation.[47] The tridiphane-glutathione conjugate concentrations *in vivo* are high enough to explain the inhibition of glutathione transferase in *Setaria*. These

* This enzyme has usually been termed "glutathione-*S*-transferase" but as glutathione and not sulfur is transferred, the IUPAC nomenclature committee has classified it "glutathione transferase".

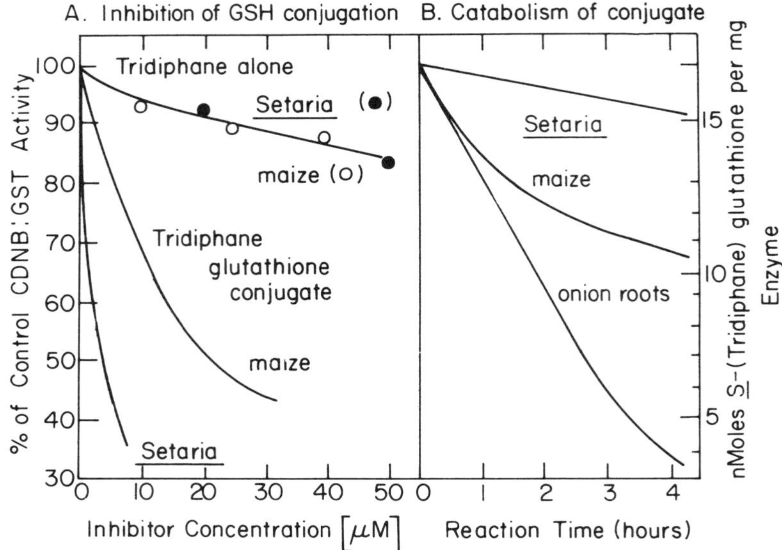

FIGURE 1. *In vitro* inhibition of glutathione-transferase activity by tridiphane-glutathione conjugate (A), and degradation of the conjugate by maize and *Setaria* enzymes (B) (replotted from data of Lamoureux and Rusness[46]). Conjugation of the artificial acceptor, CDNB, to glutathione was measured in (A), and the amount of parent conjugate remaining was measured in (B). Cell free extracts were used.

data alone are probably not enough to explain why tridiphane does not kill maize. Young maize leaves have six times more glutathione than *Setaria*,[46] which could partially explain the difference, but older maize and *Setaria* leaves have the same concentration.

The selectivity between maize and panicoid weeds is thus probably a complex series of events. These include the ability of maize to degrade much of the atrazine before tridiphane becomes an inhibitory conjugate, whereas the grasses do not degrade atrazine as quickly (Figure 1). The residual atrazine is not toxic because maize remains alive by virtue of having more glutathione and less tridiphane-glutathione conjugate than the grasses, with less activity and faster degradation of the conjugate. Thus, tridiphane must first be conjugated by the target organism. Glutathione-transferase must then be inhibited, and the tridiphane-glutathione conjugate must be stable for tridiphane to synergize atrazine or other pesticides that are conjugated to glutathione.

Tridiphane prevents degradation of EPTC and alachlor in maize.[48] The mode of synergy has not been elucidated, but these herbicides are degraded by glutathione conjugation. Surprisingly, tridiphane also synergized chlorotoluron, which is not conjugated by glutathione.[49] The effect of tridiphane is not limited to synergizing atrazine in grasses. The *Abutilon theophrasti* biotype that evolved atrazine resistance due to high levels of glutathione transferase was controlled by atrazine tridiphane mixtures.[50] Tridiphane synergized the insecticide diazinon against houseflies (*Musca domestica*).[51] Tridiphane had a 20-fold greater rate of conjugate formation with glutathione than diazinon, and diazinon does not seem to compete with this step.

B. SYNERGISTS INHIBITING MONOOXYGENASE ACTIVITIES

Many plant species possess an array of enzymes capable of oxidizing xenobiotics, including herbicides. They can confer crop/weed selectivity, as well as the differences in rates required to control different weed species. The oxidation step introduces a single oxygen from molecular oxygen atom into the herbicide molecule with the other atom of oxygen forming water. This reaction is defined as a monooxygenation mediated by a monooxygenase. There is direct evidence in many cases for such monooxygenases being NADPH dependent cytochrome-P_{450} mixed function oxidases (cyt-P_{450}).[52] These cytochrome P_{450} monooxygenases are bound to plant microsomal membranes. In bacterial herbicide-degrading systems, these enzymes are soluble.[52] Similar reactions in weeds presumably are also cyt-P_{450} mediated, but the evidence is only circumstantial.

Recently, a new type of herbicide resistance has evolved, which is ascribed to cyt-P_{450} systems. Two grass-killing herbicides, diclofop-methyl and chlorotoluron, have selected for resistant populations of *Lolium rigidum* and *Alopecurus myosuroides,* respectively.[25-28] These two new weed biotypes are also resistant to all other selective herbicides for wheat that controlled the wild type weed populations. This includes inhibitors of acetolactate synthase, acetyl CoA carboxylase, photosystem II, and tubulin polymerization.[25] A metabolic cross resistance probably evolved in these weeds using a single enzyme system. The weeds that evolved this resistance have evolved the predominant mechanism of wheat for degrading herbicides, i.e., by oxidation. All the available literature on wheat, suggests that wheat detoxifies the selective herbicides used in weed control for this crop by oxidation.[54] Wheat oxidizes these herbicides at a large number of sites, both on the rings and side chains of the herbicides. It is not clear if wheat or the grass weeds have one or many polysubstrate monooxygenases, nor is it known how the reaction(s) is/are genetically controlled. This is of less importance in our context as we should like to find monooxygenase inhibitors as synergists for particular herbicides.

Piperonyl butoxide is used to synergistically inhibit the oxidation of insecticides, especially with the very expensive pyrethroids,[55] including pyrethroid-resistant insects, by blocking the degradation of the insecticides. Piperonyl butoxide can be used to synergize herbicides in maize.[56] No evidence was presented that it acted by inhibiting monooxygenases. Piperonyl butoxide also partially suppressed the evolved cross-resistances to diclofop-methyl and chlorsulfuron in *Lolium*.[57]

Aminobenzotriazole is a well known monooxygenase inhibitor in mammalian systems. It is also an inhibitor of herbicide degradation (Figure 2). It is active in wheat, but impotent as a synergist on *Veronica persica*. This is the antithesis of what is desired from a herbicide synergist, as the crop but not the weed was controlled. Wheat degrades chlorotoluron by hydroxylation and *Veronica* degrades by *N*-dealkylation.[58] Different types of monooxygenases may be involved and they can be differentially synergized. This indicates that nuances are present and that it may be possible to find a monooxygenase inhibitor specifically affecting weeds, as was found with tridiphane. Synergism by aminobenzotriazole has often been used as proof for action of cyt-P_{450}s. This is highly speculatory as aminobenzotriazole also inhibits monooxygenases that do not contain cytochrome P_{450}.[59]

Fungicide research has centered on finding inhibitors of sterol biosynthesis, which are fungicidal but not phytotoxic.[60,61] These azole derivatives affect late monooxygenase steps leading to ergosterol. Some of them were useful plant growth retardants, due to blockage of gibberellic acid biosynthesis or other plant terpenoids, as well as sterol biosyntheses. This was traced to the blockage of kaurene oxidase, a plant monooxygenase that converts kaurene to kaurenol and then to kaurenoic acid.[61] Tetcyclasis, a known kaurene oxidase inhibitor was tested for its ability to prevent herbicide degradation in maize and cotton. It was a hundred times more active than aminobenzotriazole in these systems.[62] Tetcyclacis was also highly active in preventing the degradation of the photosystem II-inhibiting herbicide bentazon in both wheat and soybean (Figure 3). Similar differential results need to be found with less photolabile plant growth regulators. Indeed, other kaurene oxidase inhibitors such as paclobutrazol also synergize herbicides used in wheat by inhibiting their degradation by monooxygenases.

Propiconazole, a sterol biosynthesis inhibiting fungicide, was effective in synergizing difenzoquat, a specific herbicide for *Avena* spp.[65] The mechanisms of action and pathways

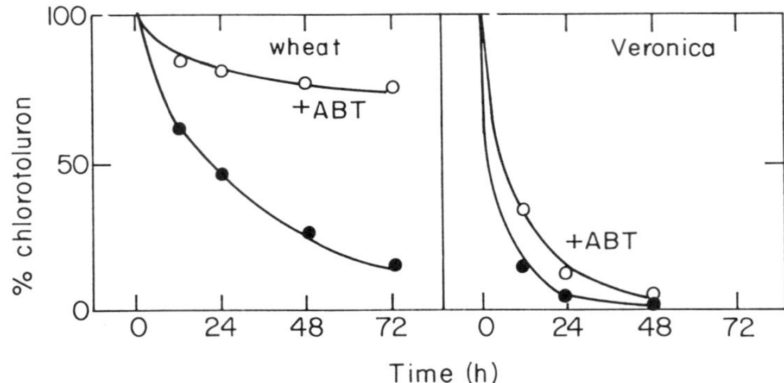

FIGURE 2. Differential synergism by prevention of monooxygenase degradation of chlorotoluron in (A) wheat, but not in (B) *Veronica* by aminobenzotriazole (ABT). The detoxification in wheat is by hydroxylation and in *Veronica* by *N*-dealkylation (modified from Gonneau et al.[58]).

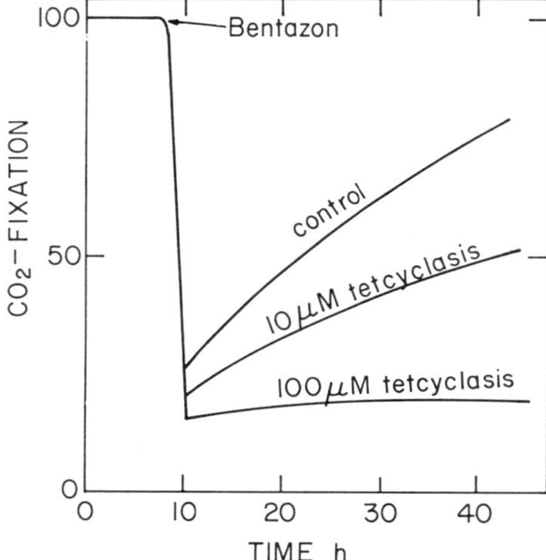

FIGURE 3. The suppression of bentazon detoxification by tetcyclasis. The prevention of loss of bentazon photosynthesis inhibition in soybean. This shows that the degradation of bentazon is prevented by tetcylasis. Similar results were achieved with wheat (redrawn from Fritsch et al.[63] and Retzlaff[64]).

of degradation of difenzoquat in wheat and *Avena* are unknown.[66] Thus, it remains to be shown that the synergy is due to monooxygenase inhibition. Clearly potential sterol biosynthesis inhibitors from fungicide development programs should be rescreened for herbicide synergism, with target weeds, and not crops.

C. SYNERGISTS SUPPRESSING THE OXIDANT DETOXIFICATION PATHWAY

Many herbicides kill plants by photogeneration of active oxygen species (Figure 4). The triazines, phenylureas, uracils, and some of the pyridazinones block electron transport at the reducing side of photosystem II of photosynthesis just before plastoquinone reduction.[18] Some phenolic herbicides act at a nearby site on photosystem II. The bipyridilium herbicides drain electrons from photosystem I, probably from ferredoxin.[69] Diverse herbicide chemistries, including the nitrodiphenylethers, cause plants to accumulate the photodynamic pigment protoporphyrin IX[70-76] by blocking the enzyme that oxidizes a precursor to protoporphyrin IX. This induces an overproduction of the precursor and its spontaneous oxidation to protoporphyrin IX.[13,77]

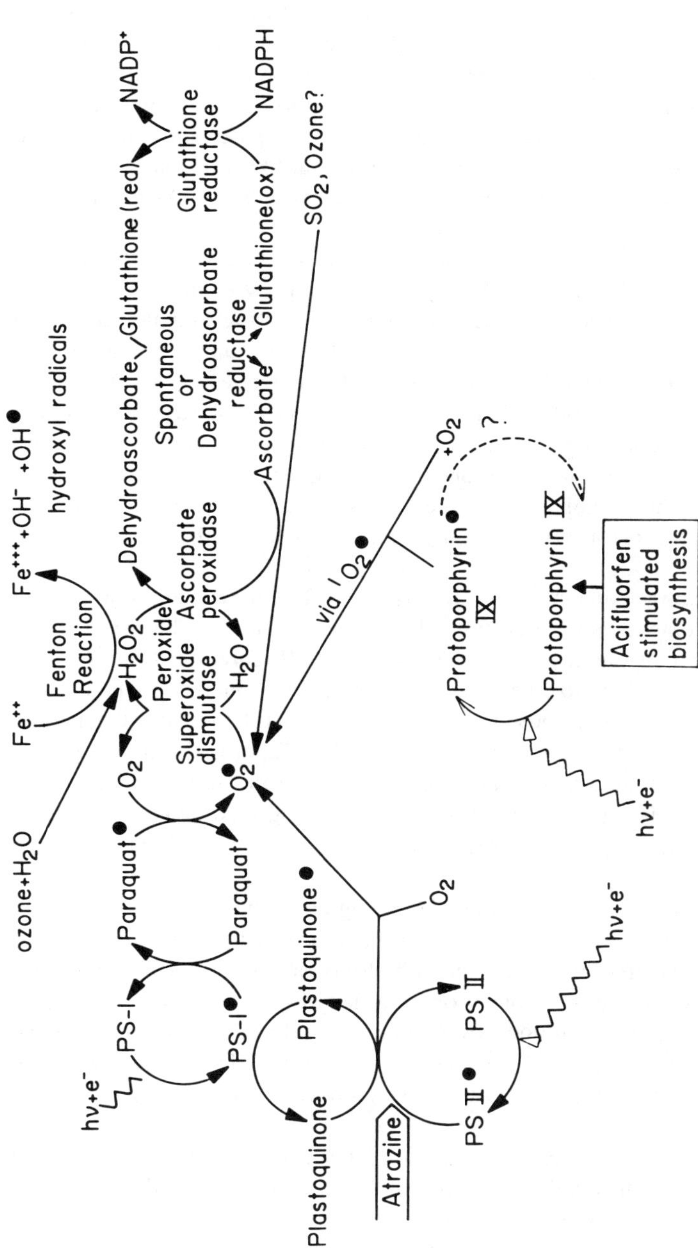

FIGURE 4. Interrelationships between oxidant generating herbicides and the pathways of oxygen detoxification. Radical species are denoted with •. Many interconversions between active species are possible, especially under physiological conditions[66] (modified from Shaaltiel and Gressel[68]).

It is not fully clear which active oxygen species is/are generated first with oxidant-generating herbicides. Membrane lipoxidation ensues when more active oxygen is generated than can be coped with by the endogenous detoxification system (Figure 4). This results in water loss, a general breakdown in the membrane-bound electron transport systems, and the transfer of solar energy to a variety of active oxygen species that can be confused with the first type generated. Chlorophyll radical, singlet oxygen, superoxide, hydroxyl radicals, etc. can be produced. This represents a self-amplifying toxic chain reaction of generation of active oxygen species; membrane lipoxidation and water leakage lead to the rapid desiccation caused by these herbicides. The stronger the light, the faster this chain reaction. Certain environmental xenobiotics (SO_2,O_3, NO_x) and some fungus-produced toxins such as cercosporin,[78,79] have similar effects. There are caveats in the literature stating that it is unlikely that singlet oxygen can be produced in leaves, where there are high levels of reductants such as NADPH, reduced glutathione, and ascorbate.[80,81]

Plants have endogenous active oxygen detoxification systems (Figure 4), which probably evolved together with photosynthesis and aerobic respiration to cope with energy "leakages" from the electron transfer chains. They include superoxide dismutase and the ability to produce and recycle oxyradical quenching agents such as glutathione and ascorbate. Plants also produce carotenes[82] and α-tocopherol,[83] which quench these active species. Herbicide toxicity requires that more radicals be produced than can be quenched by these systems before the herbicide is dissipated. Chloroplast-produced active species such as singlet oxygen and superoxide have very short diffusion distances and must be quenched before they reach membranes. The chloroplasts have their own concerted antioxidant defense mechanisms including the slowly recyclable carotenes, as well as an enzymatic pathway for oxygen detoxification.

A oxygen detoxification pathway using a series of enzymes while recycling ascorbate and glutathione has been put in context, especially by Halliwell[84] and Asada.[85] The first enzyme is superoxide dismutase, which dismutates superoxide to peroxide (Figure 4). Peroxide is less reactive than superoxide and can diffuse to other cellular compartments, but it is still potentially destructive. Peroxide can react with ferrous iron and form hydroxyl radicals (the Fenton reaction). Hydroxyl radicals are extremely damaging, more than superoxide. If peroxide is formed from superoxide, excess superoxide can react with the ferric ions, formed by the Fenton reaction, re-reducing them back to ferrous ions, and making them available to form more hydroxyl radicals. The Fenton reaction coupled with the iron recycling reaction is termed the Haber-Weiss reactions.

Chloroplasts do not contain catalase, a peroxide degrading enzyme. Chloroplasts contain ascorbate peroxidase and ascorbate, which can adequately compete with the Fenton reaction to remove the potentially destructive peroxide. The dehydroascorbate produced is recycled to ascorbate by glutathione, either spontaneously or by a dehydroascorbate reductase. The oxidized glutathione is recycled to glutathione by glutathione reductase, utilizing NADPH (Figure 4). Some NADPH must be produced to support this reaction. This system will also recycle both the glutathione and ascorbate used to quench singlet oxygen and other radical reactions.

The Halliwell-Asada pathway enzymes and reactants have been found in leaves of all species where sought. There is considerable evidence that elevated levels of the Halliwell-Asada pathway enzymes have primary responsibility for increased tolerance to some active oxygen species generating herbicides, keeping the plant alive until the herbicides can be dissipated.[86-93] This includes kinetic evidence that paraquat sprayed on leaves of the paraquat-resistant biotype of *Conyza bonariensis* rapidly but transiently inhibits photosynthesis.[87] The plants remained alive while paraquat was dissipated. The three major enzymes in the Halliwell-Asada active oxygen degradation pathway are constitutively elevated in isolated intact chloroplasts in the resistant *Conyza*.[88] Resistance was dominantly inherited under the control of a single gene. The levels of the three enzymes were elevated in the F_1 generation and high

TABLE 2
Compounds Known to Inhibit Enzymes of the Oxidant Detoxification Pathway

Enzyme	Contains	Inhibitors
Superoxide dismutase	Cu/Zn[94]	CN^-, DDC, N_3^-
Ascorbate peroxidase	Cu/Fe, thiol[95]	CN^-, DDC and thiol reagents
Dehydroascorbate reductase	Thiol[96]	IAc, PCMB, NEthMal
Glutathione reductase	Thiol[97]	IAc, PCMB, NEthMal, $ZnSO_4$

Note: DDC, diethyldithiocarbamate; IAc, iodoacetate; PCMB, *p*-chloromercurybenzoate; NEthMal, *N*-ethyl maleimide.

enzyme levels segregated with resistance in the F_2.[89] The high levels of the Halliwell-Asada pathway enzymes conferred tolerance to other herbicides and oxidant stresses.[40,90-93] Interspecific comparisons have correlated acifluorfen tolerance to increased levels of ascorbate and α-tocopherol.[83]

Blocking the Halliwell-Asada pathway and preventing regeneration of its products should decrease tolerance to oxidants. Any compound suppressing the pathway would have considerable possibilities as a synergist (Table 2). The plastid superoxide dismutase and ascorbate peroxidase are both copper-containing enzymes.[95] The former contains zinc and the latter tightly bound iron and thiol groups. The later enzymes of the pathway contain thiol groups as well.[95] Thiol binding reagents should inhibit the pathway, but they have high mammalian toxicity and would thus have little potential as synergists. There are few copper-containing enzymes and fewer yet zinc-containing enzymes in plants, and their inhibition would be more specific. Compounds that tightly bind them should serve as relatively specific synergists. Evidence is presented in Section III that indeed chelators of copper and zinc synergized oxidant generating herbicides. Conversely, iron-chelating compounds protected cells from paraquat. This is probably due to a blockage of the Haber-Weiss reactions, preventing hydroxyl radical formation.

D. INHIBITION OF GLYCOSYL TRANSFERASES

Glycosyltransferases are enzymes that conjugate a sugar moiety (usually glucose) directly to a substrate. They thus render lipophilic herbicides highly polar, as well as change their bulk configuration. Glycosylated herbicides are usually inactive. Two positions of glycosylation are common: *O*- and *N*-glycosylation, to oxygen and nitrogen groups, respectively. *O*-glycosylation reactions are usually reversible by relatively nonspecific β-glucosidases that are common in cells. Thus, if the *O*-glycosylation is directly to an active herbicide, the result is a "temporarily" inactive herbicide, which can be reactivated. Herbicides are often *O*-glycosylated at a position on the herbicide initially lacking an oxygen atom. The active herbicide was first hydroxylated, then glycosylated. Thus, it seems impossible to synergize herbicides at the level of inhibiting *O*-glycosyl transferases, although little is known about the nature and relevance of reactivated *O*-glycosylated to active herbicides. *N*-glycosylation seems to be an irreversible conjugation reaction. The easiest method to estimate whether a herbicide is *N*- or *O*-glycosylated is to react the product with a commercial β-glucosidase. The product is probably *N*-glycosylated if it remains uncleaved. There are no leads as yet on inhibitors of *N*-glycosyl transferases, but such compounds could clearly synergize many herbicides.

E. TARGET SITE RESISTANCES

Target site resistances have appeared due to mutations in herbicide binding sites that allow the weed to continue living. It should be very hard to design a synergist for such situations, except when "negative cross-resistance" is known. This is a phenomenon whereby a herbicide controls the resistant type at lower rates than it controls the wild type.[98] In some cases this is due to enhanced binding to the so-called resistant site. For example, phenolic herbicides and pyridate, (photosystem II inhibitors) are far more potent in inhibiting these reactions in triazine-resistant biotypes than sensitive biotypes.[98,99] The triazine resistance modification changed the configuration of the receptor protein (the D_1-Q_B binding protein of the photosystem II reaction center) such that it has greater affinity to these herbicides. There will be a synergistic effect if the herbicides with negative cross-resistance change the molecular configuration upon binding in such a manner that triazines will then also bind.

F. MICROBIAL DEGRADATION

Some soil-applied herbicides such as EPTC have lost effectivity against their target weeds in many locations. The weeds themselves had not evolved resistance, but the pesticides were no longer actively persistent in the soil, i.e., there are "problem soils".[32] The reason for the problem soils has been traced to an enhanced rate of microbial degradation of the pesticides.[100,101] The problem soils developed in areas with repeated use of EPTC, and EPTC was rapidly degraded in a manner yielding CO_2. EPTC problem soil was no longer problematic after sterilization, suggesting involvement of microorganisms.

Synergists have been developed as additives to soil-applied pesticides that prevent their biodegradation and enhance their persistence. Various methylcarbamates retard the degradation of propanil in the soil.[102] Dietholate has been approved to prevent degradation of EPTC[103] and other thiocarbamates (Figure 5). Mephenate is approved for use as a synergist to reduce the rate of degradation of carbamates and is formulated with propham and chlopropham. The microorganisms causing problem soils have not been isolated, and it is not totally clear how such synergists work. This requires that bioassays be directly performed with problem soils and not with isolated bacteria.

G. SYNERGIZING MYCOHERBICIDES

Mycoherbicides are biological control agents with great potential for destroying weeds. They are plant pathogenic with a host range of weeds, not crops. Conceptionally they are ideal weed control agents, i.e., natural, self-replicating herbicides that can be used wherever their host weed is a pest. They could be used to augment herbicides to control the weeds not controlled by the herbicides used in a given crop.

Mycoherbicides have so far not fulfilled their promise because they usually require dew point humidity for a minimum of 6 to 8 h, which is rare in the field. High inoculum levels (usually thousands of spores per square cm) are usually required. Part of the reason for the high inoculum threshold is the humidity requirement and part is the need to overcome the constitutive and parasite-induced defense mechanisms of the plant. Both could be overcome by synergists: adjuvants to preserve water near the spores and adjuvants to suppress the target weeds defense mechanisms. Herbicides have even been added to try to damage the leaf and facilitate physical penetration.[104] Chemicals have been "biorationally" added to modify host-pathogen biochemical interactions.

Inexpensive, simple, invert emulsions that have oil on the outside and water on the inside have been tested as synergists.[105] The droplets evaporate about 20% of their water during spraying; the rest slowly evaporates over a 2 day period giving the spores time to germinate and penetrate deep into the leaf.[105] This research was performed directly in the field. Bioassays with these emulsions are described in Section III.

Potential biorational synergists to enhance mycoherbicide activity can be divided into two groups: those that synergize the pathogen-produced toxins, and those that act as "immunosuppressants" and inhibit the plant responses. Many mycoherbicidal organisms produce toxins that help kill the plant. The cercosporins and related photodynamic dyes are produced by many *Cercospora* and other spp. and generate oxygen radicals.[78,79] The necroses appear as if droplets of paraquat had been put on leaves, due to the similar mechanisms of

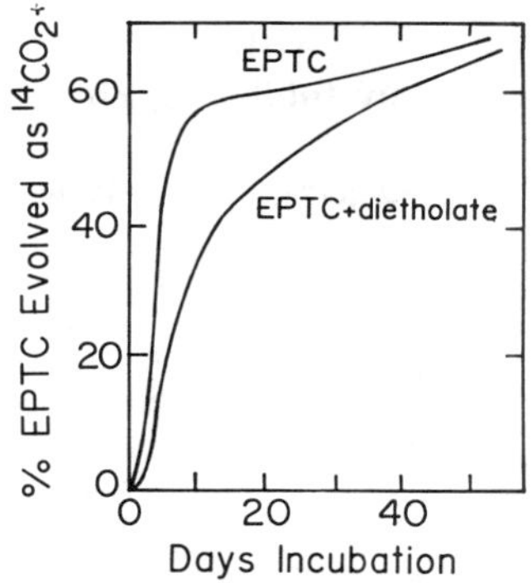

FIGURE 5. Synergistic prevention of microbial degradation of EPTC in problem soils by dietholate (replotted from Obrigawitch et al.[103]).

killing. Photodynamic fungal toxins must produce more radicals than the plant radical quenching system can detoxify. The copper and zinc chelators that synergize oxidant-generating herbicides (see Section II.C) may also synergize pathogens producing photodynamic dyes.[106] More biorational synergists can be designed as more knowledge about pathogen action and the plant responses is obtained.

Mycoherbicides can be synergized by suppressing plant defenses. Phytoalexin synthesis by the phenylpropanoid-shikimate pathway has been inhibited.[107] The inhibitors tip the balance of host-pathogen relations to the favor of the pathogens (Table 3). The activity of the first enzyme of the pathway, phenylalanine ammonia lyase, usually increases upon infection, and the products are used both to synthesize lignin and phenylpropanoid phytoalexins. The use of inhibitors of this enzyme allowed infection by tobacco mosaic virus and a noninfective strain of *Phytophthora megasperma* (Table 3). Many of these synergizing compounds are agriculturally useless as they do not normally penetrate leaves. They are fed through cut roots, stems, or petioles, or they are vacuum infiltrated.

Glyphosate acts early in the pathway to phenylpropanoids, before phenylalanine biosynthesis.[117] It depletes the levels of all of the aromatic amino acids as well as phytoalexins. Glyphosate synergized pathogenicity of bacteria as well as soil-borne and foliar-attacking fungi (Table 3). Glyphosate is truly synergistic as it is applied at sublethal levels, usually at two orders of magnitude lower concentration than is used for weed control. These low concentrations are sufficient to suppress EPSP synthase in the plant.[117] Glyphosate would not synergize *Alternaria crassa* on *Datura,* which produces sterol and not phenylpropanoid phytoalexins. Sublethal levels of glyphosate cause rapid increases of a variety of hydroxybenzoic acids in plants,[118] which may weaken them allowing fungal attack. Many other herbicides might act as synergists for mycoherbicides, either by causing local lesions for fungal penetration, or by blocking secondary metabolism at some level, preventing the appearance of phytoalexins.[119] Some of the methodologies developed for synergizing mycoherbicides are presented in the following sections.

There can also be synergistic interrelationships between mycoherbicidal pathogens. For example two pathogens alone hardly affected *Senecio vulgaris*, but their sequential treatment was lethal.[120]

III. BIOASSAY SYSTEMS

There are a few primary requirements for a bioassay system: it must be quick, easy, reproducible, and amenable for checking large numbers of potential synergists. *In vitro* bioassays for random, general herbicide activities are harder to devise than for "dedicated" assays for synergists for a particular herbicide. The fact that herbicides attack different

TABLE 3
Synergizing Pathogens by Inhibiting Phytoalexin Biosynthesis

Enzyme inhibited	Inhibitor[a]	Pathogen synergized	Ref.
Phenylalanine	AOP,AOPP	Tobacco mosaic virus	108
ammonia lyase (PAL)	R-APEP	*Phytophthora megasperma*	109
5-Enolpyruvate	Glyphosate	*Phytophthora megasperma*	110, 111
shikimate-3-phosphate		*Pseudomonas*	112
(EPSP) synthase		*Colletotrichum lindemuthianum*	113
		Alternaria cassiae	114
—	Blasticidin-S	*Pseudomonas* sp.	114
—	Phaseotoxin	*Pseudomonas* sp.	114

[a] AOP, α-aminooxyacetic acid; AOPP, L-aminooxy-β-phenylpropionic acid; OBHA, *o*-benzylhydroxylamine; R-APEP, *R*-(1-amino-2-phenylethyl)-phosphoric acid.

pathways that are not always present in all cells or all organs or all species requires considerable planning and redundancy to include different tissue types from different species to make a random general screen universal.[19,20,121]

One can, with some risk, use easy test systems in organisms that are unrelated to the weeds to be controlled. For example, phytoalexins are derived from either phenyl propanoid and sterol pathways that both are products of the mevalonate pathway. A very simple, highly sensitive system for assaying suppression of the mevalonate pathway has been developed using *Halobacterium halobium*, which makes a pigment from products of all three pathways.[122] Compounds inhibiting one of the key enzymes, 3-hydroxy-3-methylglutaryl Co-enzyme A (HMGCoA) reductase in *Halobacterium* was active in rat liver,[122] showing the possibility of extrapolating among systems.

The dedicated screen requires that the system to be synergized is well represented (i.e., weeds are used and not crops). The available techniques span from direct methods for measuring prevention of herbicide degradation to methods of estimating damage.

A. ASSESSING PREVENTION OF HERBICIDE DEGRADATION

After applying herbicide and synergist to one of the bioassay systems (Sections III.D, III.E), the efficacy must quickly be measured. It is very helpful to have radiolabeled herbicide, unless there is a specific color reaction or immunoassay available for measuring the parent herbicide or its product. Otherwise, very expensive or time consuming chemical purifications and analyses may be needed (e.g., HPLC, GC/MS, etc.).

Immunoassays are rapidly replacing direct chemical techniques for measuring pesticide residues.[123] They can assay how well potential synergists prevent herbicide degradation, but only after determining that the antibody does not artifactually detect the metabolites as well as the parent herbicide (Chapter 11).

It is not always necessary to know the precise chemical nature of the first inactive herbicide metabolite. A typical example is when the herbicide is lipophilic and the first product is polar. The general procedure is to extract the tissue with a solvent that removes both herbicide and metabolites (80% methanol), evaporate the methanol, precipitating chlorophyll, and partition the parent herbicide into an apolar solvent, leaving the metabolites in the aqueous phase. Only short term experiments should be necessary to discern synergy, as the first metabolic step is usually the rate limiting step, the one to be blocked by a synergist. Once control experiments have validated phase partitioning between parent herbicide and metabolite, large numbers of potential synergists can be tested. A secondary assay is advisable to recheck those that act well to assure that indeed the herbicide is still in the parent form. More complex assays are needed in cases where such a partitioning assay cannot be devised. The simplest and least expensive (and least quantitative) procedure for measuring inhibition of metabolite formation is TLC. Automated sample injectors and computer interfacing make HPLC, GC, and GC/MS less laborious (but more capital intensive), giving accurate results, but they are very slow for large numbers of samples. Many herbicides are racemic mixtures, with only one enantiomer biologically active, and possibly only the active one is metabolized. When enantiomers are present, the results of synergy by chemical assay must be verified by a biological test.

In summary, the chemical homework telling that synergy is feasible by preventing herbicide degradation gives the background information for devising simplified rapid specially dedicated assays that will demonstrate the capacity of a compound to be a synergist and allow QSAR of other compounds synergizing at the same site.

B. ASSESSING SYNERGIZED PHYTOTOXICITY

There is no basic difference between assessing synergized phytotoxicity in a dedicated assay (specifically designed to assay for a particular site) and just assessing herbicidal phytotoxicity, except for the added constraints and mathematical analyses of proving the "greater than additive" effect of the synergy. Mathematical analysis is unnecessary when both the potential synergist and the herbicides are coalitive, i.e., where neither are effective alone. When either or both components have to be used at a partially lethal rate, problems ensue (see Chapter 7). Partially lethal rates are hard to reproduce, requiring great uniformity of the biological system. The more "*in vitro*" one gets, using material from controlled environments, the more likely one will achieve adequate results. For example, the levels of the active oxygen detoxifying system (Section II.C) in maize and *Conyza* vary with age,[91-93] and higher levels can also be induced by various stresses.[40] The levels of glutathione transferase and glutathione can be affected by pretreatment with various xenobiotic and environmental stresses.[41] Thus, tissue from these plants from nonuniform sources will require different levels of herbicide and synergists in different experiments, complicating experiment size and design.

1. Visual Estimation

Visual estimation of damage has the advantage of being quick and inexpensive. Even when not used as the final arbiter of synergism, it is highly desirable. First, it will show up unexpected results to researchers capable of realizing when novel results are serendipitous. Second, in many biological systems, when there is no visual damage, there is no need to continue evaluating the results by other, more time consuming procedures. Conversely, when there is ample, rapid, visible evidence for synergy, why bother to analyze further? Visual estimations are often superior to other methods, especially weight determinations. Paraquat treated and synergized plants dry up within hours — in some species even without bleaching.[88] Control plants would have to be grown for at least a few days more for there to be significant dry weight differences between dead and living plants. Personnel can often visually assess damage of coded material after a minimum of training with near the same accuracy as more quantitative techniques.

2. Growth Measurements

If the herbicide to be synergized is slow to kill but slows or stops growth, the effect of the synergist can be measured as growth. This has become easier in cell culture systems with the advent of simpler techniques that can replace gravimetric methods. These include tipping cells into the side arm of special erlenmeyer flasks and measuring optical density,[124] tipping small uniform erlenmeyers at a fixed angle in a special stand and measuring the diameter of the sediment,[125,126] and using large arrays of stacked 6 cm diameter petri dishes containing thin layers of medium with cells and measuring packed volume.[127] These techniques are described at greater length in Chapter 10.

3. Estimating Membrane Damage

One of the first signs of imminent death is membrane damage. Damage to the plasmalemma leads to leakage of water, protons, and solutes to the outer medium. Damage to thylakoid membranes leads to rapid chlorophyll bleaching. As membrane damage is often due to lipoxidation, oxidation products of the membrane lipids can be measured.

a. Membrane Leakage

Herbicide and herbicide plus synergist induced-leakage can be measured in a variety of ways. The simplest is to treat leaf discs, roots, cells, etc. with the herbicide and synergist mixture for a fixed duration, wash off the material in distilled water, and incubate for a fixed time in distilled water. The proton or ion content of the water is measured by conductivity, proton extrusion with a sensitive pH meter, ion specific electrode, flame photometry, atomic absorption, or color reactions. A necessary control is to boil untreated tissue to ascertain the magnitude of the maximum possible level of ion leakage. Conductivity has the disadvantage of the nonlinearity to ion concentration, giving high sensitivity at low ionic concentrations and vastly reduced sensitivity at higher ionic strengths. This requires that the terminal incubation be done in distilled water with only nonionic additives such as sugar, etc. Many herbicides have low affinities to their sites of action and their effect is stopped by wash off. Still, conductivity is often used to assess damage. It is also possible to preload the tissue with a radiolabeled $^{86}Rb^{+}$ or dyes[128] and measure their leakage.

Measurement of transmembrane proton gradients may have utility as an *in vitro* test for herbicide protectants, as with diclofop-methyl. Both genetic and biochemical studies strongly suggest that the primary site of action of this herbicide is acetylCoA carboxylase. This enzyme activity is not antagonized by 2,4-D, yet in the field, plants are protected by their joint application. 2,4-D and diclofop-methyl interact at the level of proton extrusion, while not at the level of the enzyme.[129] Similarly, if the rapid effect of certain phosphonic esters on membrane depolarization was inhibited, there was no subsequent membrane damage.[130] The same system could be used to screen synergists; they would be compounds that either increase the magnitude or the duration of the effect.

The choice of leakage technique depends on the constraints of herbicide and synergist formulations (e.g., a potassium salt of herbicide or synergist precludes measuring potassium leakage), constraints of incubation, economics, time, and biological system.

b. Bleaching

Many herbicides cause chlorophyll bleaching, which can be used as an easy assay; the tissue is extracted and chlorophyll concentration spectrophotometrically determined. An example is given for a synergistic mixture in Table 4. The ideal solvent for such operations is not acetone, as had been used in the past,[132] except with single cell suspensions or thylakoids. Tissue must be homogenized if acetone is used. Dimethylformamide has the advantage that leaves and tissues need not be homogenized; tubes are left in the dark overnight and the chlorophyll equilibrates evenly with the solvent.[133,134] Dimethylformamide, which has a much lower vapor pressure than acetone, has the advantage that tight sealing of each extraction vessel is not required.

c. Measuring Lipoxidation Products

Membrane degradation products are often measured to assess herbicide induced cell damage, especially the formation of malonyldialdehyde and ethane.[135] When malonyldialdehyde works, it has the advantage of being a rapid and simple test. Herbicide/synergist treated cells or tissue are harvested, boiled with acid, and reacted with thiobarbituric acid, a reagent giving a color upon reaction with malonyldialdehyde and similar membrane degradation products.[135,136] Unfortunately plant pigments and some metabolites interfere with the color reading or the

TABLE 4
Synergizing Paraquat by Chelators Using Isolated *Asparagus sprengeri* Cells

	Chelator (rate) % Bleaching						
Rate of paraquat (m*M*)	No chelator	DDC (0.02%)	OxOAP (0.023%)	TEPA (0.01%)	DEHPA (0.001%)	PDC (0.005%)	8-HQ
0	0	0	16	0	3	2	0
20	10	88	74	99	84	78	0

DDC-diethyldithiocarbamate; OxOAP, oxime of 4-octyl-2-acetylphenol; TEPA, triethylene pentamine; DEHPA, di-2-ethylhexylphosphoric acid; PCD, pyridine-2,6-dicarboxylic acid; 8-HQ, 8-hydroxyquinoline. Cells were isolated according to Colman et al.[131] and incubated with mixtures for 14 h under light, and chlorophyll was then extracted. Data from Shaaltiel and Gressel.[68]

whole reaction.[137] Some putative synergists may directly interact (positively or negatively) with the color generating reagent. Thus, malonyldialdehyde estimation can be used in certain dedicated assays, but important findings should be verified by other techniques.

Ethane is easily measured after a herbicide has acted on plants, tissues, cells, or whole chloroplasts.[87,138] The reaction must be performed in closed tubes, and an aliquot of the head space is injected into a gas chromatograph. If the aliquots of headspace gasses are to be stored, glass syringes should be used, as the gases may be absorbed in plastic.[139] Care must be taken that ethane and ethylene are well resolved, as ethylene can emanate from other sources. Large scale screening can only be done with a GC having a sample changer with automatic injection, otherwise measuring ethane is too time consuming.

4. Photosynthesis

Many selective herbicides transiently inhibit photosynthesis in the "unaffected" crops (e.g., atrazine in maize) and irreversibly in target weeds, which lack a system to degrade the herbicide. A synergist that prevents degradation of such a herbicide is easy to visualize by measuring photosynthesis (Figure 3). The manometrical or IR gas measurements of CO_2 fixation are very amenable to kinetic determinations but are not easy for large scale screening, as many determinations cannot be made simultaneously. O_2 evolution can be measured with the oxygen electrode, but each determination takes time and must be performed sequentially.

$^{14}CO_2$ fixation can be inexpensive and large numbers of determinations can be made simultaneously.[87] Leaf discs, cells, plastids, or thylakoids are placed in small scintillation vials or on planchettes in racks with herbicide/synergist and incubated in the light for a predetermined time, known to show irreversible stoppage of photosynthesis. The racks are then placed in a sealed transparent chamber and $^{14}CO_2$ is injected. After a fixed time (10 to 30 minutes) the chamber is flushed with air, and the reaction stopped. Planchettes can be dried and counted directly in a geiger counter with a sample changer. More $^{14}CO_2$ must be used, but it is cheaper than scintillation fluid and vials. Samples in scintillation tubes must be bleached to prevent scintillator quenching. A simple technique using 80% acetone and bright lamps has been described.[88] Both the herbicide/synergist treatment and all measurements are performed in the same container used in measuring $^{14}CO_2$ fixation, without any transfers. This has distinct advantages in large scale synergist screening efforts.

5. Vital Staining

There are many types of measurements to assess whether cells or tissues are dying as a result of treatment. The most commonly used is fluorescein diacetate, which is enzymatically cleaved within live cells to fluorescein. Fluorescein has bright yellow fluorescence in a microscope with UV irradiation.[140] Such examination is not well suited for large scale screening. Two types of vital measurements are amenable for synergist screening: spectrophotometrically measuring whether the cells still have respiratory or photosynthetic capability with a tetrazolium dye, or measuring whether the cells can still synthesize macromolecules by incorporating radioactive monomers.

Cells or tissue pieces can be incubated overnight with 2,3,5-triphenyl-tetrazolium to measure respiration; the oxidized (formazan) color complex is extracted into alcohol, the cells sedimented, and the intensity of color is determined.[141] Nitroblue tetrazolium and a fluorescein conjugated tetrazolium have been used to measure photosynthetic capacity. Inhibitors of photosynthesis blocked the reaction.[142] In this case examination was microscopic, but the product may be extractable. If a herbicide affects any of the lipid or sterol biosynthesis pathways, the incorporation of [^{14}C] acetate can be inexpensively measured.[143] This requires a partitioning step to separate parent acetate from acetate incorporated into polar components from apolar (lipid) components.

As both protein and nucleic acid biosyntheses require large amounts of ATP, which can be produced only by healthy tissue, amino acid or nucleic acid base incorporation into macromolecules can be measured. The cheapest ^{14}C precursor is a [^{14}C]-amino acid mixture (algal protein hydrolysate), but it cannot be used when the herbicide in question prevents amino acid biosynthesis. The radioamino acids can replace missing amino acids. After incubation with tissue or cell pieces, the reaction is stopped with trichloracetic acid and/or alcohol/acetic acid, and the unincorporated precursor is removed by washing cells on a glass fiber filter or transferring tissue pieces through an alcohol series.[144] Counting of dried material can be on planchettes in a geiger counter or in a scintillation counter. Alcohol washes must remove interfering chlorophyll for scintillation.

Both the tetrazolium and incorporation type assays gave comparable results to growth assays with many herbicides.[121,144]

6. Measuring Synergized Mycoherbicide Infectivity

The efficacy of mycoherbicides is usually measured at the whole plant level as weight reduction or by visual estimation. The problems involved in estimating the intensity of pathogen infection by such visual estimation have been reviewed at length.[145,146] Human-assessed visual scales are being replaced by optical imaging. On the super macro scale, pathogenicity can be measured from satellites by infrared photography[147] — on the single leaf level by video imaging linked to microcomputers.[148] Such techniques still are time consuming and require photography, then analysis.

Two *in vitro* assays were developed to measure synergistic interactions with a mycoherbicidal preparation of *Alternaria cassiae*, which attacks the weed *Cassia obtusifolia*.[149,150] An invert emulsion[105] that preserved water around the spores allowed 100% infectivity with just one spore per droplet in a leaf microassay, whereas 20 to hundreds of spores were required in water droplets, depending on humidity.[151]

A radioimmunoassay to quantitatively detect the amount of mycelium in the leaf tissue,[150] and a fluorescence assay to detect the levels of a flavanone phytoalexin produced by the leaves,[149] were developed as *in vitro* assays for synergisms. A single flavanone phytoalexin was produced by the plant leaves, as determined by chemical and fungal bioassay. Data are shown in Table 5 using a nonphytotoxic addition of glyphosate as a synergistic inhibitor of the shikimate pathway that provides precursors for flavanone phytoalexin production. Glyphosate was added with the spores. Synergy was shown using both assay systems. The fluorescence data showed an inhibition of phytoalexin, which inversely correlated with the radioimmunoassay used as a measure of infection (Table 5). The fluorescence technique is only good to detect putative phytoalexin suppressing synergists. The invert emulsion had no effect on the level of flavonoids, and thus the fluorescence technique should be used only

TABLE 5
Synergistic Interaction in Mycoherbicide Pathogenecity: Measurement of Effects of an Invert Emulsion and a Phytoalexin Synthesis Inhibitor Using Radioimmunoassay and $AlCl_3$ Flavanone Fluorescence

Treatment	$AlCl_3$ fluorescence (mequiv naringenin per g fresh weight)	Immunoassay (mg mycelium per infection site)
Alternaria cassiae alone	3600	13
50 m*M* glyphosate alone	97	0
Invert emulsion alone	—	0
A. cassiae + emulsion	—	39
A. cassiae + glyphosate	1900	35
A. cassiae + glyphosate + emulsion	—	73

Note: 2 μl droplets containing (where stated) *A. cassiae* conidia were placed on leaflets of *Cassia obtusifolia* leaves as described previously.[151] Leaf discs (4 mm diameter) were removed at 24 h for analyses. Flavonoid appearance was shown to correlate with phytoalexin concentrations on TLC plate bioassay and was measured as Δ fluorescence with an excitation λ 340 nm and an emission at λ 472 nm using naringenin as a fluorescence standard. A polyclonal antibody was raised in rabbits against *A. cassiae* mycelia, and a standard radioimmunoassay technique was used for measurement, using glass-glass homogenized mycelium for calibrating absolute amounts. Data are from different experiments.[149,150,152]

for one type of synergist, the radioimmunoassay is amenable for all synergists. Immunological techniques can be applied to other organisms by raising specific antibodies. Specific chemical assays need be found for quickly detecting suppression of production of other phytoalexins.

C. ASSAYING SYNERGISM WITH ENZYME PREPARATIONS

Measuring the relative effects of putative synergists is probably most accurate at the level of purified enzymes, but crude enzyme preparations can often be used. If a precursor to a pathway is used as a substrate for a bioassay, it may be possible to screen the effect of putative inhibitors of the whole pathway. Extraneous penetration and translocation factors are obviated, kinetics are easy to perform, and accurate K_i values can be determined. Partial purification of an enzyme preparation can remove cellular components that interfere with measurements. With luck, an enzyme preparation can be stabilized and stored for extended periods, allowing use of a single preparation for a long duration. The first glimmers of progress with various pathways are described below.

1. Direct Inhibition of Glutathione Transferases

Because tridiphane itself is not directly a glutathione transferase inhibitor but must first be conjugated (Figure 1), purified enzyme preparations may miss some inhibitors of this enzyme. Additionally, there is some information that tridiphane is capable of inhibiting more pathways than just glutathione transferase.[48,49]

Whereas there have been many publications on compounds protecting plants by increasing glutathione transferase, there seem to have been only a few for measuring synergy at the enzyme level.[46,51]

2. Direct Inhibition of Monooxygenases

Some natural resistances,[153] and some recently evolved resistances to herbicides,[154] are due to high levels of monooxygenases that can be inhibited by synergists. Isolated microsomes were used to assay the effects of putative cytochrome P_{450} inhibitors. The levels of cytochrome P_{450} monooxygenases in plant systems are usually far lower than animal systems. This led many researchers to use tissues having high monooxygenase activity such as avocado pericarp

TABLE 6
Bioassaying Synergists Inhibiting Monooxygenases at Varying Levels

Process inhibited	Species	Potential synergists	Ref.
Microsomal fractions			
Bentazon hydroxylation	Maize	Tetcyclasis	155
Bentazon hydroxylation	Sorghum	Unstated	159
Chlorotoluron hydroxylation	Wheat	Aminobenzotriazole, menadione	160
Diclofop hydroxylation	Wheat	Tetcyclasis[a]	53
Furanocoumarin synthesis	*Amni majus*	Tetcyclasis, ancymidol	161
Cell suspension cultures			
Furanocoumarin synthesis	*Amni majus*[b]	Tetcyclasis, ancymidol	162
Chlorotoluron hydroxylation	Cotton, maize	Aminobenzotriazole, 3(2,4-dichlorophenoxy-1-propyne, tetcyclasis	62
Chlorotoluron hydroxylation	*Alopecurus*	Prochloraz	163
Chlorotoluron hydroxylation	Wheat	Prochloraz	163
Whole plants or parts			
MCPA metabolism	Potato	Aminobenzotriazole	164
Chlorotoluron hydroxylation	Maize, cotton	Aminobenzotriazole	62
Chlorotoluron hydroxylation	Wheat	Aminobenzotriazole	165, 166
Chlorotoluron hydroxylation	*Bromus sterilis*	Aminobenzotriazole	167
Chlorotoluron hydroxylation	*Alopecurus* (R)	Aminobenzotriazole	168
Chlorotoluron hydroxylation	*Galium aparine*	Aminobenzotriazole	
Bentazon metabolism	Soybean	Tetcyclasis, BAS110	169
Bentazon hydroxylation	*Amaranthus* etc.	Tetcyclasis, BAS111	170
Isoproturon and others	*Alopecurus*	Tridiphane	49
Diclofop-methyl	*Lolium*	Piperonylbutoxide	57

[a] Also by herbicides, especially chlorsulfuron and haloxyfop, which may compete for binding sites on monooxygenases.
[b] See Matern et al.[171] for many more examples.

or *Helianthus tuberosus* slices. These are possibly excellent systems for studying monooxygenases per se, but have little direct relevance to herbicide-degradation. The activity of herbicide-degrading cytochrome P_{450}s could be enhanced over 20-fold by pretreating seeds with naphthalic anhydride,[155-157] and even more so by ethanol.[158] The affinity of a maize P_{450} preparation for different herbicides varies, and the competitiveness for the site(s) on P_{450} is roughly correlated with the affinity.[158] This would support a hypothesis that there is a single P_{450} or a family of P_{450}s with similar broad substrate specificities but with different substrate affinities. Aminobenzotriazole and tetcyclasis inhibited bentazon degradation in crop microsomal preparations (Table 6). This correlates with the prevention of recovery of inhibition of photosynthesis in whole tissue by tetcyclasis (Figure 3). Not all oxidative degradations of herbicides by microsomal preparations are by cytochrome P_{450}s. The *N*-demethylation of chlorotoluron by *Veronica* was not inhibited by carbon monoxide (and thus does not meet the criteria of a P_{450}) unlike the methyl hydroxylation.[160]

It should be possible to spectrophotometrically or metabolically measure the synergism by inhibition of cytochrome P_{450} activity. The spectrophotometric method is fastest but may require a highly purified enzyme preparation, as the levels are low and plant pigments interfere. The synergist should prevent the spectral shift from a peak of 450 nm to 430 nm that occurs when the herbicide binds to the enzyme. This would not be useful in cases where the synergist acts by suicidally binding to the P_{450}, causing the same shift in spectrum. So far researchers have preferred to measure the inhibition of appearance of degradation products of the herbicide (Table 6). Secondary or further metabolites of the herbicide are often found in crude preparations, as enzymes are present that attack the primary metabolites upon their appearance. This could be precluded (in theory) by a few washes of the microsomal pellet to remove soluble enzymes as well as soluble substrates such as sugars needed for further conjugation.

The microsomal extract should be from the weeds that need more adequate control because of specificity. In greenhouse tests there was a tenfold synergy of bentazon by tetcyclasis in *Amaranthus retroflexus*, but some species were unaffected and others had only a two- to fourfold synergy.[170] There was no comparison between these whole plant studies and *in vitro* studies.

3. Direct Inhibition of the Oxidant Detoxification Pathway

Many compounds can inhibit the enzymes of this pathway (Table 2). Some, such as cyanide, azide, and the thiol reagents are hardly suitable for consideration as synergists because of their human toxicity. A concept and initial data on using chelators of copper and zinc to synergize oxidant generating herbicides has been described.[68,172]

Superoxide dismutase is assayed in a negative-negative reaction.[173] Superoxide is either enzymatically generated (xanthine-xanthine oxidase) or light generated with riboflavin. Superoxide dismutase degrades the superoxide before it can interact with a tetrazolium dye, preventing color formation. Kinetic parameters such as K_m and V_{max} cannot be measured.

As chloroplast-derived superoxide dismutase is commercially available, it would seem to be an ideal mechanism for measuring potential synergism. Alas, many of the potential synergists that we tested directly interact with the color generating system; they prevent color formation without superoxide dismutase.[174] Perhaps a polarographic method that has been described[175] could be useful.

Ascorbate peroxidase activity is spectrophotometrically measured by following the oxidation of ascorbate to dehydroascorbate at 290 nm.[176] This can easily be performed by first isolating intact chloroplasts, breaking them, and using a crude enzyme preparation of the soluble supernatant.[88] There has been no screening for potential synergists at this level, to the best of our knowledge.

Glutathione reductase is easy to measure as the decrease in absorbance at 340 nm due to the oxidation of NADPH by oxidized glutathione.[177] Other than the toxic thiol reagents, there are no known specific inhibitors. This would be an excellent target for synergy, due to the key role of this enzyme (Figure 4).

4. Direct Inhibition of Glycosyltransferases

The activity of these enzymes is easy to measure using radioactive uridine diphosphate-[^{14}C]-glucose as a substrate and then either a partitioning or a TLC system. Many of these enzymes have been isolated to homogeneity, but the more interesting *N*-glycosyltransferases have been harder to purify, but their activity can be measured in crude preparations.[178] No specific inhibitors of this group of enzymes are known to the best of our knowledge.

5. Detection of Negative Cross-Resistance at the Target Site

The possibility that some herbicides may synergistically prevent or delay herbicide resistance by exerting greater control of resistant biotypes than sensitive ones is described in Section II.E. Such negative cross-resistance in triazine-resistant weeds has been measured at the level of isolated thylakoids by various researchers (Table 7). Whole plant data are needed to show that the thylakoid data can be extrapolated to the higher plant.[98] Indeed, in one case it has been found that bentazon exerts negative cross-resistance on triazine-resistant *Amaranthus retroflexus* (Figure 6), but has (positive) cross-resistance with triazine-resistant *Chenopodium album* and *Solanum nigrum*.

TABLE 7
Enhanced Efficiency (Negative Cross Resistance) of Some Herbicides in Controlling Weeds That Evolved Resistance

Primary resistance	Enhanced sensitivity	Enhancement factor[a]	Number of species	Ref.
Atrazine	Dinoseb	3.3	5	179, 180
	Dinoterb	14.2	1	99
	Medinoterb	5.0	1	99
	DNOC	2	2	181, 182
	Chlorpropham	2.2	1	183
	Pyridate[b]	4.8	1	99
	Ioxynil	2.2	1	99
	Bromoxynil	1.4	1	99
Mecoprop	Benazolin	1.9	1	184
MSMA-DSMA	Paraquat	2.0	1	185
	Bentazon	1.5	1	185
Chlorsulfuron	Imazaquin[c]	33.3	1	186
Paraquat	Glufosinate	3.8	1	187

[a] Enhanced factor calculated as the I_{50} of the sensitive divided by the I_{50} of the resistant biotype. Most of the data for atrazine are from isolated thylakoids and not from plant studies. Values are averages from various reports.
[b] Experiments were performed with the active metabolite of pyridate.
[c] Most of the chlorsulfuron resistant mutants isolated in this study were (positively) cross-resistant to imazaquin.

One case of a 30-fold negative cross-resistance by imazaquin was reported with a chlorsulfuron-resistant line of *Datura*.[186] It has not yet been reported whether this follows through to the level of acetolactate synthase, nor is it relevant to weed control, as there are field cases of positive cross-resistance among sulfonylurea and imidazolinone herbicides.

6. Others

Enzyme assays for synergisms can be elaborated only as the enzymes responsible for loss of herbicide activity become known or at least become detectable.

D. ASSAYING SYNERGISM AT THE CELLULAR LEVEL

The cell culture/callus/protoplast systems have the many advantages cited in Chapter 10. Such systems, are even more advantageous for seeking synergists because they can be dedicated to a specific herbicide or herbicide group and have greater predictive value. False-negative results are reduced because the assay is "dedicated". Once it is ascertained that the specific pathway to be blocked is operative in the system, reliability is rather assured. Few advantages can be foreseen from using callus or protoplast systems. Calli grow slowly and are less amenable to liquid additions, and callus-cells are not equally bathed in solution. The diffusion time for a herbicide varies from the outside to the center of a callus piece, whereas even with suspension cultures made up of clusters of 8 to 16 cells, diffusion time is minimal. It is harder to aportion equal amounts of calli than cells.

Calli should be used in metabolism studies where metabolism to lignin is expected.[189] Calli are more differentiated than cell suspensions and often produce lignin-containing vascular elements. Protoplasts are harder to prepare than cell cultures, and must be freshly made for each experiment. Protoplasts could be useful to measure synergies specific to herbicides such as dichlobenil, which inhibit cellulose biosynthesis. It is easy to follow inhibition of resynthesis of cellulose on protoplasts with fluorescent dyes such as calcofluor. Protoplasts could also be useful for any compounds affecting the plasmolemma, where "popped" protoplasts could be a good assay. Leakage from cell suspensions has been used for kinetically measuring

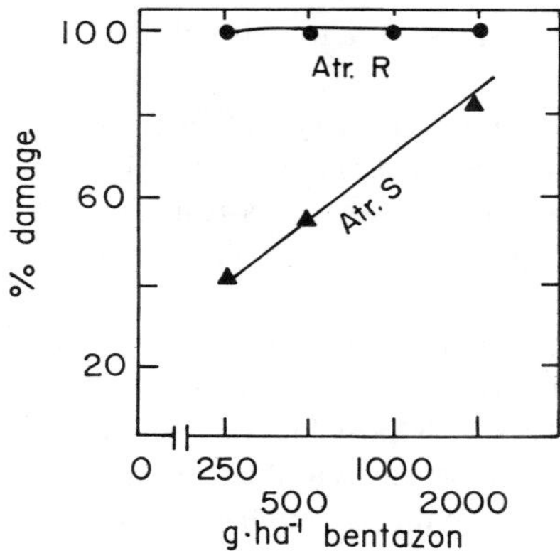

FIGURE 6. Negative cross-resistance of bentazon with atrazine-resistant *Amaranthus retroflexus*, an example of synergistically managing resistance. Bentazon was treated post-emergence in pot experiments in the greenhouse. Unfortunately, *Chenopodium album* and *Solanum nigrum* resistant biotypes showed positive cross-resistance to bentazon at use rates lower than those normally applied (plotted from data kindly provided by Würzer, BASF, AG, Limburgerhof[188]).

membrane effects.[128] Cell cultures can be continuously transferred, but the precise optimal stage of growth must be fixed to obtain uniformity; many herbicides have different effectivity on different growth stages.[190]

Primary cultures, used just after isolation have many uses, especially when photosynthetically active cultures are needed. The easily isolated cells from florist asparagus (*A. sprengerii)*[131,191] and *A. officinalis* [192] are ideal. These species contain only spongy mesophyll, and whole cells are easily banged out of finely minced leaves. Enough cells can be prepared and cleaned in a few hours for hundreds of assays, and the isolated cells can be stored. Algae have many potential uses because they are photosynthetic, although they have not been extensively used for such studies. Fern protonema may become useful; it is possible to easily isolate herbicide-resistant cell lines in ferns[193,194] and it may well be possible to use the resistant lines for synergy studies.

1. Glutathione Transferase in Cell Cultures

Glutathione[195] and glutathione transferase[196] levels in maize cultures can be raised with herbicide protectants. Conversely, compounds such as tridiphane have not been tested to ascertain whether they act as synergists at the cellular level. If so, an easy, highly reproducible quick assay system could be developed.

2. Monooxygenase Synergists in Cell Cultures

Cell culture systems have been widely used to show oxidative degradation of herbicides.[197-202] Fewer studies have shown synergistic depressions of monooxygenase activities in cell cultures. Cultures have the advantage over microsomal preparations in that they do not need to be prepared and purified, and they are "stable" compared to enzyme preparations. There is no need to exogenously provide cofactors. Cytochrome P_{450}s have been measured in cell cultures.[203] Still, it is not known yet whether herbicide binding can be measured as shifts in absorbancy at 450 nm. The possibility may be improved with pretreatments with P_{450} enhancers such as naphthalic anhydride[155-157] and ethanol,[158] coupled with sophisticated multiple scanning and computer-linked sensitive spectrophotometry to allow measurement of spectral shifts.

Many compounds synergistically prevent the monooxygenase type degradation of a variety of herbicides (Table 6). Clearly cell cultures have much to offer in investigating potential synergists of rapidly degraded herbicides. Because of species variations in specific monooxygenase activities that carry over to cell cultures,[72] the target weed for a synergist should be used.

None of the data on antimonooxygenase synergists have dealt with 2,4-D, yet many studies with cell cultures of many species have shown that 2,4-D is degraded in cultures.[197-202,204] As 2,4-D resistance has appeared, with cross-resistance to a monooxygenase-degraded sulfonylurea,[205] the time is ripe to look for synergists to suppress this resistance.

3. Cell Assays for Inhibitors of the Oxidant Detoxification Pathway

The oxidant-degradation pathway is primarily a chloroplast pathway, requiring cell cultures with active chloroplast systems. It is possible, but not easy, to maintain green steady state suspension cultures. These are quite green to the eye, especially when compared to etiolated cultures, but have far less chlorophyll per gram tissue than leaves. Green steady state cultures are selectively affected by inhibitors of photosynthesis (but etiolated cultures are unaffected), and they do fix some CO_2.[206] The etiolated cultures are more affected than green cultures by protoporphyrin IX generating herbicides, probably due to a lack of chloroplasts containing the oxidant detoxifying pathway.[72,127]

Green cell culture systems are needed for synergy assays that have normal chloroplasts, and normal levels of the oxidant detoxification pathway. These cannot be obtained from the steady state dividing green cell lines available, but primary isolations of leaf cells, can be used.[131] These generalizations are not with exceptions; nongreen callus systems have been used to isolate mutants with higher levels of resistance to photooxidants.[207,208] Regenerated plants from such calli had cotolerance to sulfur dioxide,[207] suggesting that the photooxidant system is also operative in nongreen cells.

Herbicides can be strongly synergized by chelators of copper and zinc using the *Asparagus* system (Table 4). Thus, the system can detect putative protectants. 8-Hydroxyquinoline protected the cells from herbicide damage. Hydroxquinoline preferentially chelates iron and probably prevents the generation of hydroxy radicals from hydrogen peroxide by the iron catalyzed Fenton reaction (Figure 4).

The *Asparagus* cell system is actually too sensitive for a good dedicated assay. All compounds that strongly chelate copper and zinc were active synergists (Table 4) as shown by $^{14}CO_2$ fixation or chlorophyll bleaching after a fixed time in the light with herbicide and/or putative synergist. The results with both techniques were similar, although bleaching is the simplest assay. Additionally, it should be possible to directly determine synergist enhanced production of hydroxyl radicals, with a recently developed assay using dimethyl sulfoxide as a sink for hydroxyl radicals.[209] Those chelators that did synergize herbicides in cells had to be used at much higher concentrations in whole plants. This is probably because they do not easily traverse the cuticle on the leaf epidermis, especially not at the speed of paraquat, which inhibits intact leaf photosynthesis as quickly as can be measured.[87] Thus, a whole tissue or whole plant assay is really needed once it can be shown that there is a potential for synergy.

4. Glycosyl Transferases

Glycosylated metabolites of herbicides have often been found in cell suspension culture systems,[178,201] but no specific inhibitors of glycosylation are known. Cell cultures might provide a system to look for them.

5. Target Site Resistances

Cells isolated from triazine-resistant *Chenopodium album* plants are resistant in green suspension cultures, while the wild type *C. album* cells are inhibited.[210] Such a system could be utilized to assay for herbicides showing negative cross-resistance, that would synergize against evolution of resistance. As thylakoids are easy to isolate, they are preferable over plants for a prescreen.

6. Synergizing Against Problem Soils at the Cell Level

No single bacterial or fungal culture has been isolated and found responsible for problem soils. Thus, one cannot assay for synergy with pure cultures. Even mixed, isolated cultures have not been used. Small, well mixed samples of problem soils have been used to compare the rate of herbicide degradation using herbicides and/or putative synergists. As the soil is an "inert support", such systems are *in vitro* assays using mixed cultures of unknown organisms.

7. Synergizing Mycoherbicides — Measurement at the Cell Level

The elicitation of high levels of antifungal phytoalexins has been shown in cell culture.[161,211-213] Many inhibitors of phytoalexin biosynthesis pathways have been shown to operate at the level of isolated cells (Table 4). Dedicated screens for such phytoalexin synthesis-suppressors could be designed. Some compounds such as citrate suppressed elicited phytoalexin biosynthesis, *in situ* and at the cell level.[212] The reasons for elicitation are unclear, but suggest that a random screen may also be designed.

E. ASSAYS WITH TISSUE PIECES

Leaf discs, roots, coleoptile and stem sections, excised seedlings, as well as storage tissue pieces from carrot, beet, potato, avocado, and Jerusalem artichoke slices have often been used for *in vitro* assays of herbicides and other plant growth regulators. As each single tissue cannot universally assay all herbicides, such systems are not amenable to all herbicides, but could be developed into dedicated systems. Roots, unlike the other organs listed, can be cultured axenically in continuous culture. Many herbicides specifically act on roots, allowing design of systems for assaying synergy. Still, the effects on root-acting herbicides are mimicked by nongreen callus and cell suspension systems.[206,214]

Coleoptile and stem sections could be easily used to develop *in vitro* assays for compounds that synergize herbicides by speeding translocation. These would be modifications of the widely used auxin translocation systems. Leaf disc assays have the advantage over whole plants by allowing a microassay with careful control of light, temperature and humidity, allowing greater reproducibility. The presence of a cuticle more closely approaches the situation of plants.

Multicellular but miniature whole plants can also be cultured *in vitro*. *Arabidopsis thaliana* (Cruciferae) has been widely used to obtain herbicide-resistant mutants[186] as have ferns.[193,194] Free living duckweeds *(Lemnaceae)* have been very useful in herbicide screening.[19] The solutions can be sterile and microassays can be set up in microtiter dishes.[20] The fronds lack a cuticle and rapidly take up organic materials, and they are affected by photosynthesis-inhibiting herbicides. They have been used as an assay for herbicide protectants,[20,215] but not, to the best of our knowledge, for synergists.

1. Assays for Potential Inhibitors of Monooxygenases

Synergy has been found using stem, coleoptile and leaf segments, for inhibiting metabolism of various herbicides (Table 6). In many cases it was clearly shown that a monooxygenase type reaction was inhibited. Such assays could be used to show synergy in screening for potential synergists, but tissue pieces from the target weeds should be used and not the crops listed in Table 6.

2. Leaf Assays for Synergists to Photooxidant-Generating Herbicides

Leaf discs of various plant species have been either floated on or dipped in solutions containing herbicides and/or putative synergists and placed on wet filter paper. The petri dishes were then placed under strong lights to generate the photooxidants. The synergy was measured either as visual estimation, photosynthetic $^{14}CO_2$ fixation, or as chlorophyll bleaching.

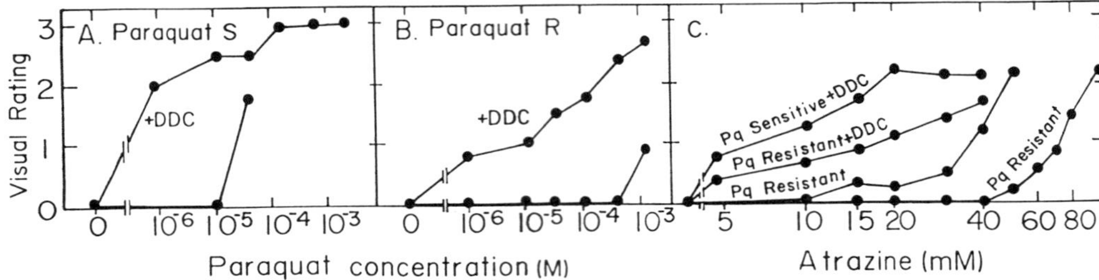

FIGURE 7. Synergizing oxidant generating herbicides: visual estimation of effects on leaf discs of paraquat resistant and sensitive *Conyza bonariensis* plants were cultured to the young rosette stage. Leaf discs 7 mm diameter were excised and floated on 0.1% Tween 20 in distilled water containing herbicide, and where stated 89 m*M* (A,B) and 222 m*M* (C) diethyldithiocarbamate. A and B were incubated for 6 h and C for 24 h in 0.85 mEin m^{-2} s^{-1} light. Visual estimations were scored by at least two people as 0: no damage to slight burn; 1: moderate burn but eventually recovers; 2: severe burn, eventually dying; 3: dead. (From Shaaltiel, Y. and Gressel, J., unpublished.)

TABLE 8
Use of an *In Vitro* Bean and *Conyza* Leaf Disc Assay to Estimate Synergy of Oxidant-Generating Herbicides by Metal Chelators[a]

Herbicide	Chelator	Metal chelated	Synergy factor
Bean			
Paraquat	9 m*M* Diethyldithiocarbamate	Cu^{++}	>180
	0.5 m*M* Di-2-ethylhexylphosphate	Zn^{++}	35
Acifluorfen	1.2 m*M* Oxime of 4-octyl-2-acetylphenol	Cu^{++}	31
	0.5 m*M* Di-2-ethylhexylphosphate	Zn^{++}	22
Conyza			
(Paraquat sensitive)			
Acifluorfen	22 m*M* Diethyldithiocarbamate	Cu^{++}	>100
(Paraquat resistant)			
Acifluorfen	22 m*M* Diethyldithiocarbamate	Cu^{++}	10

Note: Leaf discs (7 mm diameter) of beans or paraquat resistant and sensitive *Conyza bonariensis* were floated on water with 0.1% Tween 20 containing herbicide and chelator and incubated under 0.45 mEin m^{-2} s^{-1} (PAR) light, for 4 h (beans) or 14 h (*Conyza*) in the different experiments shown. Chlorophyll content was measured according to Moran.[133, 134] The synergy factor is the ratio of the interpolated I_{50} of nonsynergized to synergized.

[a] Collated and calculated from data in Gressel and Shaaltiel.[172]

Visual observations showed that the amount of paraquat required to control paraquat-sensitive (Figure 7A) and paraquat-resistant (Figure 7B) *Conyza* could be decreased by a copper chelator. The visual data also show that the paraquat-resistant *Conyza* is more resistant to atrazine than the wild type, and resistance was abolished by a copper chelator (Figure 7C). Other herbicides could be synergized by chelators, and the assay is also effective with bean discs (Table 8). Similar studies were performed as ion leakage, from the maize inbreds that have paraquat and drought co-tolerance due to elevated levels of the oxidant detoxification pathway.[92]

3. *In Vitro* Assays for Measuring Synergy of Mycoherbicides

Leaf discs used to measure the effects of both adjuvants to keep water around the spores and inhibitors of phytoalexin production are described in Table 5 and Sections II.G and III.B.6.

IV. CONCLUDING REMARKS

In vitro systems for bioassaying synergies have been verified at the higher plant level and could possibly be used in the field. *In vitro* systems have a clear place in performing highly sensitive pre-screening and QSAR studies on synergists when just the primary activity need be measured, without the limitations placed by a nonuniform, whole plant system. The systems are amenable to modification to develop dedicated screens that are special for any herbicide insufficiently active on whole plants due to degradation of the herbicide or its toxic products. *In vitro* systems are also amenable to finding synergists for mycoherbicides, especially those with the sensitive immunoassays for infection, and fluorescence assays for certain types of phytoalexins.

The many levels of bioassay systems and the simple methods of measuring effects should be more widely used. Synergists are needed to make herbicides more cost-effective and to allow less environmental load of herbicides.

ACKNOWLEDGMENTS

The authors are grateful to Dr. A. Warshawsky for advice and collaboration on chelator studies. The authors' research on synergizing photooxidant-generating herbicides was supported in part by the Yeda Fund and the research on synergizing mycoherbicides by the Yeda Fund and the National Council for Research and Development, Jerusalem. J. G. has the Gilbert de Botton Professorial Chair of Plant Sciences.

REFERENCES

1. **Anon.,** *Alternative Agriculture,* National Acad. Press, Washington, D.C., 1989, 448.
2. **Le Baron, H. M. and Gressel, J.,** *Herbicide Resistance in Plants,* Wiley, New York, 1982, 401.
3. **LeBaron, H. M. and McFarland, J.,** Herbicide resistance in weeds and crops: an overview and prognosis, in *Managing Resistance to Agrochemicals,* Green, M. B., LeBaron, H. M., and Moberg, W. K., Eds. American Chemical Society, Washington, D.C., 1990, 336.
4. **Gressel, J.,** Why get resistance? It can be prevented or delayed, in *Herbicide Resistance in Weeds and Crops,* Caseley, J. C., Cussans, G. W., and Atkin, R. K., Eds., Butterworth, Oxford, 1991, 1.
5. **Gressel, J. and Segel, L. A.,** The paucity of genetic adaptive resistance of plants to herbicides: possible biological reasons and implications, *J. Theor. Biol.,* 75, 349, 1978.
6. **Gressel, J. and Segel, L. A.,** Interrelating factors controlling the rate of appearance of resistance: the outlook for the future, in *Herbicide Resistance in Plants,* LeBaron, H. M. and Gressel, J., Eds., Wiley, New York, 1982, 325.
7. **King, R. P., Lybecker, D. W., Schweizer, E. E., and Zimdahl, R. L.,** Bioeconomic modeling to simulate weed control strategies for continuous corn *(Zea mays), Weed Sci.,* 34, 972, 1986.
8. **Streibig, J. C., Andreasen, C., and Fredshavn, J.,** A computer programme for economic thresholds, Proc. EWRS Symp., 1990, 459.
9. **Cousens, R., Doyle, C. J., Wilson, B. J., and Cussans, G. W.,** Modelling the economics of controlling *Avena fatua* in winter wheat, *Pestic. Sci.,* 17, 1, 1986.
10. **Beyer, E. M., Duffy, M. J., Hay, J. V., and Schlueter, D. D.,** Sulfonylureas, in *Herbicides, Chemistry, Degradation and Mode of Action,* Vol. 3, Kearney, P. C. and Kaufman, D. D., Eds., Marcel Dekker, New York, 1988, 118.
11. **Duke, S. O.,** Glyphosate, in *Herbicides, Chemistry, Degradation and Mode of Action,* Vol. 3, Kearney, P. C. and Kaufman, D. D., Eds., Marcel Dekker, New York, 1988, 2.
12. **Rendina, A. R., Felts, J. M., Beaudoin, J. D., Craig-Kennard, A. C., Look, L. L., Paraskos, S. L., and Hagenah, J. A.,** Kinetic characterization, stereoselectivity, and species selectivity of the inhibition of plant acetyl-CoA carboxylase by the aryloxy-phenoxypropionic acid grass herbicides, *Arch. Biochem. Biophys.,* 265, 219, 1988.
13. **Matringe, M., Camadro, J. M., Labbe, P., and Scalla, R.,** Protoporphyrinogen oxidase as a molecular target for diphenyl ether herbicides, *Biochem. J.,* 260, 231, 1989.
14. **Anon.,** *Herbicide Handbook,* 6th ed., Weed Science Society of America, Champaign, IL., 1989, 301.
15. **Bocion, P. F.,** Personal communication, 1990.

16. **Gressel, J.,** Synergizing herbicides, *Rev. Weed Sci.,* 5, 49, 1990.
17. **Blumhorst, M. R. and Kapusta, G.,** Mefluidide as an enhancing agent for post emergence broadleaf herbicides in soybeans *(Glycine max), Weed Technol.,* 1, 149, 1987.
18. **Gressel, J.,** Herbicide tolerance and resistance: alternation of site of activity, in *Weed Physiology,* Duke, S. O., Ed., C.R.C. Press, Boca Raton, FL, 1985, 159.
19. **Gressel, J.,** Plant tissue culture systems for screening of plant growth regulators: hormone, herbicides and natural phytotoxins, in *Advances in Cell Cultures,* Vol. 3, Maramorosch, K., Ed., Academic Press, New York, 1984, 93.
20. **Gressel, J.,** *In vitro* plant cultures for herbicide pre-screening, in *Applications of Biotechnology to Agricultural Chemistry,* LeBaron, H. M., Mumma, R. O., Honeycutt, R. C., and Duesing, J. H., Eds., American Chemical Society, Washington, D.C., 1987, 41.
21. **Stoltenberg, D. E., Gronwald, J. W., Burton, J. D., and Wyse, D. L.,** Effect of sethoxydim and haloxyfop on acetyl-CoA carboxylase in tolerant and susceptible fescue species, *Weed Sci. Soc. Am.,* 179 (Abstr.), 1988.
22. **Secor, J. and Cseke, C.,** Inhibition of acetyl-CoA carboxylase activity by haloxyfop and tralkoxydim, *Plant Physiol.,* 86, 10, 1988.
23. **Burton, J. L., Gronwald, J. W., Somers, J. A., Connally, J. A., Gegenbach, B. G., and Wyse, D. L.,** Inhibition of plant acetyl CoA carboxylase by the herbicides sethoxydim and haloxyfop, *Biochem. Biophys. Res. Comm.,* 148, 1030, 1987.
24. **Lichthethaler, H. K.,** Mode of action of herbicides affecting acetyl CoA carboxylase, *Z. Naturforsch.,* 45c, 521, 1990.
25. **Gressel, J.,** Multiple resistances to wheat selective herbicides: new challenges to molecular biology, *Oxford Surv. Plant Molec. Cell. Biol.,* 5, 195, 1988.
26. **Heap, I. M.,** Herbicide cross-resistance in a population of annual ryegrass *Lolium rigidum,* in *8th Aust. Weeds Conf.,* Lemerle, D. and Leys, A. G., Eds., 1987, 1140.
27. **Powles, S. B., Holtum, J. A. M., Matthews, J. M., and Liljegren, D. R.,** Multiple herbicide resistance in annual ryegrass *(Lolium rigidum):* the search for mechanism, in *Fundamental and Practical Approaches to Combating Resistance,* Green, M. B., LeBaron, H. M., and Moberg, M. K., Eds., American Chemical Society, Washington, D.C., 1990, 394.
28. **Moss, S. R.,** Herbicide resistance in black-grass *(Alopecurus myosuroides), Brit. Crop Prot. Conf. Weeds,* 1987, 879.
29. **Andersen, R. N. and Gronwald, J. W.,** Non-cytoplasmic inheritance of atrazine tolerance in velvetleaf *(Abutilon theophrasti), Weed Sci.,* 35, 496, 1987.
30. **Gronwald, J. W., Andersen, R. N., and Yee, C.,** Atrazine resistance in velvetleaf *(Abutilon theophrasti)* due to enhanced atrazine detoxification, *Pestic. Biochem. Physiol.,* 34, 149, 1989.
31. **Beurat, E. and Maigre, D.,** La resistance des Amaranthus dans les mais du Tessin n'est pas une resistances chloroplastique, *13e' Conference du Columa,* Paris, 1/II p. 23, 1986.
32. **Kaufman, D. D., Katan, Y., and Edwards, D. F.,** Microbial adaptation and metabolism of pesticides, in *Agricultural Chemicals and the Future,* Hilton, J. L., Ed., Rowan and Allanheld, Totawa, 1984, 437.
33. **Mannervik, B. and Danielson, U. H.,** Glutathione transferases-structure and catalytic activity, *CRC Crit. Rev. Biochem.,* 23, 283, 1988.
34. **Mozer, T. J., Tiemeier, D. C., and Jaworski, E. G.,** Purification and characterization of corn glutathione *S*-transferase, *Biochemistry,* 22, 1068, 1983.
35. **Shimabukuro, R. H., Frear, D. S., Swanson, H. R., and Walsh, W. C.,** Glutathione conjugation, an enzymatic basis for atrazine resistance in corn, *Plant Physiol.,* 47, 10, 1971.
36. **O'Connell, K. M., Breaux, E. J., and Fraley, R. T.,** Different rates of metabolism of two chloroacetanilide herbicides in Pioneer 3320 corn, *Plant Physiol.,* 86, 359, 1988.
37. **Shah, D. M., Hironaka, C. M., Wiegand, R. C., Harding, E. I., Krivi, G. G., and Tiemeier, D. C.,** Structural analysis of a maize gene coding for glutathione-*S*-transferase involved in herbicide detoxification, *Plant Molec. Biol.,* 6, 203, 1986.
38. **Wiegand, R. C., Shah, D. M., Mozer, T. J., Harding, E. I., Diaz-Collier, J., Saunders, C., Jaworski, E. G., and Tiemeier, D. C.,** Messenger RNA encoding a glutathione-*S*-transferase responsible for herbicide tolerance in maize is induced in response to safener treatment, *Plant Molec. Biol.,* 7, 235, 1986.
39. **Timmerman, K. P.,** Molecular characterization of corn glutathione *S*-transferase isozymes involved in herbicide detoxication, *Physiol. Plant,* 77, 465, 1989.
40. **Burke, J. J., Gamble, P. E., Hatfield, J. L., and Quisenberry, J.,** Plant morphological and biochemical responses in field water deficit. Responses of glutathione reductase activity and paraquat sensitivity, *Plant Physiol.,* 79, 415, 1985.
41. **Jachetta, J. J. and Radosevich, S. R.,** Enhanced degradation of atrazine by corn *(Zea mays), Weed Sci.,* 29, 37, 1981.
42. **Dean, J. V., Gronwald, J. W., and Eberlein, C. V.,** Induction of glutathione *S*-transferase isozymes in sorghum by herbicide antidotes, *Plant Physiol.,* 92, 467, 1990.

43. **Zorner, P. S. and Olsen, G. L.,** The effect of DOWCO 356 on atrazine metabolism in giant foxtails, *Proc. North Cent. Weed Control Conf.,* 36, 115, 1981.
44. **Zorner, P. S. and Sheppard, B. R.,** Factors regulating the performance of tridiphane and triazine in controlling panicoid grasses, *Weed Sci. Soc. Am.,* 290(Abstr.), 1985.
45. **Ezra, G., Dekker, J. H., and Stephenson, G. R.,** Tridiphane as a synergist for herbicides in corn *(Zea mays)* and proso millet *(Panicum milaceim), Weed Sci.,* 33, 287, 1985.
46. **Lamoureux, G. L. and Rusness, D. G.,** Tridiphane [2-(3,5-dichlorophenyl)-2-(2,2,2-trichloroethyl)oxirane] an atrazine synergist: enzymatic conversion to potent glutathione-*S*-transferase inhibitor, *Pestic. Biochem. Physiol.,* 26, 323, 1986.
47. **Zorner, P. S.,** Personal communication of unpublished results, 1989.
48. **Ray, P. G., Markley, L. D., Snelling, J., and Coleman, S.**, Synergism of a variety of herbicidal chemistries by tridiphane, *Weed Sci. Soc. Am.,* 290(Abst.), 1990.
49. **Caseley, J. C., Copping, L., and Mason, D.,** Control of herbicide resistant black-grass with herbicide mixtures containing tridiphane, in *Herbicide Resistance in Weeds and Crops,* Caseley, J. C., Cussans, G. W., and Atkin, R. K., Eds., Butterworth, Oxford, 1991, 429.
50. **Gronwald, J. W.,** Personal communication of unpublished results, 1990.
51. **Lamoureux, G. L. and Rusness, D. G.,** Synergism of diazinon toxicity and inhibition of diazinon metabolism in the house fly by tridiphane: inhibition of glutathione-*S*-transferase activity, *Pestic. Biochem. Physiol.,* 27, 318, 1987.
52. **O'Keefe, D., Romesser, J. A., and Leto, K. J.,** Plant and bacterial cytochromes P_{450}, involvement in herbicide metabolism, in *Phytochemical Effects of Environmental Compounds,* Saunders J. A., Channing, L. K., and Conn, E. E., Eds., Plenum Press, New York, 1987, 151.
53. **McFadden, J. J., Frear, D. S., and Mansager, E. R.,** Aryl hydroxylation of diclofop by a cytochrome P_{450} dependent monooxygenase from wheat, *Pestic. Biochem. Physiol.,* 34, 92, 1989.
54. **Gressel, J.,** *Wheat Herbicides: The Challenge of Emerging Resistance,* Biotechnology Affiliates, Checkendon/ Reading, U.K., 1988, 247.
55. **Metcalf, R. L. and McKelvey, J. J.,** *Insecticide Synergism in the Future for Insecticides. Needs and Prospects,* Wiley, New York, 1979.
56. **Varsano, R.,** Enhancement of herbicide activity by piperonyl butoxide, Ph.D. thesis, Hebrew University, Jerusalem, Israel, 1988, 118.
57. **Powles, D.B. and Liljegren, D.,** A grass weed (*Lolium rigidum)* biotype displaying cross-resistance to herbicides with different modes of action: first studies on the mechanism of cross resistance, *Weed Sci. Soc. Am.,* 187(Abst.), 1988.
58. **Gonneau, M., Pasquette, B., Cabanne, F., Scalla, R., and Durst, F.,** Metabolism of chlorotoluron in tolerant species: possible role of cytochrome P_{450} monooxygenases, *Weed Res.,* 28, 19, 1988.
59. **Blee, E. and Durst, F.,** Hydroperoxide dependent sulfoxidation catalyzed by soybean microsomes, *Arch. Biochem.,* 254, 43, 1987.
60. **Belai, I. and Matolcsy, G.,** Inhibition of insect cytochrome P_{450} by some metyrapone analogues and compounds containing a cyclopropylamine moiety and their evaluation as inhibitors of juvenile hormone biosynthesis, *Pestic. Sci.,* 24, 205, 1988.
61. **Vanden-Bossche, H. , Marichal, P., Gorrens, J., Bellens, D., Verhoeven, H., Coene, M.-C., Lauwers, W., and Janssen, P. A. J.,** Interaction of azole derivatives with cytochrome P_{450} isozymes in yeast, fungi, plants and mammalian cells, *Pestic. Sci.,* 21, 289, 1987.
62. **Cole, D. J. and Owens, W. J.,** Influence of monooxygenase inhibitors on the metabolism of the herbicides chlortoluron and metolachlor in cell suspension cultures, *Plant Sci.,* 50, 13, 1987.
63. **Fritsch, H., Rademacher, W., and Retzlaff, G.,** Inhibition of plant growth, gibberellin biosynthesis and cinnamate-4-monooxygenase by selected growth regulators, *Proc. Int. Bot. Congr. Berlin,* 2(Abstr.), 113b, 1987.
64. **Retzlaff G.,** Personal communication of unpublished results, 1989.
65. **Kudsk, P., Thonke, K. E., and Streibig, J. C.,** The phytotoxicity of difenzoquat to wild oat as influenced by other pesticides, in *Adjuvants and Agrochemicals,* Chow, P. N. P., Grant, C. A., and Hinshalwood, A. M., Eds., CRC Press, Boca Raton, FL, 1989, 175.
66. **Shaner, D. L.,** Mode of action of difenzoquat, in *Wild Oats Symposium Proceedings,* Vol. 2, Smith, A. E. and Hsaio, A. I., Eds., Agriculture Canada Research Station, Regina, 1984, 49.
67. **Grimes, H. D., Perkins, K. K., and Boss, W. F.,** Ozone degrades into hydroxyl radical under physiological conditions, *Plant Physiol.,* 72, 1016, 1983.
68. **Shaaltiel, Y. and Gressel, J.,** Biochemical analysis of paraquat resistance in *Conyza* leads to pinpointing synergists for oxidant generating herbicides, in *Pesticide Science and Biotechnology,* Greenhalgh, R. and Roberts, T. R., Eds., Blackwell Scientific, London, 1987, 183.
69. **Summers, L. A.,** *The Bipyridilium Herbicides,* Academic Press, New York, 1980, 449.
70. **Duke, S. O., Lydon, J., and Paul, R. N.,** Oxadiazon activity is similar to that of *p*-nitrodiphenylether herbicides, *Weed Sci.,* 321, 152, 1989.

71. **Lydon, J. and Duke, S. O.,** Porphyrin synthesis is required for photobleaching activity of the *p*-nitro substituted diphenylether herbicides, *Pestic. Biochem. Physiol.,* 31, 74, 1988.
72. **Matringe, M. and Scalla, R.,** Studies on the mode of action of acifluorfen-methyl in nonchlorophyllous cells: accumulation of tetrapyrroles, *Plant Physiol.,* 86, 619, 1988.
73. **Becerril, J. M. and Duke, S. O.,** Protoporphyrin IX content correlates with activity of photobleaching herbicides, *Plant Physiol.,* 90, 1175, 1989.
74. **Yanase, D. and Andoh, A.,** Porphyrin synthesis involvement in diphenyl ether-like mode of action of TNPP-ethyl, a novel phenylpyrazole herbicide, *Pestic. Biochem. Physiol.,* 35, 70, 1989.
75. **Sandmann, G. and Böger, P.,** Accumulation of protoporphyrin IX in the presence of peroxidizing herbicides, *Z. Naturforsch.,* 43c, 699, 1988.
76. **Nicolaus, B., Sandmann, G., Watanabe, H., Wakabayashi, K., and Böger, P.,** Herbicide-induced peroxidation: influence of light and diuron on protoporphyrin IX formation, *Pestic. Biochem. Physiol.,* 35, 192, 1989.
77. **Witkowski, D. A. and Halling, B. P.,** Inhibition of plant protoporphyrinogen oxidase by the herbicide acifluorfen-methyl, *Plant Physiol.,* 90, 1239, 1989.
78. **Hartmann, P. E., Dixon, W. J., Dahl, T. A., and Daub, M. E.,** Multiple modes of photodynamic action by cercosporin, *Photochem. Photobiol.,* 47, 5, 1988.
79. **Hartman, P. E., Suzuki, C. K., and Stack, M. E.,** Photodynamic production of superoxide *in vitro* by altertoxins in the presence of reducing agents, *Appl. Environ. Microbiol.,* 55, 7, 1989.
80. **Martin, J. P. and Logsdon, N.,** Oxygen radicals are generated by dye-mediated intracellular photooxidations: a role for superoxide in photodynamic effects, *Arch. Biochem. Biophys.,* 256, 39, 1987.
81. **Martin, J. P. and Logsdon, N.,** The role of oxygen radicals in dye-mediated photodynamic effects in *Escherichia coli* B, *J. Biol. Chem.,* 262, 7213, 1987.
82. **Ben-Amotz, A., Gressel, J., and Avron, M.,** Massive accumulation of phytoene induced by norflurazon in *Dunaliella bardawil* prevents recovery from photoinhibition, *J. Phycol.,* 23, 176, 1987.
83. **Finckh, B. F. and Kunert, K. J.,** Vitamins C and E, an antioxidative system against herbicide-induced lipid peroxidation in higher plants, *J. Agric. Food Chem.,* 33, 574, 1985.
84. **Halliwell, B. and Gutteridge, J.,** *Free Radicals in Biology and Medicine,* Oxford Science Publishers, Oxford, 1985, 346.
85. **Asada, K. and Takahashi, M.,** Production and scavenging of active oxygen in photosynthesis, in *Photoinhibition,* Kyle, D. J., Osmond, C. B., and Arntzen, C. J., Eds., Elsevier, Amsterdam, 1987, 228.
86. **Harvey, B. M. R. and Harper, D. B.,** Tolerance to bipyridylium herbicides, in *Herbicide Resistance in Plants,* LeBaron, H. M. and Gressel, J., Eds., Wiley, New York, 1982, 251.
87. **Shaaltiel, Y. and Gressel, J.,** Kinetic analysis of resistance to paraquat in *Conyza*: evidence that paraquat transiently inhibits leaf chloroplast reactions in resistant plants, *Plant Physiol.,* 85, 869, 1987.
88. **Shaaltiel, Y. and Gressel, J.,** Multienzyme oxygen radical detoxifying system correlated with paraquat resistance in *Conyza bonariensis, Pestic. Biochem. Physiol.,* 26, 22, 1986.
89. **Shaaltiel, Y., Chua, N.-H., Gepstein S., and Gressel, J.,** Dominant pleiotropy controls enzymes co-segregating with paraquat resistance in *Conyza bonariensis, Theor. Appl. Genet.,* 75, 850, 1988.
90. **Shaaltiel, Y., Glazer, A., Bocion, P. F., and Gressel, J.,** Cross tolerance to herbicidal and environmental oxidants of plant biotypes tolerant to paraquat, sulfur dioxide and ozone, *Pestic. Biochem. Physiol.,* 31, 13, 1988.
91. **Jansen, M. A. K., Shaaltiel, Y., Kazzes, D., Canaani, O., Malkin, S., and Gressel, J.,** Increased tolerance to photoinhibitory light in paraquat-resistant *Conyza bonariensis* measured by photoacoustic spectroscopy and $^{14}CO_2$ fixation, *Plant Physiol.,* 91, 1174, 1989.
92. **Malan, C., Greyling, M. M., and Gressel, J.,** Correlation between antioxidant enzymes, SOD and glutathione reductase, and environmental and xenobiotic stress tolerance in inbred maize, *Plant Sci.,* 69, 157, 1990.
93. **Jansen, M. A. K., Malan, C., Shaaltiel, Y., and Gressel, J.,** Mode of evolved photo-oxidant resistance to herbicides and xenobiotics, *Z. Naturforsch.,* 45c, 463, 1990.
94. **Sawada, Y., Ohyama, T., and Yamazaki, I.,** Preparation and physicochemical properties of green pea superoxide dismutase, *Biochim. Biophys. Acta,* 268, 305, 1972.
95. **Asada, K., Nakano, Y., and Hossain, M. A.,** Scavenging system of hydrogen peroxide in chloroplasts: ascorbate peroxidase and monodehydroascorbate and dehydroascorbate reductase, in *Oxidative Damage and Related Enzymes,* Rotilio, G. and Bannister, J. V., Eds., Harwood, Chichester, 1984, 342.
96. **Hossain, M. A. and Asada, K.,** Purification of dehydroascorbate reductase from spinach and its characterization as a thiol enzyme, *Plant Cell Physiol.,* 25, 85, 1984.
97. **Halliwell, B. and Foyer, C. H.,** Properties and physiological function of a glutathione reductase purified from spinach leaves by affinity chromatography, *Planta,* 139, 9, 1978.
98. **Gressel, J. and Segel, L. A.,** Negative cross-resistance; a possible key to atrazine resistance management: a call for whole plant data, *Z. Naturforsch.,* 45c, 470, 1990.
99. **Durner, J., Thiel, A., and Böger, P.,** Phenolic herbicides: correlation between lipophilicity and increased inhibitor sensitivity of thylakoids from a higher plant mutant, *Z. Naturforsch.,* 42c, 881, 1986.

100. **Obrigawitch, T., Wilson, R. G., Martin, A. R., and Roeth, F. W.,** The influence of temperature, moisture and prior EPTC application on the degradation of EPTC in soils, *Weed Sci.,* 30, 175, 1982.
101. **Moorman, T. B.,** Populations of EPTC-degrading microorganisms in soils with accelerated rates of EPTC degradation, *Weed Sci.,* 36, 96, 1988.
102. **Kaufman, D. D., Blake, J., and Miller, D. E.,** Methylcarbamates affect acylanilide herbicide residues in soil, *J. Agric. Food Chem.,* 19, 204, 1971.
103. **Obrigawitch, T., Roeth, F. W., Martin, A. R., and Wilson, R. G.,** Addition of R-33865 to EPTC for extended herbicide activity, *Weed Sci.,* 30, 417, 1982.
104. **Hodgson, R. J., Wymore, L. A., Watson, A. K., Snyder, R. H., and Collette, A.,** Efficacy of *Colletotrichum coccoides* and thiadiazuron for velvet leaf (*Abutilon theophrasti*) control in soybean (*Glycine max*), *Weed Technol.,* 2, 473, 1988.
105. **Quimby, P. C., Fulgham, F. E., Boyette, C. D., and Connick, W. J.,** An emulsion replaces dew in biocontrol of sicklepod — a preliminary study, in *Pesticide Formulation and Application Systems,* Hoved, D. A. and Beestman, G. B., Eds., American Society Testing Materials, Philadelphia, 1990, 264.
106. **Gressel, J. and Shaaltiel, Y.,** Synergists for herbicidal composition, *Israel Patent Application,* 77, 817, 1990.
107. **Bailey, J. A. and Mansfield, J. W., Eds.,** *Phytoalexins,* Wiley, New York, 1982.
108. **Massala, R., Legrand, M., and Fritig, B.,** Comparative effects of competitive inhibitors of phenylalanine ammonia-lyase on the hypersensitive resistance of tobacco to tobacco mosaic virus, *Plant Physiol. Biochem.,* 25, 217, 1987.
109. **Waldmuller, T. and Grisebach, H.,** Effects of R-(1-amino-3-phenylethyl)phosphonic acid on glyceollin accumulation and expression of resistance to *Phytophthora megasperma* f. sp. *glycinea* in soybean, *Planta,* 172, 424, 1987.
110. **Ward, E. W. B.,** Suppression of metalaxyl activity by glyphosate: evidence that host defense mechanisms contribute to metalaxyl inhibition of *Phytophthora megasperma.* f. sp. *glycinea* in soybeans, *Physiol. Plant Pathol.,* 25, 381, 1984.
111. **Keen, N. T., Holliday, M. J., and Yoshikawa, M.,** Effects of glyphosate on glyceollin production and the expression of resistance to *Phytophthora megasperma* f. sp. *glycinea, Phytopathology,* 72, 1467, 1982.
112. **Holliday, M. J. and Keen, N. T.,** The role of phytoalexins in the resistance of soybean leaves to bacteria, effect of glyphosate on glyceollin accumulation, *Phytopathology,* 72, 1470, 1982.
113. **Johal, G. S. and Rahe, J. E.,** Glyphosate hypersensitivity and phytoalexin accumulation in the incompatible bean anthracnose host-parasite interaction, *Physiol. Mol. Plant Pathol.,* 32, 267, 1988.
114. **Sharon, A., Amsellem., Z., and Gressel, J.,** Glyphosate suppression of an elicited defense response: increased susceptibility of *Cassia obtusifolia* to a mycoherbicide, *Plant Physiol.,* 98, 654, 1992.
115. **Keen, N. T., Ersek, T., Long, M., Bruegger, B., and Holliday, M. J.,** Inhibition of HR of soybean leaves to incompatible *Pseudomonas* spp. by blasticidin S, streptomycin and elevated temperature, *Physiol. Plant Pathol.,* 18, 325, 1981.
116. **Patil, S. S. and Gnanamanickam, S. S.,** Suppression of bacterially induced hypersensitive reaction and phytoalexin accumulation in bean by phaseotoxin, *Nature,* 259, 486, 1976.
117. **Amrhein, N.,** Specific inhibitors as probes into the biosynthesis and metabolism of aromatic amino acids, in *The Shikimic Acid Pathway,* Conn, E. E., Ed., Plenum Press, New York, 1986, 83.
118. **Lydon, J. and Duke, S. O.,** Glyphosate induction of elevated levels of hydroxybenzoic acids in higher plants, *J. Agric. Food Chem.,* 36, 813, 1988.
119. **Lydon, J. and Duke, S. O.,** Pesticide effects on secondary metabolism of higher plants, *Pestic. Sci.,* 25, 361, 1989.
120. **Hallett, S. G., Paul, N. D., and Ayres, P. G.,** *Botrytis cinerea* kills groundsel (*Senecio vulgaris*) infected by rust *(Puccinia lagenophorae), New Phytol.,* 114, 105, 1990.
121. **Davis, D. G., Stolzenberg, R. L., and Dusky, J.,** A comparison of various growth parameters of cell suspension cultures to determine phytotoxicity of xenobiotics, *Weed Sci.,* 32, 235, 1984.
122. **Fitch, M. E., Mangels, A. R., Altmann, W. A., Hawary, M. E., Qureshi, A. A., and Elson, C. E.,** Microbiological screening of mevalonate-suppressive minor plant constituents, *J. Agric. Food Chem.,* 37, 687, 1989.
123. **Van Emon, J., Seiber, J. N., and Hammock, B. D.,** Immunoassay techniques for pesticide analysis, in *Analytical Methods for Pesticides and Plant Growth Regulators: Advanced Analytical Techniques and Specific Applications,* Vol. 17, Sherma, J., Ed., Academic Press, New York, 217, 1989.
124. **Sung, Z. R.,** Turbidimetric measurement of plant cell culture growth, *Plant Physiol.,* 57, 460, 1976.
125. **Gilissen, L. J. W., Hänisch ten Cate, C. H., and Keen, B.,** A rapid method of determining growth characteristics of plant cell populations in batch suspension culture, *Plant Cell Rep.,* 2, 232, 1983.
126. **Lewinsohn, E. and Gressel, J.,** Unpublished results, 1985.
127. **Gaba, V., Cohen, N., Shaaltiel, Y., Ben-Amotz, A., and Gressel, J.,** Light requiring acifluorfen action in the absence of bulk photosynthetic pigments in plant cell cultures, *Pestic. Biochem. Physiol.,* 31, 1, 1988.
128. **Zilkah, S. and Gressel, J.,** Cell suspensions for kinetic studies of inhibitor action, *Planta,* 145, 273, 1979.

129. **Shimabukuro, R. H., Walsh, W. C., and Wright, J. P.,** Effect of diclofop-methyl and 2,4-D on transmembrane proton gradient: a mechanism for their antagonistic interaction, *Physiol. Plant.*, 77, 107, 1989.
130. **Linsel, G., Dahse, I., and Müller, E.,** Electrophysiological evidence for the herbicidal mode of action of phosphonic acid esters, *Physiol. Plant.*, 73, 77, 1988.
131. **Colman, B. B., Mawson, T., and Espie, G. S.,** The rapid isolation of photosynthetically active mesophyll cells from *Asparagus cladophylls, Can. J. Bot.*, 57, 1505, 1979.
132. **Arnon, D. I.,** Copper enzymes in isolated chloroplasts: polyphenoloxidase in *Beta vulgaris, Plant Physiol.*, 4, 1, 1949.
133. **Moran, R. and Porath, D.,** Chlorophyll determination in intact tissues using *N,N*-dimethylformamide, *Plant Physiol.*, 65, 478, 1980.
134. **Moran, R.,** Formulae for determination of chlorophyllous pigments extracted with *N,N*-dimethylformamide, *Plant Physiol.*, 69, 1376, 1982.
135. **Girotti, A. W.,** Mechanisms of lipid peroxidation, *J. Free Radic. Biol. Med.*, 1, 87, 1985.
136. **Schmedes, A. and Holmer, G.,** A new thiobarbituric acid (TBA) method for determining free malondialdehyde (MDA) and hydroperoxides selectively as a measure of lipid peroxidation, *J. Amer. Oil Chem. Soc.*, 66, 6, 1989.
137. **Kostka, P. and Kwan, C.-Y.,** Instability of malondialdehyde in the presence of H_2O_2: implications for the thiobarbituric acid test, *Lipids*, 24, 545, 1989.
138. **Kunert, K. J. and Böger, P.,** The bleaching effect of the diphenyl ether oxyfluorfen, *Weed Sci.*, 29, 169, 1981.
139. **Blankenship, S. M. and Hammett, L. K.,** Changes in oxygen, carbon dioxide, and ethylene mixtures stored in plastic or glass syringes, *Hort. Sci.*, 22, 278, 1987.
140. **Widholm, J. M.,** The use of fluorescein diacetate and phenosafranine for determining viability of cultured plant cells, *Stain Technol.*, 17, 189, 1972.
141. **Towill, L. E. and Mazur, P.,** Studies on the reduction of 2,3,5-triphenyltetrazolium chloride as a viability assay for plant tissue cultures, *Can. J. Bot.*, 53, 1097, 1975.
142. **Robertson, D. and Earle, E. D.,** Nitro-blue tetrazolium: a stain for photosynthetic activity in protoplasts, *Plant Cell Rep.*, 6, 70, 1987.
143. **Ezra, G., Gressel, J., and Flowers, H. M.,** Effects of the herbicide EPTC and the protectant DDCA on incorporation and distribution of (2-^{14}C) acetate into major lipid fractions of maize cell suspension cultures, *Pestic. Biochem. Physiol.*, 19, 224, 1983.
144. **Zilkah, S. and Gressel, J.,** The estimation of cell death in suspension cultures evoked by phytotoxic compounds: differences among techniques, *Plant Sci. Lett.*, 12, 305, 1978.
145. **Horsfall, J. G. and Cowling, E. B.,** Pathometry: the measurement of plant disease, in *Plant Disease: An Advanced Treatise*, Vol. 2, Horsfall, J. G. and Cowling, E. B., Eds., Academic Press, New York, 1978, 119.
146. **Hebert, T. T.,** The rationale for the Horsefall-Barratt plant disease assessment scale, *Phytopathology*, 72, 1269, 1982.
147. **Toler, R. W., Smith, B. D., and Harlan, J. C.,** Use of aerial color infrared photography to evaluate crop disease, *Plant Dis.*, 65, 24, 1981.
148. **Lindow, S. E. and Andersen, G. L.,** Microcomputer measurements of pathogen injury to weeds, *Weed Sci.*, Suppl. 34, 38, 1986.
149. **Sharon, A. and Gressel, J.,** Elicitation of a flavanoid phytoalexin accumulation in *Cassia obtusifolia* by a mycohderbicide: determination by $AlCl_3$ spectrofluorimetry, *Pestic. Biochem. Physiol.*, 41, 142, 1991.
150. **Sharon, A., Amsellem, Z., and Gressel, J.,** Detection and quantificaion of *Alternaria cassiae* infection by leaf immunoblot and radioimmunosorbent assay, *J. Phytopathology*, 1992, in press.
151. **Amsellem, Z., Sharon, A., Quimby, P. C., and Gressel, J.,** Complete abolition of inoculum threshold of two mycoherbicides (*Alternaria cassiae* and *A. crassa)* when applied in invert emulsion, *Phytopathology*, 80, 925, 1990.
152. **Amsellem, Z., Sharon, A., and Gressel, J.,** Abolition of selectivity of two mycoherbicidal organisms and enhanced virulence of avirulent fungi by an invert emulsion, *Phytopathology*, 81, 985, 1991.
153. **Hutchison, J. M., Shapiro, R., and Sweetser, P. B.,** Metabolism of chlorsulfuron by tolerant broadleaves, *Pestic. Biochem. Physiol.*, 22, 243, 1984.
154. **Cotterman, J. C. and Saari, L. L.,** Rapid metabolic inactivation is the basis for cross resistance to chlorsulfuron in diclofopmethyl-resistant annual ryegrass (*Lolium rigidum* Gaud.), *Pestic. Biochem. Physiol.*, 43, 182, 1992.
155. **McFadden, J. J., Gronwald, J. W., and Eberlein, C. V.,** *In vitro* hydroxylation of bentazon by microsomes from naphthalic anhydride-treated corn shoots, *Biochem. Biophys. Res. Commun.*, 168, 206, 1990.
156. **Frear, D. S. and Mardaus, M. C.,** 1,8-Naphthalic anhydride-induced microsomal oxidation of diclofop, chlorsulfuron and triasulfuron in wheat, *Am. Chem. Soc. 199th National Meeting*, 18(Abstr.), 1990.
157. **Maxson, J. M. and Barrett, M.,** Naphthalic anhydride induction of corn microsome cytochrome P_{450} and imazethapyr metabolism, *Weed Sci. Soc. Am.*, 243(Abstr.), 1990.

158. **Frear, D. S.,** Personal communication of unpublished results, 1990.
159. **Moreland, D. E. and Corbin, F. T.,** Metabolism of bentazon by excised tissues and a microsomal preparation from grain sorghum seedlings, *Weed Sci. Soc. Am.,* 179(Abstr.), 1990.
160. **Mougin, C., Cabanne, F., Canivenc, M.-C., and Scalla, R.,** Hydroxylation and *N*-demethylation of chlorotoluron by wheat microsomal enzymes, *Plant Sci.,* 66, 195, 1990.
161. **Hamerski, D. and Matern, U.,** Biosynthesis of psoralens. Psoralen 5-monooxygenase activity from elicitor-treated *Ammi majus* cells, *FEBS Lett.,* 239, 263, 1988.
162. **Hamerski, D. and Matern, U.,** Elicitor-induced biosynthesis of psoralens in *Ammi majus* L. suspension cultures, *Eur. J. Biochem.,* 171, 369, 1986.
163. **Cabanne, F., Canivenc, M. -C., Gaudry, J. C., and Scalla, R.,** Differential effects of prochloraz on metabolism of chlorotoluron in wheat and black-grass, two tolerant species, *Weed Sci. Soc. Am.,* 283(Abstr.), 1990.
164. **Cole, D. J. and Loughman, B. C.,** Factors affecting the hydroxylation and glycosylation of (4-chloro-2-methylphenoxy) acetic acid in *Solanum tuberosum* tuber tissue, *Physiol. Veg.,* 23, 879, 1985.
165. **Cabanne, F., Huby, D., Gaillardon, P., Scalla, R., and Durst, F,** Effect of the cytochrome P_{450} inactivator 1-aminobenzotriazole on the metabolism of chlorotoluron and isoproturon in wheat, *Pestic. Biochem. Physiol.,* 28, 371, 1987.
166. **Gaillardon, P., Cabanne, F., Scalla, R., and Durst, F.,** Effect of mixed function oxidase inhibitors on toxicity of chlorotoluron and isoproturon to wheat, *Weed Res.,* 25, 397, 1985.
167. **Gonneau, M., Pasquette, B., Cabanne, F., Scalla, R., and Loughman, B. C.,** Transformation of phenoxyacetic acid and chlortoluron in wheat, barren brome, cleavers and speedwell. Effects of an inactivator of monooxygenases, *Brit. Crop Protection Conf. — Weeds,* 1987, 29.
168. **Kemp, M. S. and Caseley, J. C.,** Synergistic effects of 1-aminobenzotriazole on the phytotoxicity of chlorotoluron and isoproturon in a resistant population of blackgrass, *Brit. Crop Prot. Conf. —Weeds,* 1987, 895.
169. **Leah, J. M., Worrall, T. L., and Cobb, A. H.,** Metabolism of bentazone in soybean and the influence of tetcyclasis, Bas 110 and Bas 111, *Brighton Crop Prot. Conf. — Weeds,* 433, 1989.
170. **Moore, B. A., Zorner, P. S., Carlson, D. R., and Winkler, V. W.,** Increased weed control through inhibition of bentazon metabolism with tetcyclacis, *Weed Sci. Soc. Am.,* 59(Abstr.), 1990.
171. **Matern, U., Wendorff, H., Hamerski, D., Pakusch, A. E., and Kneusel, R. E.,** Elicitor-induced phenylpropanoid synthesis in *apiaceae* cell cultures, *Bull. Liaison Groupe Polyphenols,* 14, 173, 1988.
172. **Gressel, J. and Shaaltiel, Y.,** Biorational herbicide synergists, in *Biotechnology in Crop Protection,* Hedin, P. A. , Menn, J. J., and Hollingworth, R. M., Eds., American Chemical Society Washington, D. C., 1988, 4.
173. **Beauchamp, C. and Fridovich, I.,** Superoxide dismutase. Improved assays and an assay appplicable to acrylamide gel, *Anal. Biochem.,* 44, 276, 1971.
174. **Shaaltiel, Y. and Gressel, J.,** Unpublished results, 1987.
175. **Rigo, A. and Rotilio, G.,** Electrochemical assay, in *Handbook of Methods for Oxygen Radical Research,* Greenwald, R. A., Ed., CRC Press, Boca Raton, FL, 1985, 221.
176. **Nakano, Y. and Asada, K.,** Hydrogen peroxide is scavenged by ascorbate specific peroxidase in spinach chloroplasts, *Plant Cell Physiol.,* 22, 867, 1981.
177. **Foyer, C. H. and Halliwell, B.,** The presence of glutathione and glutathione reductase in chloroplasts: a proposed role in ascorbic acid metabolism, *Planta,* 133, 21, 1976.
178. **Shimabukuro, R. H.,** Detoxification of herbicides, in *Weed Physiology* Vol. 2, Duke, S. O., Ed., CRC Press, Boca Raton, FL., 1985, 215.
179. **Fuerst, E. P., Arntzen, C. J., Pfister, K., and Penner, D.,** Herbicide cross-resistance in triazine-resistant biotypes of four species, *Weed Sci.,* 34, 344, 1986.
180. **Salhoff, C. R. and Martin, A. R.,** *Kochia scoparia* growth response to triazine herbicides, *Weed Sci.,* 34, 40, 1986.
181. **Oettmeier, W., Masson, K., Fedtke, C., Konze, J., and Schmidt, R. R.,** Effect of different photosystem II inhibitors on chloroplasts isolated from species either susceptible or resistant toward *s*-triazine herbicides, *Pestic. Biochem. Physiol.,* 18, 357, 1982.
182. **Lehoczki, E., Laskay, G., Pölös, E., and Mikulas, J.,** Resistance to triazine herbicides in horseweed *(Conyza canadensis), Weed Sci.,* 32, 669, 1984.
183. **Bulcke, R., Verstraete, F., Van Himme, M., and Stryckers, J.,** Biology and control of *Epilobium ciliatum* Rafin., in *Weed Control on Vine and Soft Fruits,* Cavalloro, R. and Robinson, D. W., Eds., 1987, 57.
184. **Lutman, P. J. W. and Snow, H. S.,** Further investigations into resistance of chickweed (*Stellaria media), Brit. Crop Prot. Conf. — Weeds,* 1987, 901.
185. **Haigler, W. E., Gossett, B. J., Harris, J. R., and Toler, J. E.,** Resistance of common cocklebur (*Xanthium strumarium)* to the organic arsenical herbicides, *Weed Sci.,* 36, 24, 1988.
186. **Saxena, P. K. and King, J.,** Herbicide resistance in *Datura innoxia, Plant Physiol.,* 86, 683, 1986.

187. **Pölös, E., Mikulas, J., Szigeti, Z., Laskay, G., and Lehoczki, E.,** Cross-resistance to paraquat and atrazine in *Conyza canadensis, Brit. Crop. Prot. Conf.— Weeds,* 1987, 909.
188. **Würzer, B.,** Unpublished results, 1989.
189. **Sandermann, H., Scheel, D., and Trenck, T. V. D.,** Use of plant cell cultures to study the metabolism of environmental chemicals, *Ecotox. Environ. Safety,* 8, 167, 1985.
190. **Zilkah, S. and Gressel, J.,** Differential inhibition of dikegulac of dividing and stationary cells in *in vitro* cultures, *Planta,* 142, 281, 1978.
191. **Espie, G. S. and Colman, B.,** The effect of pH, O_2 and temperature on the CO_2 compensation point of isolated *Asparagus* mesophyll cells, *Plant Physiol.,* 83, 113, 1987.
192. **Hills, M. J.,** Photosynthetic characteristics of mesophyll cells isolated from cladophylls of *Asparagus officinalis* L., *Planta,* 169, 38, 1986.
193. **Hickok, L. G., Warne, T. R., and Slocum, M. K.,** *Ceratopteris richardii*: applications for experimental plant biology, *Am. J. Bot.,* 74, 1304, 1987.
194. **Carmi, N. , Zilberstein, A., and Gressel, J.,** Unpublished results, 1990.
195. **Ezra, G. and Gressel, J.,** Rapid effects of a thiocarbamate protectant on macromolecular synthesis and glutathione levels in maize cell cultures, *Pestic. Biochem. Physiol.,* 17, 48, 1982.
196. **Edwards, R. and Owen, W. J.,** Regulation of glutathione-*S*-transferases of *Zea mays* in plants and cell cultures, *Planta,* 17, 99, 1988.
197. **Smith, A. E. and Oswald, T. H.,** Degradation of phenoxyalkylcarboxylic acids by white clover (*Trifolium repens)* cell suspensions, *Weed Sci.,* 27, 389, 1979.
198. **Feung, C. S., Hamilton, R. H., and Mumma, R. O.,** Metabolism of 2,4-dichlorophenoxyacetic acid. X. Identification of metabolites in rice root callus tissue cultures, *Agric. Food Chem.,* 24, 1013, 1976.
199. **Bristol, D. W., Ghanuni, A. M., and Oleson, A. E.,** Metabolism of 2,4-dichlorophenoxyacetic acid by wheat cell suspension cultures, *Agric. Food Chem.,* 25, 1308, 1977.
200. **Feung, C. S., Loerch, S. L., Hamilton, R. H., and Mumma, R. O.,** Comparative metabolic fate of 2,4-dichlorophenoxyacetic acid in plants and plant tissue culture, *Agric. Food Chem.,* 26, 1064, 1978.
201. **Davis, D. G., Hodgson, R. H., Dusbabek, K. E., and Hoffer, B. L.,** The metabolism of the herbicide diphenamid (*N-N-dimethyl*-2,2-diphenyl-acetamide) in cell suspensions of soybean (*Glycine max*), *Physiol. Plant.,* 44, 87, 1978.
202. **Scheel, D. and Sandermann, H.,** Metabolism of 2,4-D in cell suspension cultures of soybean and wheat. I. General considerations, *Planta,* 152, 248, 1981.
203. **Enkhardt, U. and Pommer, U.,** Bestimmung des cytochrom P_{450}-gehaltes in intakten zellen aus pflanzlichen suspensionskulturen, *Biochem. Physiol. Pflanzen,* 184, 183, 1989.
204. **Montague, M. J., Enns, R. K., Siegel, N. R., and Jaworski, E. G.,** Inhibition of 2,4-dichlorophenoxyacetic acid conjugation to amino acids by treatment of cultured soybean cells with cytokinins, *Plant Physiol.,* 67, 701, 1981.
205. **Harrington, K. C.,** Distribution and cross-tolerance of MCPA-tolerant nodding thistle, *Proc. 42nd N.Z. Weed and Pest Control Conf.,* 1989, 39.
206. **Zilkah, S. and Gressel, J.,** Correlations in phytotoxicity between white and green calli of *Rumex obtusifolius, Nicotiana tabacum* and *Lycopersicon esculentum, Pestic. Biochem, Physiol.,* 9, 334, 1978.
207. **Tanaka, K., Furusawa, I., Kondo, N., and Tanaka, K.,** SO_2 tolerance of tobacco plants regenerated from paraquat-tolerant callus, *Plant Cell Physiol.,* 29, 743, 1988.
208. **Furasawa, I., Tanaka, K., Thanuton, P., Mizuguchi, M., and Asada, Y. K.,** Paraquat resistant tobacco calluses with enhanced superoxide dismutase activity, *Plant Cell Physiol.,* 25, 1247, 1984.
209. **Babbs, C. F., Pham, J. A., and Coolbaugh, R. C.,** Lethal hydroxyl radical production in paraquat-treated plants, *Plant Physiol.,* 90, 1267, 1989.
210. **Blein, J. P. ,** Mise en culture de cellules de jeunes plantes de *Chenopodium album* L. sensibles ou resistantes a l'atrazine, *Physiol. Veg.,* 18, 703, 1980.
211. **Hahlbrock, K., Lamb, C. J., Purwin, C., Ebel, J., Fautz, E., and Schäfer, E.,** Rapid response of suspension-cultured parsley cells to the elicitor from *Phytophthora megasperma* var. *sojae, Plant Physiol.,* 67, 768, 1981.
212. **Apostol, I., Low, P. S., Heinstein, P., Stipanovic, R. D., and Altman, D. W.,** Inhibition of elicitor-induced phytoalexin formation in cotton and soybean cells by citrate, *Plant Physiol.,* 84, 1276, 1987.
213. **Wendorff, H. and Matern, U.,** Differential response of cultured parsley cells to elicitors from two non-pathogenic strains of fungi, *Eur. J. Biochem.,* 161, 391, 1986.
214. **Zilkah, S., Bocion, P. F., and Gressel, J.,** Target tissue for napropamide inhibition: effects on green and white calli and seedlings, *Weed Sci.,* 26, 711, 1978.
215. **Lewinsohn, E. and Gressel, J.,** Benzyl viologen mediated counteraction of diquat and paraquat phytotoxicities, *Plant Physiol.,* 76, 125, 1984.

Chapter 13

CONCLUDING REMARKS

Jens C. Streibig and Per Kudsk

In the course of history of chemical weed control, bioassay use has played and still will play an important role to appraise the three key factors in herbicide development and use: biological activity, selectivity, and environmental fate.

There is a dichotomy in herbicide research with bioassays: to detect bioactivity of new leads, the influence of adjuvants, herbicide mixtures, and synergists (Chapters 2 to 8, 10 and 12); and to measure residues in various media such as soil, water, plant, etc. (Chapters 9 and 11).

An important issue is to quantitatively assess the efficacy of herbicides. One of the best ways to summarize herbicide dose-response curves is to use regressions that can present results in a parsimonious way and also can tell us about the quality of data. Regression requires some explicit assumptions about the response curves that by proper statistical analysis can be tested (Chapter 3). Additional tests having biological meanings can be done, e.g., whether the curves of assayed compounds are similar (parallel curves) and how adjuvants, formulation product, and herbicide mixtures affect plant responses, etc.

Bioassays on whole plants will still play an important role in the development of herbicides because whole plant activity is the ultimate aim. A proper description of the dose-response relationship will help giving us a better understanding of the activity of a compound, the effect of formulations, adjuvants, application techniques, environmental factors, etc. In the light of the political demands for reduction of pesticide use in many countries, a more precise understanding of the optimal environment for herbicide action will enable us to better predict herbicide performance in the field.

Detection of herbicidal activity in living matter still is an important area of bioassay work. In the future, screening programs on whole plants will be further developed parallel to new and perhaps more precise *in vitro* test systems. In many respects the different test systems are complementary to each other. The research front will probably include parallel studies of well defined biochemical and more complex whole plant systems to correlate the results with physical and chemical properties of herbicides. A more comprehensive understanding of the links among different test systems of different complexity and the relationships with molecular shape, lipophilicity, electron charge distribution, etc. of herbicides will make a sound basis for utilizing methods currently available for the study of quantitative structure activity relationship (QSAR). In the future this may reduce the laborious trial and error screening for new herbicides and optimize use of herbicides already in the market place.

The development of new herbicides now is so costly that other avenues of research and development are exploited, such as biorational approaches. Biorational approaches are based on proper assay systems and intelligent use of current chemical, biochemical, and biological knowledge, and will undoubtedly improve in the future.

We can distinguish between various levels of complexity in herbicide research. The description of the dose-response relationship at all levels in Table 1 can give us a better understanding of the intrinsic selectivity of a compound. In some instances, metabolic studies can explain most of the selective nature of a certain compound at dose ranges used in the field.

In other instances there are too many unknown factors, which make it virtually impossible to explain selectivity. The worst case is when we look at herbicides with several sites of action

0-8493-6603-8/93/$0.00+$.50

TABLE 1
Classification of Various Systems to Study Herbicide Action

System	Data relevance	Type of experiments
Super complex	Growth	Anatomy, morphology, dry matter production
Complex	Metabolism	Tracer studies, gas related exchange, cell component studies, cell and tissue culture
Defined	Site related	*In vitro* studies with enzymes, binding studies

Note: After Fedkte.[1]

and where absorption, translocation, and metabolism form an integral part of the whole picture. To some extent this is still the case for the auxin herbicides. First, when we can quantify the selectivity as a delicate balance between dose, uptake, translocation, and metabolism, we will be able to determine the real nature of selectivity.

> *"In order to obtain simple systems, suitable for a more direct, 'exact' approach, the living organism has to be disintegrated. The further this disintegration goes, the more 'exact' will be our approach, but less life will remain in the biological system. As soon as we really approach exactitude, life will be gone."*
>
> Ariëns, Simonis, and van Rossum[2]

In the early days, herbicide bioassays were often a substitute for quantitative chemical analysis of residues. Progress in analytical chemistry lowered the detection limits for residual herbicides and simplified hitherto quite complex analytical procedures. Whatever the precision of modern chemical analysis, the link between the actual measurement of a residue must inevitably be related to the biological effect of that residue. Bioassay of soil residues of sulfonylureas is still quicker, cheaper, and more meaningful than chemical analysis. In other instances chemical analysis does not always indicate biological availability for such compounds as paraquat and glyphosate.

The new low rate sulfonylurea herbicides did put the significance of whole plant bioassay into a new perspective, and probably accelerated the development of immunological assays for herbicides in soil plant and environment. It can be anticipated that these immunochemical techniques will play an important role in the future, particularly where costly chemical hardware is unavailable.[3]

As time goes by, the unwanted side-effects of 45 years of modern herbicide use have become a general concern. This has increased the interest for the ecotoxicological aspects in herbicide research. Ecotoxicology is a fairly new discipline that, broadly speaking, links a dose or a residue to their biological effect in organism or whole ecosystems. In the present context, we could restrict the ecotoxicological question to be an appraisal of the subtle effect of low dose rates of herbicides, whereas phytotoxicity in relation to weed control in crops is concerned with the effect of higher doses. In this kind of research, bioassay test systems are important. The no-observable-effect level (NOEL) and related topics, for example QSAR studies, require properly defined bioassays linked with chemical measurement of herbicides.

For the farmer and the public at large, it is a matter of efficient weed control, safe crop rotation, and minimizing environmental risks.

REFERENCES

1. **Fedtke, C.,** *Biochemistry and Physiology of Herbicide Action,* Springer-Verlag, Berlin, 1982, 6.
2. **Ariëns, E. J., Simonis, A. M., and van Rossum, J. M.,** Drug receptor interaction. Interaction of one or more drugs with one receptor system, in *Molecular Pharmacology* Vol. 1, Academic Press, New York, 1964, 119.
3. **Stanker, L. H, Watkins, B. E., and Vanderlaan, M,** Environmental monitoring by immunoassays, in *Pesticide Chemistry,* Proc. 7th Int. Congr. Pesticide Chemistry (IUPAC), Hamburg, Frehse, H., Ed., VCH, Weinheim, 397, 1990.

INDEX

A

B

C

D

E

J

K

L

M

T

U

V

W

X

Z